Polymer Coatings

Scrivener Publishing
100 Cummings Center, Suite 541J
Beverly, MA 01915-6106

Publishers at Scrivener
Martin Scrivener (martin@scrivenerpublishing.com)
Phillip Carmical (pcarmical@scrivenerpublishing.com)

Polymer Coatings

Technology and Applications

Edited by

Inamuddin, Rajender Boddula, Mohd Imran Ahamed and Abdullah M. Asiri

Scrivener Publishing

WILEY

This edition first published 2020 by John Wiley & Sons, Inc., 111 River Street, Hoboken, NJ 07030, USA and Scrivener Publishing LLC, 100 Cummings Center, Suite 541J, Beverly, MA 01915, USA

For more information about Scrivener publications please visit www.scrivenerpublishing.com.

Wiley Global Headquarters
111 River Street, Hoboken, NJ 07030, USA

For details of our global editorial offices, customer services, and more information about Wiley products visit us at www.wiley.com.

Library of Congress Cataloging-in-Publication Data

ISBN 978-1-119-65499-5

Cover image: Pixabay.Com
Cover design by Russell Richardson

Set in size of 11pt and Minion Pro by Manila Typesetting Company, Makati, Philippines

Printed in the USA

10 9 8 7 6 5 4 3 2 1

Contents

Preface

Polymer coatings are thin polymer films that are applied to flat surfaces or irregular objects. Protective and decorative layers can be served by these coatings. They can be used as functional coatings with corrosion inhibitors or for decorative purposes like in paints. Polymeric coatings are known to be made of organic materials. However, they may contain metallic or ceramic grains to enhance endurance, properties or appearance. Polymeric coatings can be obtained using natural and synthetic rubber, urethane, polyvinyl chloride, acrylic, epoxy, silicone, phenolic resins or nitrocellulose, etc. There is a wide range of fabrication methods to design and construct polymer-coated materials. Compared to conventional coatings, they offer efficient and cost-effective coatings, facile fabrication methods with excellent properties such as corrosion, wear, and heat resistance, higher mechanical strength, and additional benefits, including good chemical and blocking resistance, and excellent scratch/abrasion resistance. Besides which, high gloss to matt looks, soft-touch effect, no color chage after UV exposure, excellent adhesion on metal and plastics, short drying time, fast hardness development, and easy formulation are other advantages of these coatings. Polymer coatings have various applications in the field of painting, storage media, semiconductors, optical devices, fluorescent devices, etc., and interest in them has increased due to their applications in areas such as electronics, defense, aeronautical and automotive industries.

This edition of *Polymer Coatings: Technology and Applications* explores the cutting-edge technology of polymer coatings. It discusses fundamentals, fabrication strategies, characterization techniques, and allied applications in fields such as corrosion, food, pharmaceutical, biomedical systems and electronics. It also discusses a few new innovative self-healing, antimicrobial and superhydrophobic polymer coatings. Subsequently, current industrial applications and possible potential activities are also discussed. This book is an invaluable reference guide for engineers, professionals, students and faculty members working in areas such as coatings, polymer chemistry, and materials science and engineering. Based on thematic topics, this edition contains the following eighteen chapters:

Chapter 1 provides an up-to-date account of fabrication methods for polymer coatings from the basic science to the latest innovations. The techniques which are described and discussed include blade coating, dip coating, spray coating, thermal spray coating, pulsed laser deposition, plasma polymerization, flow coating, spin coating, sol-gel and grafting.

Chapter 2 includes the different fabrication methods of organic/inorganic coating, namely, sol-gel method, cold spray technique, chemical vapor deposition, physical vapor deposition, thermal spray coating, electroplating deposition and electroless deposition. The classification of different coating methods for various organic/inorganic matrices and nanofillers are reported in detail.

Chapter 3 describes various eco-friendly dry powder coating techniques explored in the formulation and development of dry powder inhalers. Additionally, the chapter also includes a segment detailing the process analytical technology techniques, force controlling agents, implications in inhaler device coating and use of computational fluid dynamics in coating technology.

Chapter 4 first introduces the growth of bioinspired superhydrophobic coatings. Then, several theoretical backgrounds are discussed briefly. Afterwards, various methods are considered relating to the importance of creating chemical and physical textures on the surface. Additionally, the development of superhydrophobic and self-cleaning coatings with added nanoparticles are also presented.

Chapter 5 first investigates the nature (substrate-substance) and applications of superhydrophobic coatings. Afterwards, superhydrophobic coating applications are divided into three major categories of restrictive attributes, self-cleaning and smart attributes. All applications of hydrophobic coatings which have been examined in several studies are discussed.

Chapter 6 provides a brief overview of adsorptive polymer coatings, their techniques and a comprehensive comparison. Moreover, adsorptive polymer coating applications in various fields are also discussed. Furthermore, a future perspective of existing challenges provides a better direction and understanding for overcoming these challenges in coming days.

Chapter 7 deals with the formulations and chemistry of polyurethane (PU) coatings, and also provides an insight into the development of PU over the conventional coatings. A detailed discussion of the advantages of PU coatings and their future scope in industry is also presented.

Chapter 8 emphasizes a unique type of polymer coatings based on electroactive material. Fabrication, essential characteristics, and potential applications of electroactive polymer coatings are discussed.

Chapter 9 deals with the importance of conducting polymer coatings in the field of corrosion resistance.

Chapter 10 discusses the main objectives, materials and techniques used for encapsulating food components or coating food surfaces such as fruits and vegetables, meat and meat products, eggs, cheese, nuts, and fried food. Biopolymers including polysaccharides, proteins, and waxes are the main ingredients used for this purpose.

Chapter 11 discusses the scope of biopolymers as edible coating in food products. The chapter emphasizes the various types of raw materials used for preparing edible coating. The role of edible packaging in microbial spoilage, mechanical damage, and consumer acceptance of food is discussed along with its advantages and limitations and selection criteria as edible coating for different varieties of food products.

Chapter 12 addresses the wider aspects of pharmaceutical coatings using different types of polymers and their applications in the development and manufacturing of conventional and modified release drug delivery systems. A historical perspective on pharmaceutical coatings along with their physical attributes and characterization are also discussed, which will guide researchers and pharmaceutical manufacturers to their appropriate selection.

Chapter 13 summarizes the critical characteristics of self-healing polymeric coatings. Progress in existing self-healing coating methods and realistic frameworks of polymeric coatings are presented. Surface self-regeneration and anti-corrosive protective layer fractures are discussed. Issues related to the transition from laboratories to valid industrial application of these self-healing technologies are addressed.

Chapter 14 describes various methods that have revolutionized the role of this fascinating strategy in biological science especially in biomedical applications, notably infectious therapy, drug delivery system for therapeutic agent and protective layer for implants and biomedical devices. The major focus is given to some key applications which are trendsetting for surface functionalization of implants and biomedical materials.

Chapter 15 describes the role of polymers against various microorganisms such as bacteria, protozoans and fungi. These polymers mimic the action of antimicrobial peptides which are utilized by immune systems of such living organisms to kill the microorganism. The main purpose of antimicrobial coating is to combat antimicrobial resistance and infections.

Chapter 16 discusses in detail the various processes and techniques that are most commonly used for the coating of polymers which protect active pharmaceutical ingredient (API) against environmental hazards and bodily fluids, protecting the body from adverse effects of API and modifying the release of API.

Chapter 17 discusses the different conducting polymer coatings used over metal surfaces for corrosion protection along with the role of conducting polymers and various coating techniques. Additionally, this chapter summarizes the performance improvement and bulk modifications of conducting polymers and extensive studies on the protective coating of conductive polymer materials are discussed.

Chapter 18 presents extensive research studies reported by worldwide scientists and specialists in the area of polymer coatings for industrial applications. New and emerging industrial applications are discussed, including microsystems, oil and gas industries, electronics, biomedical systems, pipeline, automotive industries, micro bit storage systems, anti-corrosion and antibacterial coatings.

Chapter 19 discusses recent advancements in the usage of polymers for coating of different dosage forms such as tablets, capsules, implants, nanoparticles, and liposomes. Additionally, mechanisms of polymeric film formation and applications of polymer coatings in the different areas of biomedicine are clearly explained as are the application of different polymers in various coating functions.

Editors
Inamuddin
Rajender Boddula
Mohd Imran Ahamed
Abdullah M. Asiri
February 2019

1

Fabrication Methods for Polymer Coatings

Hüsnügül Yilmaz Atay

İzmir Katip Çelebi University, Department of Material Science and Engineering, Çiğli İzmir, Turkey

Abstract

Polymer coatings mean the top layer applied on any substance for purposes like protection and decoration. It is possible to apply to synthetic materials as well as metals and ceramics. They are resistant to high temperatures, such as up to about 280°C. The polymeric coating process comprises applying a polymeric material onto a supporting substrate and coating the substrate surface. Polymeric coatings can be obtained using natural and synthetic rubber, urethane, polyvinyl chloride, acrylic, epoxy, silicone, phenolic resins or nitrocellulose, etc. There are a wide range of fabrication methods to design and construct polymer-coated materials. In this chapter, the techniques are described and discussed including blade coating, spray coating, thermal spray coating, pulsed laser deposition, plasma polymerization, flow coating, spin coating, sol–gel, dip coating, and grafting. The key point is provided to highlight current methods and recent advances in polymer coating fabrication techniques.

Keywords: Polymer coatings, fabrication methods, blade coating, spray coating, thermal spray coating, pulsed laser deposition, plasma polymerization, flow coating, spin coating, sol–gel, dip coating, and grafting

1.1 Introduction

Polymer coatings are thin polymer films that are applied to flat surfaces or irregular objectives. Protective and decorative layers can be served by these coatings [1]. They can be functional coatings such as adhesives or photographic films. They can be used as corrosion inhibitors or for decorative purposes like paints. Moreover, for modifying the surfaces, they can be utilized such as paper coatings or hydrophobic coatings.

Email: hgulyilmaz@gmail.com

Inamuddin, Rajender Boddula, Mohd Imran Ahamed and Abdullah M. Asiri (eds.) Polymer Coatings: Technology and Applications, (1–20) © 2020 Scrivener Publishing LLC

Polymeric coatings are known to be made of organic materials. However, they may contain metallic or ceramic grains to enhance endurance, property, or appearance [2]. They offer various properties and additional benefits, for instance, very good chemical resistance, very good blocking resistance, and excellent scratch and abrasion resistance. Besides, high gloss to matt looks, soft touch effect, noncoloring after UV exposure, excellent adhesion on metal and plastics, short drying time, fast hardness development, and easy formulation are other acquirements of those coatings [3].

In general, polymer coatings are architected to manufacture a film of a kind of polymer. The process should be as fast as possible. The thickness is typically 1–100 m. The type of coating method varies according to the thickness of the desired covering, the rheology of the running, and the velocity of the web [2].

1.1.1 Starting Liquid Types

Before passing through to the coating methods, it is better to explain starting liquid types to obtain an impermeable and indiscrete polymer coating deposit. Three different types of starting liquids can be used to achieve this output. These are indicated as: polymer solutions, monomer liquids, and polymer latexes [2].

1.1.1.1 Polymer Solutions

It is necessary to decrease the viscosity of the polymer to make it a stickable fluid. For this purpose, the polymer is decomposed in a dissolvent. The fluidity property of the solution is regulated by varying the amount of solvent in the solution. The resulting fluid is covered onto the substrate. The dissolvent should then be removed by a drying operation. The glass transition temperature of the dispersion rises with removal of the solvent. If the drying temperature is smaller than the glass transition temperature, the coating passes to the solid phase. However, when the drying temperature is higher than room temperature, it is seen that solidification or hardening continues during the cooling of the coating. On the other hand, some of the polymers can crystallize when the dissolvent is removed. While some are cooling, they form semi-crystalline final polymer coatings [2].

Generally, most polymers are insoluble in water and organic solvents are used for dissolution. The solvent is selected in terms of both its ability to dissolve the polymer and its influence on the drying step. Due to the need to add different additives and reduce the cost, it may be necessary to use more than one volatile solvent [2].

The use of coatings produced with polymer solution is favored as they can be applied to a wide variety of polymers at the processing site specifications and formulated according to adjustable properties to produce evaluated properties in the last product. The quantity of polymeric material soluble in a solvent is relatively small. The drying requirement therefore appears as a function of the unit thickness of the coating. On the other hand, there are environmental and safety concerns due to complications related to solvent use. Solvent recycling is another important problem. It is also another disadvantage that flammable solvents need to be captured by expensive driers [2].

1.1.1.2 Liquid Monomers

Many monomers have fluid properties at room temperature. Therefore, there is no need to decrease their flow resistance at the coating process temperature. Also, they can be covered directly without adding any dissolvent. Oligomeric precursors can be said to be in this category. Without the need for a drying operation, the monomer liquids are allowed to solidify by serial curing reactions. Meanwhile, the molecular weight of the covering material rises in the period from progression of curing to the formation of a solid polymer layer. Hardening reactions are initiated by exposing them to energetic sources, for instance, ultraviolet light or electron beams [2].

The most widely used and popular coating material is epoxy in the field. They are not monomers, yet they are formed by a chemical reaction of oligomeric resins with its hardeners. The liquids can also be produced with dissolvent to enhance interoperability. Acrylates as liquid monomers are widely used for ultraviolet curing [2].

Monomer fluids do not require much drying step because they contain very little solvent. Therefore, they are quite attractive ways for coatings. The final coating properties (e.g., density of crosslinking) can be managed at the curing stage using parameters such as temperature, ultraviolet density, or the resin chemistry. Nevertheless, in some cases, the materials used in functional polymer solution coatings may be less expensive than monomers and initiating agents. Besides, due to the high degree of crosslinking, the final product can sometimes be brittle [2].

1.1.1.3 Polymer Latex

A latex can be defined as the dissipation of polymeric grains in water. In the case of lower water solubility of the polymeric materials having functional properties, the latex paths supply an environmentally suitable solution for forming enduring covering. Grains varying in size from ~ 10 nm to 1 umm can be manufactured from various polymeric chemicals by emulsion polymerization. It is easy to use, especially because they are synthesized in dispersion form and can be stabilized in the process. For some special applications, latex may also be formulated with other phases, such as ceramic grains [2].

The drying of the latex suspensions appears to be slightly different from the drying of the hard colloidal grains. This process is known as "film formation" depicted in Figure 1.1. Because water is removed, the grains go into the "consolidation" step and become more concentrated in suspension. When the drop time is over, surface tension, capillary, and van der Waals forces begin to pull the grains toward each other. Those forces must be powerful sufficiently to allow the grains to flatten at the grain–grain contact points. Consequently, the pores between the particles become smaller. This stage is called "compression." The final stage is the "union" stage. Here, the polymer chains boil the particles together and transcend the boundaries between the grains. With that process, a finalized covering is formed that lacks gaps that were once between the individual particles [2].

Water is a liquid medium that can be used in latex coatings. Thus, monomer and solvent may be an eco-friendly alternative to other coatings used.

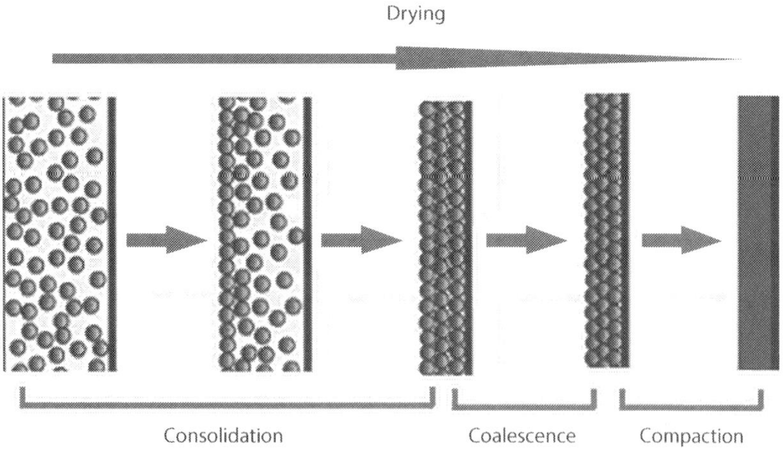

Figure 1.1 Latex film formation stages [2].

Various coatings, such as paints and varnishes, seem to start as latex dispersions. On the other hand, it is pricey to transport and purchase latexes as raw material on a commercial scale. Besides, drying of water is an operation that requires more energy [2–4].

1.1.2 Polymer Coating Methods

Coatings made of polymeric materials can provide many different surfaces: metallic, ceramic, or synthetic materials, using a number of different techniques [5]. They must adhere well to the substrate. They should also not be readily susceptible to moisture, salt, heat, or different kind of chemicals. Generally, the following properties are required for a good coating film [3]:

- Water-based resins. Low- or zero-volatile organic compounds (VOC)
- Very good stain and chemical resistance
- Very good blocking resistance
- Excellent scratch and abrasion resistance
- High gloss to matt looks
- Soft touch effect
- Nonyellowing after UV exposure
- Excellent adhesion on metal and plastics
- Carbodiimides for 2K systems

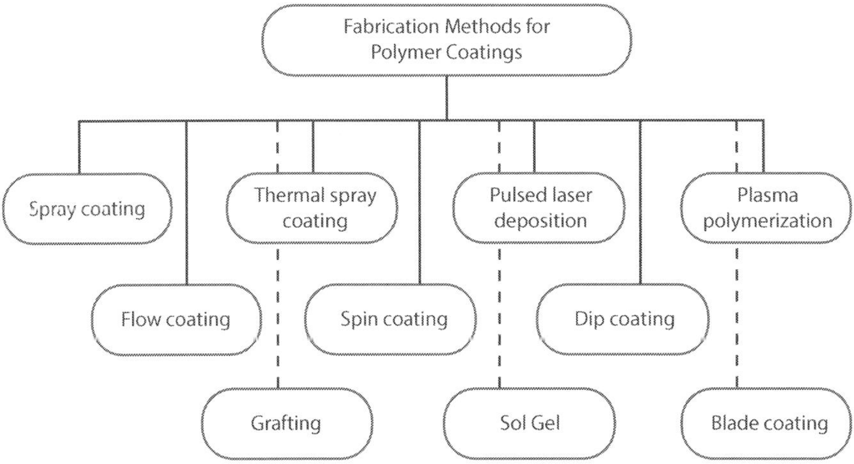

Figure 1.2 Fabrication techniques for polymer coatings.

- Short drying time, hence fast hardness development
- Easy formulation

Applied coating methods can affect the product quality, and thus the coating methods are important to obtain desired properties. Different fabrication methods are demonstrated in Figure 1.2.

1.1.2.1 Blade Coating

The blade coating can be defined as a process in which a certain amount of covering material is applied to the underside and the excess is removed by a measuring blade to obtain the desired coating thickness [6, 7]. This coating method has several advantages for obtaining a good coating film. Homogeneity of the coating area, small amount of material waste, prevention of intermediate layer melting, roll-to-roll production compliance, and economic use of the material [8–10]. In this method, fast drying process will prevent the slowing of the manufacturing process by solvent annealing [10]. Control of the thickness can be adjusted by controlling manufacturing conditions such as sol concentration, blade gap, and blade covering velocity [8].

1.1.2.2 Spray Coating

Spray coating technique is a process method in which the printing material (ink) is constrained through a nozzle and thereby forming a thin aerosol [11]. In this process, the performance of polymer solar cells seems to be limited by certain disadvantages, for instance, isolated droplets, non-uniform surfaces, and holes at some points. Regarding the process parameters, the flow rate, the pressure, the substrate temperature, the density of the mixing dissolution, the spraying time, the distance between the sample, and the air brush can be listed [6].

1.1.2.2.1 Nozzle-to-Substrate Distance

The distance between the nozzle and the surface is considered to be one of the process parameters, since it has a big effect on the morphology of the deposited part in the spray coating. Many studies were performed to examine and achieve the best distance of the nozzle and surface for the active coating. Vak *et al.* [12] found three areas between the air brush nozzle and the substrates, which were "wet," "intermediate," and "dry." They then concluded that the perfect linear control distance was in the "intermediate-region." This result is described as the spray time function.

Susanna *et al.* [13] observed that, with the same material, the deposited material remained wet when the distance between the sample and the air brush was less than 15 cm. They reported that they manufactured dry and powdered coatings over 20 cm from the substrate. In this study, the "intermediate" region was 17 cm.

1.1.2.2.2 Solvent and Mixed Solvents Effect

There are many different solvents used by scientists to investigate the effects of dissolvent on coating quality and efficiency, for example, chlorobenzene, dichlorobenzene, trichlorobenzene, p-xylene, toluene, etc. The choosing of solvent affects the choice of nozzle–substrate distance for achieving the suitable thickness and the covering morphology. It is therefore important in all spray coating methods. Fundamentally, the principle behind selecting a dissolvent should be to select a quick-drying dissolvent to inhibit droplets from redissolving the substrates. However, this velocity should not be high enough to permit a homogenous and pinhole-free coating to be formed [14].

Pin-hole thin films are not desirable in that those films impair the efficiency of the coating. The quantity of fluid sprayed on the surface must be higher than at least a threshold to produce films without pin-hole. In this way, it may be possible for the droplets placed on the surface to join into a fully wet region. [6].

1.1.2.2.3 Substrate Temperature and Annealing Treatment

Green *et al.* [15] conducted an extensive work examining the performance of spray-coated devices over the effect of annealing temperature. In this study, it was seen that the efficiency of the devices amended with annealing temperature. The change in the intensity of the short-circuit current causes this to happen. Lee *et al.* [16] showed that the morphological surfaces of spray-coated equipment were not affected by thermal annealing at a micrometer scale at 150°C. Though, conflicting results have been reported by Dang *et al.* [17]. It was found that annealing affected the morphological texture of the active regions of the rotary casting and improved toward phase separation.

1.1.2.3 Thermal Spray Coating

Thermal spraying can be defined as a method of improving the surface of a solid object. This method can be used to coat a variety of materials to supply resistance to abrasion, erosion, cavitation, corrosion, abrasion, or heat. It can also be applied to supply electrical conductivity or insulation, lubricity, high or low friction, victim wear, chemical resistance, and many other desired features [18].

The application area of thermal spray is very wide. Extending the life of new components and repairing and reconstructing worn or damaged components may be some of them. In general, all thermal spraying methods include the adhesion of little softened grains to a dirt-free and conditioned surface to form a well coating. The combination of thermal energy and kinetic energy results in the flattening of the particles on the surface and on top of one another, with the successive layers forming an adhesive coating. Some advantages of thermal spray are shown below [18]:

- Metallurgical cold processing.
- No heat input, no deterioration.
- A mechanical bonding process required.
- Many different materials can be sprayed such as: steels, stainless steels, nickel alloys, copper, bronzes, molybdenum, ceramics, tungsten carbides, etc.
- Thickness can be applied between 100 and 750 microns, but more can be provided.
- Line of sight process.

Some of different types of thermal spray methods are used in the industry. In flame spray method, gas and oxygen are used for melting the wires before spray process with compressed air. Powder spray uses acetylene and oxygen for softening or melting the powders with the gas flow accelerating the grains to the substrate. In arc spray method, direct current (DC) electrical power is used for melting the wires before spray operation with compressed air. Plasma spray uses electric arc in an inert gas for creating a plasma that can soften the spray powders. The gas stream can project the powder onto the substrate for low oxide situation. High velocity oxygen fuel (HVOF) can use an accelerated oxygen or fuel flame for softening the spray powders and projecting them onto the surface with high levels of kinetic energy. All of the five thermal sputtering processes described above can be used to amend the substrate properties. Thanks to these engineering coatings, properties such as improved wear resistance, thermal barriers, electrical and thermal conductivity, hard chrome exchange, and insulation can be obtained on the material surfaces [18].

1.1.2.4 Pulsed Laser Deposition

The word laser is an abbreviation of light magnification with excited radiation emission. It has precious unique features such as narrow frequency bandwidth, consistency, and elevated power density. Because the light beam is too dense, it may be possible for vaporizing the hardest and heat-resistant materials. In addition, it has features such as high precision, reliability, and spatial resolution. Therefore, it has utility in the processing of thin coatings for the materials modification, heat treatment of the material surface, welding, and micro-modeling industry [19, 20].

In view of the pulsed laser deposition principle, it is generally seen that a pulsed laser beam is focused on the surface of a solid material. Electromagnetic radiation is strongly absorbed by the solid surface, which allows quick evaporation of target materials. When the content of the vaporized materials is examined, it is seen that it consists of highly excited and ionized species [21].

The pulsed laser deposition technique makes it particularly easy to store materials with complex stoichiometry. Therefore, it has attracted great interest of late years. The coating of the YBa2Cu3O7 thin film, a superconductor, was the first such process. Since then, the material, which is particularly difficult to obtain by multi-element oxides, has been successfully applied with this technique. The main advantages of this technique are [21]:

- *Conceptually quite simple:* Using a laser beam to evaporate the target surface and produce a coating of the same content as the target.
- *Versatile:* A wide range of materials can be coated in different gases over a wide gas pressure range.
- *Cost-effective:* A single laser can work with multiple vacuum systems.
- *Fast:* Good quality products can be obtained in 10 or 15 min.
- *Scalable:* Scaling can be achieved by advancing complex oxides toward volume production.

1.1.2.5 Plasma Polymerization

Plasma polymerization can be described that, in this process, organic and inorganic polymers can be deposited from a monomer vapor using an electron beam, ultraviolet radiation, or radiation discharge [22]. For low pressure plasma coating, it is ensured that gas or liquid monomers are processed, which are polymerized under the action of plasma. In general, the coating thicknesses obtained are in the range of one micrometer. The adhesion of the coatings to the surface is quite good [23].

When it is examined to activate and degrease, the process is much more complex. For example, barrier coatings are produced in fuel tanks, scratch-resistant coatings in headlights, and hydrophobic coatings. There are three coating methods used in large-scale applications [23]: hydrophobic coatings, PTFE (polytetrafluoroethylene)-like coatings, and hydrophilic coatings.

Coating of metals by plasma polymerization allows various effects such as continuous activation and functional coatings for one to several weeks. Process parameters should be set to take into account the characteristics of some equipment, such as generator type and power, electrode assembly, and material properties of the workpiece.

Plastics can be simply coated by plasma polymerization. In this way, a scratch-resistant coating can be made to CDs and DVDs without damaging their quality. If the products have low friction, PTFE-like coatings can be used to increase this. It may also be possible to bind functional groups such as amino groups used for bioanalytical applications to the plastic surface.

The difficulty in coating materials such as glass and ceramics is to prepare the surface accordingly. Once this problem is solved, there is no other element to prevent the application of the coatings. Check the adhesion of the covering to the surface in all cases. Where "mismatch" occurs between the coating and the substrate material, it may be ensured that the intermediate layers are applied as a binder. It is known that textile materials are very well-coated in plasma. However, the difficulty here lies in the long-term protection of coatings against surfactants. If hydrophobic coating is desired, fluorine-containing gases or monomers may be used [22, 23].

1.1.2.6 Flow Coating

The flow coating technique is a suitable form of coating to form polymer coating thickness gradients in the submicron regime. The device has a fixed blade that is fixed in a gap above a moving stage. The height of this gap can vary from tens of microns to hundreds of microns. The surface to be covered is firmly fixed to the stage, and the grains of a polymer dissolution are deposited between the blade and the surface and compacted. Next, the blade is accelerated relative to the surface. In the flow coating technique, the polymer is drawn through the substrate under forces caused by capillary forces holding between the fixed blade and the substrate, and by friction on the same solution as the blade. This process is in principle similar to the other measured flows for instance dip coating and blade coating.

In the fluid coating process, the capillary forces ensure that the polymer dissolution is kept under the knife at the first state at zero speed. In progress of time, the volume gradually decreases because of the evaporation of the dissolvent from the edges. At lower speeds, frictional pull may cause material to escape under the blade, although the capillary forces are intended to hold the material between the surface and the blade. This material then dries by evaporation of the solvent and is left in the shape of a wet coating [24–26].

1.1.2.7 Spin Coating

The spin coating method is a very common technique for applying thin coatings to the surfaces. It is seen that this technique is used in many different industries and technology branches. The most significant benefit of spin coating is the ability to manufacture very smooth coatings quickly and easily. The thickness of these films can range from several nanometres to several microns. The use of spin coatings is particularly common in organic electronics and nanotechnology. This production is based on many methods that can be utilized in different semiconductor industries [27, 28].

1.1.2.7.1 General Theory

In the spin coating process, it can be expressed as applying the desired material by pouring it onto the surface of a substrate and coating it evenly along its surface as it rotates in a solvent (an "ink") (Figure 1.3). As the coating liquid is dropped to the center of the target, the stage begins to rotate. As the rotation accelerates, the resulting centrifugal force allows the liquid to spread over the entire surface to form the coating layer. Factors such as the viscosity of the fluid, the rotational speed of the target, the acceleration of rotation, and the aeration that affect the drying rate affect the thickness of the film formed [29].

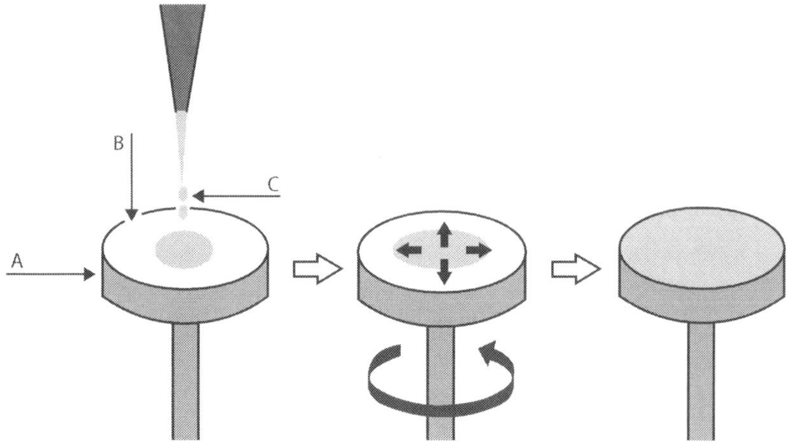

Figure 1.3 A coating specimen obtained by spin coating a small molecule in solution by using a static dispersion (A. Rotating stage. B. Target surface. C. Coating fluid.) [29].

1.1.2.7.2 Applications

Spin coatings can vary greatly when viewed as a field of application. They can be used to cover surfaces smaller than a few square millimeters used in the state of the art, as well as to flat panel TVs with a diameter of 1 m or more. When evaluated as a coating material, it is possible to coat substrates such as photoresists, insulators, organic semiconductors, synthetic metals, nanomaterials, metal and metal oxide precursors, transparent conductive oxides, and many other different materials.

1.1.2.7.3 Advantages and Limitations

Combined with a thin and homogeneous coating, the greatest advantage in the spin coating process can be shown as the simplicity and relative ease of installation of a process. The high air flow caused by the high rotational speeds causes rapid drying. This helps ensure high stability in macroscopic or nano-length scales.

A single substrate is used for spin coating. This is the limitation of this process. Therefore, the yield seems lower than roll-to-roll operations. Certain nanotechnologies require time for self-assembly and/or crystallization. Fast drying times in spin coating can therefore lead to poor performance.

1.1.2.7.4 Spin Speed

The current rotational speed ranges are important because they define the thickness range that can be obtained from a particular solution. By rotating the coating, it is possible to produce homogeneous films relatively easily from about 1000 rpm. However, high quality coatings can be obtained with caution up to 500 or 600 rpm and sometimes lower than this.

1.1.2.7.5 Coating Duration

In general, standard spin coatings allow the substrate to spin until the film is completely dry. However, the boiling point and vapor pressure of the dissolvent utilized, as well as the environmental parameters in which the extrusion coating is made, such as temperature and humidity, affect this. A spin coating time of 30 s is recommended as a starting point for most processes, since this is considered to be sufficient [28].

1.1.2.8 Sol–Gel

Sol–gel technology can be defined as the use of liquid solutions to produce solid films with a wide chemical composition structure on various substrates at relatively low temperatures. Sol–gel-coated coatings are very diverse in function, for example, electrical conductivity, superconductivity, ferroelectric behavior, corrosion resistance, selective barrier for wear resistance and gas permeability, etc. There are many interdependent or independent factors that determine the manufacturing and specialities of sol–gel coatings [30].

Historically, the first synthesis of silica gel goes back to 1846. However, it is seen that sol–gel chemistry has been comprehensively researched ever since the 1970s. In the sol–gel method, chemical reactions of the basic precursor, which is usually an organometallic compound, are used in alcoholic dissolutions. These reactions produce various inorganic networks that can be constituted from a metal alkoxide dissolution [31, 32].

The sol–gel method has three steps for the production of the latest metal oxide protocols: these are hydrolysis, condensation, and drying. The operation begins by subjecting the metallic precursor to rapid hydrolysis to manufacture the metal hydroxide solution. In the following step, condensation occurs, leading to the constitution of a three-dimensional gel. After drying the obtained gel, the product obtained can be quite easily transformed to

Xerogel or Airgel depending on the drying mode. The structure of metal precursors and solvents in the sol–gel process has an effective role in metal oxide synthesis. The sol–gel method can be classified in two ways, depending on the nature of the solvent used: aqueous sol–gel and anhydrous sol–gel method. If water is used as the reaction media, the aqueous solution is the method. However, if organic solvent is utilized as the reaction media, it is called anhydrous sol–gel. Figure 1.4 shows the reaction steps used for the manufacturing of metal oxide nanostructures in the sol–gel technique [33, 34].

The production steps of a radar-absorbing covering material manufactured by the sol–gel coating technique are shown in Figure 1.5. In this study, barium hexaferrite particles were synthesized by sol–gel [35, 36]. The precursors used were barium nitrate (Ba (NO3) 2) and ferric citrate monohydrate (C6H5FeO7.H2O), chelating agent citric acid monohydrate (C6H8O7.H2O), and pH regulator ammonium hydroxide (NH40H). Firstly, the process began by dissolving the ferric citrate and barium nitrate separately in citric acid. Using a magnetic stirrer, the solutions were vigorously stirred until a clear dissolution was achieved. Ammonium hydroxide was inserted to the dissolution until the pH reached 7 to ensure homogeneous suspension and stable pH. Next, the dissolution was held in a water bath at 80°C for 15 h in air to remove the water slowly from the solution. In this way, a high-viscosity wet gel was obtained. The wet gel obtained to prepare the dry gel was kept in an oven at 180°C for 15 h. Finally, in the

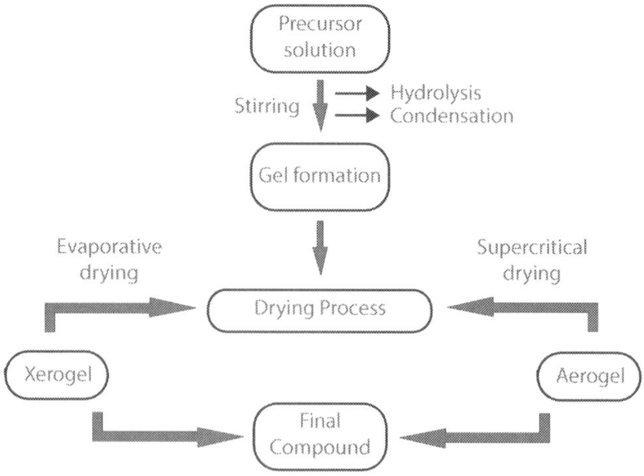

Figure 1.4 Reaction pathway for production of metal oxides in the sol–gel [33].

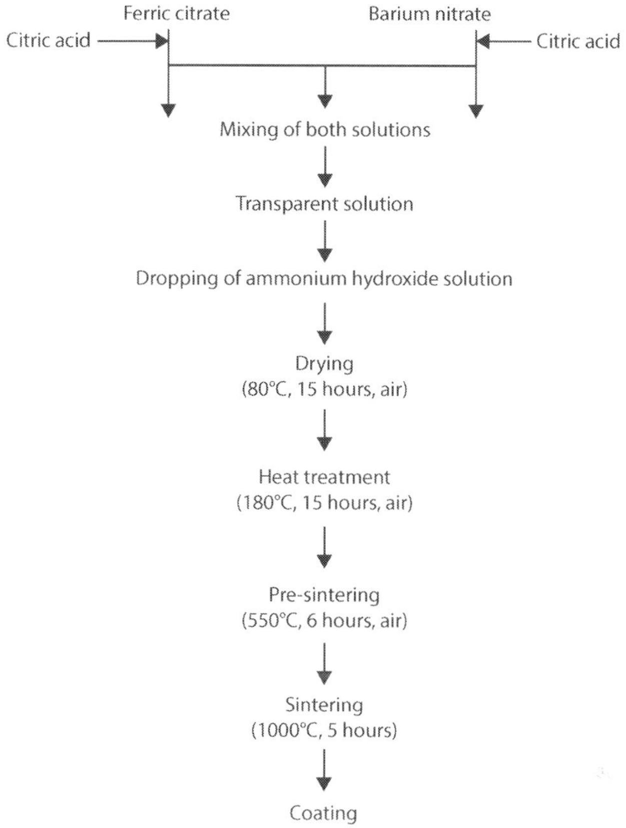

Figure 1.5 Coating is produced by sol–gel method [37].

sintering step, sintering was first performed at 550°C for 6 h to evaporate the impurities of the dry gel. The process was then completed by sintering in a tube oven at 1000°C for 5 h [37, 38].

1.1.2.9 Dip Coating

Immersion coating is considered one of the earliest mercantile available processes. The primitive patent in that field was given to Jenaer Glaswerk Schott in 1939 for sol–gel reproduced silica films [39, 40]. Today, it is seen that this technique is used for various applications such as sol–gel derived coatings, ferroelectrics, dielectrics, sensors and actuators, membranes, superconducting layers, protective coatings, and passivation layers. The process is described by three important technical steps:

1. *Immersion and dwell time:* At this stage, it is ensured that the substrate is kept in the coating solution for sufficient wetting for sufficient interaction. For this purpose, the surface is immerged in the precursor dissolution at a stable rate and then allowed to stand for a certain period of time.
2. *Deposition and drainage:* At this stage, film deposition is made. By pulling the substrate upward with a stable velocity, a thin precursor solution layer is retained. Excess fluid is discharged from the surface.
3. *Evaporation:* The dissolvent vaporizes from the liquid in this step. The deposited thin film may be supported by heated drying.

This is the fluid that forms it. Heat treatment is then carried out to burn the residual organics. In addition, the heat treatment carried out in this step may provide to crystallize the functional oxides.

Although that covering technique appears to be quite simple, microscopic processes need to be understood in more detail during immersion coating to adapt the final coatings. There is an important connection between the construction of the dissolution or the formed sol and the microstructure of the deposited coating. This is the coating process that establishes the connection. In the standard operation, the substrate is drawn upright from the solution tank with a stable velocity [41]. With the movement of the moving substrate, the liquid is dragged through a mechanical boundary layer based on the current line [42].

The flux moving upward is compensated by evaporation. As a result, the position of the film and the film remain constant relative to the surface of the coating bath. Upon evaporation and discharge of the solvent, the film has an approximately wedge-like shape terminating in a drying line. The vapor–solid–solid three-phase limit herein is defined as the drying line. Above this line, the nonvolatile species form the deposited part, which may be exposed to further curing. After gradual concentration of the inorganic species by evaporation, agglomeration and gelation occur to formalize a kind of dry gel or xerogel layer. The process ends with the final drying [39].

1.1.2.10 Grafting

This method is a capable tool for achieving surface grafting, surface modification, and functionality of polymer brushes on a solid surface. The final functionalized polymer chains can be grafted onto the solid surface, which is referred to as grafting. This grafting reaction may proceed with surface polymerization. In this way, a thin polymer brush layer is formed on the solid surface, which adjudicates the surface features. By combining surface roughness with mixed brushes, interchangeable ultrahydrophobic surfaces can be obtained [43].

The composition of surfaces and interfaces has a very important status in determining the performance of the materials. For example, areas of critical importance include the control of surface and interface properties, colloid stabilization, adhesion, lubrication, rheology, immobilization of catalysts, and the production of multiphase materials [44]. In addition to these classic areas, surface modification has lately played an important role in bioengineering, nonlinear optics, (bio) sensors, nano patterning, molecular recognition, waveguides, and electronic microcircuit operations. Accumulation of surface-bound thin polymer layers with convenient physical and chemical features for optimizing surface properties is one of the first operations [45].

The process is due to the elevated surface density, the precise placement of the chain on the surface, and the easy and controllable insertion of polymer chains with prolonged durability of the grafted layers. The most important benefit is the fact that several different polymers are attached to the same substrate. Prevention of macro-phase separation is achieved by chemical grafting. The features of the coating are determined by the consolidation of particular functional components. Polymer brushes are formed when the linked polymer chains contacted to the solid substrate with a chain end reach a relatively high grafting density. This can be clearly noticeable from other grafted polymer layers. Because of the excluded volume effect, the brush-like layers are formed when the surface is completely coated with a comparatively dense grafted chain monolayer normally stretched to the support. Parameters such as graft density, chain length, and chemical content of the chains can control the brush features [43, 46].

References

1. Ching, Y.C. and Liou, N.S., Effects of high temperature and ultraviolet radiation on polymer composites, in: *Durability and Life Prediction in Biocomposites, Fibre-Reinforced Composites and Hybrid Composites*, 2019.
2. Francis, L.F., Stadler, B.J.H., Roberts, C.C., *Materials Processing: A Unified Approach to Processing of Metals, Ceramics and Polymers*, 2016.
3. Bakker, R., Polymers Coatings: Technology and Applications. CH - Fabrication methods for polymer coatings. https://www.stahl.com/polymers. Retrieved in 15.12.2019.

4. National Research Council, *Polymer Science and Engineering: The Shifting Research Frontiers*, The National Academies Press, Washington, DC, 1994, https://doi.org/10.17226/2307.
5. Kausar, A., Polymer coating technology for high performance applications: Fundamentals and advances. *J. Macromol. Sci. Part A*, 55, 440–448, 2018.
6. Farhana, A. and Ahmad, İ., Spray coating methods for polymer solar cells fabrication: A review. *Mater. Sci. Semicond. Process.*, 39, 29–39, 2019.
7. Shim, E., 10 - Bonding requirements in coating and laminating of textiles, in: *Woodhead Publishing Series in Textiles, Joining Textiles*, I. Jones and G.K. Stylios (Eds.), pp. 309–351, Woodhead Publishing, Cambridge, U.K., 2013.
8. Chang, Y.H., Tseng, S.R., Chen, C.Y., Meng, H.F., Chen, E.C., Horng, S.F., Hsu, C.S., Polymer solar cell by blade coating. *Org. Electron.*, 10, 5, 741–746, 2009.
9. Schneider, A., Traut, N., Hamburger, M., Analysis and optimization of relevant parameters of blade coating and gravure printing processes for the fabrication of highly efficient organic solar cells. *Sol. Energy Mater. Sol. Cells*, 126, 149–154, 2014.
10. Tsai, P.T., Tsai, C.Y., Wang, C.M., Chang, Y.F., Meng, H.F., Chen, Z.K., Lin, H.W., Zan, H.W., Horng, S.F., Lai, Y.C., Yu, P., Bulk Heterojunction Photovoltaic Cells with Triphenylamine-Based Amorphous Polymer and Non-Halogenated Solvent Processing Provide Reproducible Performance. *Org. Electron.: Phys. Mater. Appl.*, 15, 4, 893–903, 2014.
11. Krebs, F.C., Fabrication and processing of polymer solar cells: A review of printing and coating techniques. *Sol. Energy Mater. Sol. Cells*, 93, 4, 394–412, 2009.
12. Vak, D., Kim, S.S., Jo, J., Oh, S.H., Na, S.I., Kim, J., Kim, D.Y., Fabrication of organic bulk heterojunction solar cells by a spray deposition method for low-cost power generation. *Appl. Phys. Lett.*, 91, 8, 1102–1103, 2007.
13. Susanna, G., Salamandra, L., Brown, T.M., Di Carlo, A., Brunetti, F., Reale, A., Airbrush spray-coating of polymer bulk-heterojunction solar cells. *Sol. Energy Mater. Sol. Cells*, 95, 7, 1775–1778, 2011.
14. Nie, W., Coffin, R., Liu, J., MacNeill, C.M., Li, Y., Noftle, R.E., Carroll, D.L., Exploring Spray-Coating Techniques for Organic Solar Cell Applications. *Int. J. Photoenergy*, 2012, Article ID 175610, 7 pages, https://doi.org/10.1155/2012/175610, 2012.
15. Green, R., Morfa, A., Ferguson, A.J., Kopidakis, N., Rumbles, G., Shaheen, S.E., Performance of bulk heterojunction photovoltaic devices prepared by airbrush spray deposition. *Appl. Phys. Lett.*, 92, 3, 3301–3303, 2008.
16. Lee, J.H., Sagawa, T., Yoshikawa, S., Thickness dependence of photovoltaic performance of additional spray coated solar cells. *Org. Electron.*, 12, 12, 2165–2173, 2011.

17. Dang, M.T., Wantz, G., Bejbouji, H., Urien, M., Dautel, O.J., Vignau, L., Hirsch, L., Role of the casting solvent. *Sol. Energy Mater. Sol. Cells*, 95, 12, 3408–3418, 2011.
18. Metalisation Ltd. Pear Tree Lane. Dudley, https://metallisation.com/applications/thermal-spray-engineering-applications/. Retrieved in 22.12.2019.
19. Metev, S.M. and Veiko, V.P., *Laser Assisted Microtechnology*, Springer, Berlin, Heidelberg, 1994.
20. Chrisey, D.B. and Hubler, G.K., *Pulsed Laser Deposition of Thin Film*, John Wiley & Sons, Inc., New York, 1994.
21. Pulsed Laser Deposition (PLD), Abstract. http://groups.ist.utl.pt/rschwarz/rschwarzgroup_files/PLD_files/PLD.htm. Retrieved in 05.01.2020.
22. Martin, P.M. (Ed.), Chapter 1 - *Deposition Technologies: An Overview*, in: *Handbook of Deposition Technologies for Films and Coatings*, Third Edition, pp. 1–31, William Andrew Publishing, Norwich, NY, 2010.
23. https://www.plasma.com/en/plasmatechnik/low-pressure-plasma/plasma-polymerisation/&gclid=EAIaIQobChMIzPSOqaWe4wIVQ6qaCh1Dx-QB5EAAYASAAEgI_nPD_BwE
24. Stafford, C.M., Roskov, K.E., Epps, T.H. III, Fasolka, M.J., Generating thickness gradients of thin polymer films via flow coating. *Rev. Sci. Instrum.*, 77, 023908, 2006.
25. Weinstein, S.J. and Ruschak, K.J., Coating flows. *Annu. Rev. Fluid Mech.*, 36, 29, 2004.
26. Ruschak, K.J., Coating flows. *Annu. Rev. Fluid Mech.*, 17, 65, 1985.
27. Spin Coating Theory, University of Louisville, Micro/Nano Technology Center, Louisville, Kentucky, Ocotober 2013, https://louisville.edu/micronano/files/documents/standard-operating-procedures/SpinCoatingInfo.pdf.
28. Spin Coating: A Guide to Theory and Techniques, https://www.ossila.com/pages/spin-coating. Retrieved in 02.01.2020.
29. Coating and Dispensing Technology, Spin Coating, https://www.keyence.com/ss/products/measure/sealing/coater-type/spin.jsp. Retrieved in 12.01.2020.
30. John, D.M. and Eric, B., Physical Properties of Sol-Gel Coatings. *J. Sol-Gel Sci. Technol.*, 19, 19–25, 2000.
31. Righini, G.C. and Chiappini, A., Glass optical waveguides: A review of fabrication techniques. *Opt. Eng.*, 53, 071819, 2014.
32. Sol-Gel Principles, http://www.tn.ifn.cnr.it/facilities/sol-gel-room/sol-gel-principles. Retrieved in 09.01.2020.
33. Ruohong, S. and Paul, C., Synthesis of Metal Oxide Nanostructures by Direct Sol-Gel Chemistry in Supercritical Fluids. *Chem. Rev.*, 112, 3057–3082, 2012.
34. Livage, J., Basic Principles of Sol-Gel Chemistry, in: *Sol-Gel Technologies for Glass Producers and Users*, M.A. Aegerter and M. Mennig (Eds.), Springer, Boston, MA, 2004.

35. Aksit, A., Onar, N., Ebeoglugil, M.F., Birlik, I., Celik, E., Ozdemir, I., Structural, electrical, and electromagnetic properties of cotton fabrics coated with polyaniline and polypyrrole. *J. Appl. Polym. Sci.*, 113, 1, 358, 2009.
36. Mendoza-Suarez, G., Rivas-Vazquez, L.P., Fuantes, A.F., Escalante-Garcia, J.I., Ayala-Valenzuela, O.E., Valdez, E., Investigation of Structural, Thermal and Magnetic properties of Strontium substituted Barium Hexaferrite Synthesized via co-precipitation Method. *Mater. Lett.*, 57, 868–872, 2002.
37. Yılmaz Atay, H. and Çelik, E., Barium Hexaferrite Reinforced Polymeric Dye Composite Coatings for Radar Absorbing Applications. *Polym. Compos.*, 35, 3, 602–610, 2014.
38. Yılmaz Atay, H., A Comparison on Radar Absorbing Properties of Nano and Micro-scale Barium Hexaferrite Powders Reinforced Polymeric Composites. *Mugla J. Sci. Technol.*, 2, 1, 88–92, 2016.
39. Brinker, C.J., Dip Coating Chapter 10, in: *Chemical Solution Deposition of Functional Oxide Thin Films*, T. Schneller (Eds.), Springer-Verlag, Wien, 2013.
40. Geffcken, W. and Berger, E., Verfahren zur Ä nderung des Reflexionsvermögens optischer Gläser. Deutsches Reichspatent, assigned to Jenaer Glaswerk Schott & Gen, Jena. 736, 411, 1939.
41. Brinker, C.J., Hurd, A.J., Frye, G.C., Schunk, P.R., Ashley, C.S., Sol-gel thin film formation. *J. Ceram. Soc. Jpn.*, 99, 862–877, 1991.
42. Scriven, L.E., Physics and application of dip-coating and spin-coating, in: *Materials Research Society symposium proceedings*, Better ceramics through chemistry III, vol 121, C.J. Brinker, D.E. Clark, D.R. Ulrich (Eds.), Materials Research Society, Pittsburgh, PA, pp. 717–729, 1988.
43. Minko, S., Grafting on Solid Surfaces: "Grafting to" and "Grafting from" Methods, in: *Polymer Surfaces and Interfaces*, M. Stamm (Ed.), Springer, Berlin, Heidelberg, 2008.
44. Ratner, B., New ideas in biomaterials science – A path to engineered biomaterials. *J. Biomed. Mater. Res.*, 27, 837, 1993.
45. Zhao, B. and Brittain, W.J., Polymer brushes: Surface-immobilized macromolecules. *Prog. Polym. Sci.*, 25, 677, 2000.
46. Uyama, Y., Kato, K., Ikada, Y., Surface modification of polymers by grafting. *Adv. Polym. Sci.*, 137, 1, 1998.

2

Fabrication Methods of Organic/ Inorganic Nanocomposite Coatings

Anandraj Mohan Kumar[1]*, Rajasekar Rathanasamy[1], Gobinath Velu Kaliyannan[1], Moganapriya Chinnasamy[1] and Sathish Kumar Palaniappan[2]

[1]*Department of Mechanical Engineering, Kongu Engineering College, Erode, India*
[2]*Department of Mining Engineering, Indian Institute of Technology Kharagpur, Kharagpur, India*

Abstract

In recent years, nanocomposite coatings have displays increasing interest due to its applications like electronic, defense, aeronautical and automotive industries. The nanocomposite coatings offer efficient and cost effectiveness coatings with excellent properties such as corrosion, wear, heat resistive properties, and higher mechanical strength. Compare to the conventional methods of composite coatings, it provides higher performance with ease of fabrication methods. In this chapter different fabrication techniques have discussed with appropriate schematic diagrams. The different organic/inorganic matrices and nanofillers have listing in table. The various nanocomposite coatings has discussed namely, sol-gel method, cold spray technique, chemical vapor deposition, physical vapor deposition, thermal spray coatings, electroplating deposition and electroless deposition methods.

Keywords: Nanocomposite, sol-gel method, cold spray technique, chemical vapor deposition, physical vapor deposition, thermal spray coatings, electroplating deposition, electroless deposition

Abbreviations

O/I	-	organic/inorganic
PEG	-	Polyethylene Glycol

**Corresponding author*: amohanmech2006@gmail.com

Inamuddin, Rajender Boddula, Mohd Imran Ahamed and Abdullah M. Asiri (eds.) Polymer Coatings: Technology and Applications, (21–40) © 2020 Scrivener Publishing LLC

PBZ	-	Rubber-modified polybenzoxazine
TMOS	-	Polymers containing reactive trimethoxysilyl
PVDF	-	Polyvinylidene fluoride
ETFE	-	Ethylene tetrafluoroethylene
PANi	-	Polyaniline
PPy	-	Polypyrrole
PVA	-	Polyvinyl Alcohol
CNT	-	Carbon Nanotubes
PEO	-	Poly ethylene oxide
PTFE	-	Polytetrafluoroethylene

2.1 Introduction

The blending of organic and inorganic constitutions is a most general approach to obtain surface coatings with optimal properties. The two elements with different or same essential properties are mixed at micro, nano, or molecular level. The composite coatings have appeared like general coatings with an opaque appearance. OINC stands for organic-inorganic nanocomposite coatings, which is nano scale hybrid coatings depend on organic matrix. In the earlier 15 years, research on organic-inorganic nano composite coatings has been increasing randomly in both academic and industrial fields. Many researchers published more papers related to nano composites coatings. The nanocomposite comprises two phases of materials, namely, matrix and fillers. Fillers are disseminated in matrix material [1]. Classification of nanocomposite is based on different types of filler materials and matrix materials. The nano structured fillers are classified into three types, namely, zero-dimensional, one-dimensional, and two-dimensional nano fillers. Zero-dimensional nano structured fillers are in the form of nano particles, and one-dimensional and two-dimensional nano structures are in the form of nano tubes and nano layers, respectively.

The matrix materials are categorized into two main groups, namely, organic and inorganic. The four major classifications of nanocomposite coatings are as follows: organic/inorganic (O/I) nano composite coating, organic/organic (O/O) nano composite coating, inorganic/organic nano (I/O) composite coating, and inorganic/inorganic (I/I) nano composite coatings.

The other name of organic based nano composite coatings called polymer based nano composite coatings. The different polymer materials are listed in Table 2.1. Various nano fillers utilized for nano composite coatings are listed in Table 2.2. Different organic nanocomposite materials are listed in Table 2.3.

Table 2.1 Different polymer-based matrix materials.

Matrix	**References**
Epoxy	[2–7]
Polyethylene Glycol (PEG)	[8–11]
Polyurethane	[12, 13]
Polyamic Acid and polyimide	[14]
Chitosan	[15, 16]
Rubber-modified polybenzoxazine (PBZ)	[17]
Polymers containing reactive trimethoxysilyl (TMOS)	[18]
Polyvinylidene fluoride (PVDF)	[19]
Pullulan	[20]
Fluoroacrylic polymer	[21, 22]
Ethylene tetrafluoroethylene (ETFE)	[23]
Polyaniline (PANi)	[24–26]
Polyacrylate	[3]
Polycarbonate	[27]
Polypyrrole (PPy)	[27–30]
Polyester	[31]
Polystyrene	[32]
Poly(n-vinyl carbazole)	[33]
Fluorinated Polysiloxane	[34]
Polyacrylic	[35]
UV-curable polymers	[36]
Polyamide	[37]
Polydimethylsiloxane	[38]
Polyvinyl Alcohol (PVA)	[39]

Table 2.2 Different inorganic nano fillers used in nanocomposite coatings.

Inorganic nano fillers	References
Clay	[4, 31]
Carbides	[40]
Borides	[41]
Metallic particles	[24, 42]
Nitrites	[43–50]
Oxides	[4, 51, 52]
Nano-diamond	[53, 54]
Carbon Nanotubes (CNT)	[55]

Table 2.3 Different organic nano fillers used in nanocomposite coatings.

Organic nano fillers	References
Poly ethylene oxide (PEO)	[56]
Polyaniline (PANi)	[23]
Polytetrafluoroethylene (PTFE)	[57–59]
Nano-cellulose and Cellulose Nano-crystal	[25, 36]

2.1.1 Transparency of Organic/Inorganic Nanocomposites

Equation (2.1) can be used to calculate transmissions of light through non-homogeneous coatings like organic-inorganic nano composite coatings. This equation is based on Rayleigh scattering theory, where, L denotes thickness of the coatings, r_p is the radius of the scattering element, ϕ_p volume fraction, λ denotes wavelength of the light, n_p indicates refractive indices of the inorganic phase and nm denotes refractive indices of the polymer matrix. Transparency of the organic inorganic nano composite is high since size of the inorganic phase is lesser compared to wavelength of light.

$$T = \exp\left[-\frac{3\phi_p L r_p^3}{4\lambda^4}\left(\frac{n_p}{n_m}-1\right)\right] \quad (2.1)$$

2.2 Fabrication Methods

Different fabrication methods of nanocomposite coatings are briefly discussed below; it includes sol-gel method, cold spray technique, chemical vapor deposition, physical vapor deposition, thermal spray coatings, electroplating deposition, and electroless deposition methods.

2.2.1 Sol-Gel Method

Sol-gel method is one of physical deposition methods. The nano composite coating is very much appropriate to attain good quality of coatings upto micro-scale level. This methods exhibit few draw backs that include crack formation and thickness restrictions. Further, heat treatment is also critical. If coating thickness is greater than the critical thickness value, tensile stresses may induce in coating material, and it leads to crack formation. The sol-gel chemistry has employed to generate I/O nanocomposite coatings. The nano composite coatings have formed in the form of inorganic nano fillers dispersed in an organic matrix. An extensive variety of oligomers and low molecular weight of organic compounds are used as organic matrix materials. Silanes and organic molecules have combined and form nanocomposite coatings which comprise nanoparticles. Inorganic and organic nano composites have been interlinking by coupling agents.

In sol-gel technology, different coatings methods exist, namely, flow coating, spray coating, spin coating, and dip coating technologies. The different stages of dip coatings are shown in Figure 2.1. The dip coating technique has performed under controlled atmosphere. First step of dip coating technique is dipping the substrate into the initial solution. At controlled speed, withdrawn of substrate has performed. The inner structure and film thickness of coating has influenced by withdrawal speed and evaporation conditions. The solution is uniformly spreading on the surface of the substrate by joined effect of capillarity and viscous drag. The effect of characteristic of the films has applied more on post heat treatment of the coated substrates.

2.2.2 Cold Spray Technique

Figure 2.2 shows the principle of cold spray technique for preparation of nano composite. Copper and CNT, nano composite coatings, have fabricated on the copper plate of thickness 0.3 mm. The cold spray technique system consist of different parts, namely, pressure gas supply, heater, powder feeder, de laval-type nozzle, xy axis movable table, and

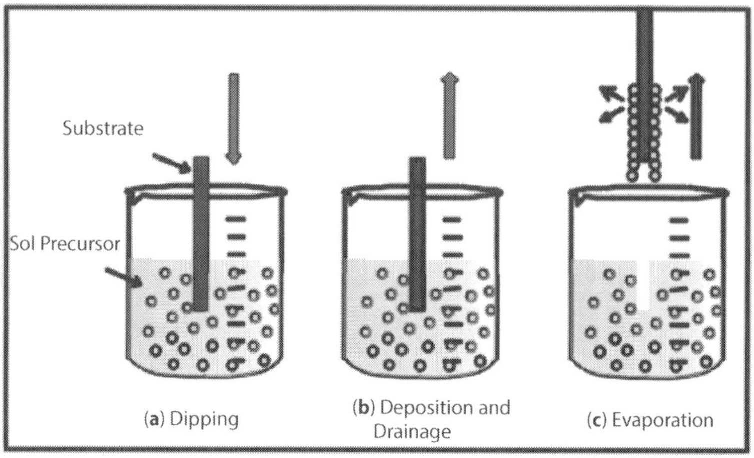

Figure 2.1 Sequential stages of the dip-coating technique for thin film deposition: (a) the substrate has dipped and immersed in the sol precursor, (b) the substrate has withdrawn at constant speed rate, and (c) solvent evaporation yields to the gelation of the layer [60].

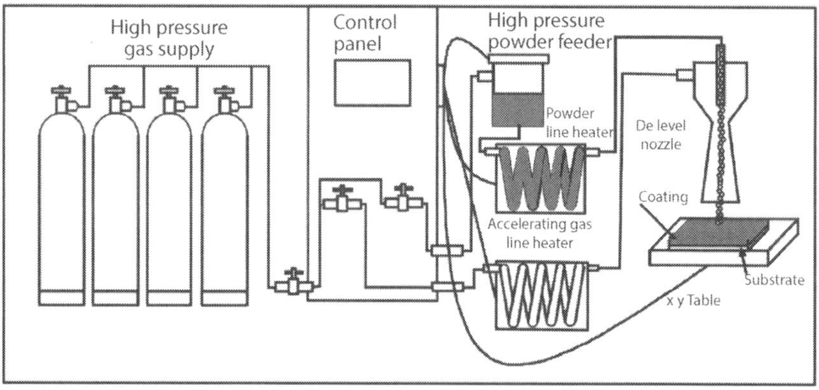

Figure 2.2 Principle of cold spray technique for preparation of nanocomposite [62].

control panel. The feed disc speed 7.5 rpm and mass expended in feeder disc determine feed rate of powder. This method of fabrication of coating materials has performed with low melting temperature of sprayed materials. The cold spray method of fabrication of coating material is not a conventional thermal spray coating methods like plasma, gas flame, and detonation spraying. Since the process has performed at low temperature, it avoids oxidation, decomposition, and phase changing during the process. The attained coatings having low oxygen concentration and low porosity compared to conventional spray coating methods. In other words, this coating method has high strength and higher interfacial bonding strength. This method is also applicable for the metallic matrix materials like copper [41, 55], aluminum [53, 61], cobalt [40], or alloy matrix [43–45], and nanofillers, namely, carbide, carbon nano tubes, and diamond [40, 41, 43–45, 53, 55, 61].

2.2.3 Chemical Vapor Deposition

The chemical vapor deposition process initiates with materials in tanks (red, blue, and cyan colors). These materials are utilizing to building blocks of the required nanocomposite coatings on the surface of substrate. The materials are coated by melting or reducing the pressure and then transferred to vacuum chamber, which containing materials to be coated on substrate. Speed up the process of converting the monomers to polymers by linking up in chains on the surface of materials. Figure 2.3 shows principle of chemical vapor deposition method. This method can be used for inorganic to inorganic nano composite coatings [63–65] and organic to inorganic nano composite coatings (platinum (II) hexafluoroacetylacetonate as precursors) [66]. Aerosol assisted chemical vapor deposition

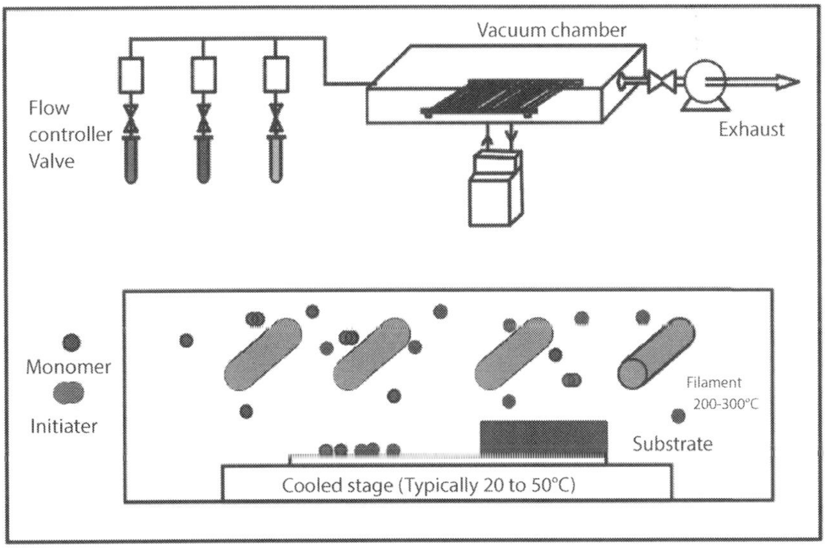

Figure 2.3 Principle of chemical vapor deposition method.

method [67] can be used to increase quality of coatings. A specific advantage of this method is used to fabricate high quality films.

2.2.4 Physical Vapor Deposition

The physical vapors deposition method includes thermal evaporation [68], ion implantation [69, 70], laser ablation [71], and laser assisted deposition [72]. This technique is suitable for inorganic and inorganic nanocomposite coatings [73–78] and organic and inorganic nanocomposite coatings. Schild [72] has introduced the nanocomposite coatings by aerosol assisted plasma deposition. Figure 2.4 shows basic principle of physical vapor deposition method. The physical vapor deposition process has utilized to fabricate nanocomposite coatings in the range from nanometers to micrometers. This method comprising three basic steps has listed below:

- ✓ Evaporation of the materials from a solid phase has been performing by gaseous plasma.
- ✓ Vacuum or partial vacuum has transferred to the surface of the substrate.
- ✓ Fabricate thin film on the surface of substrate by condensation.

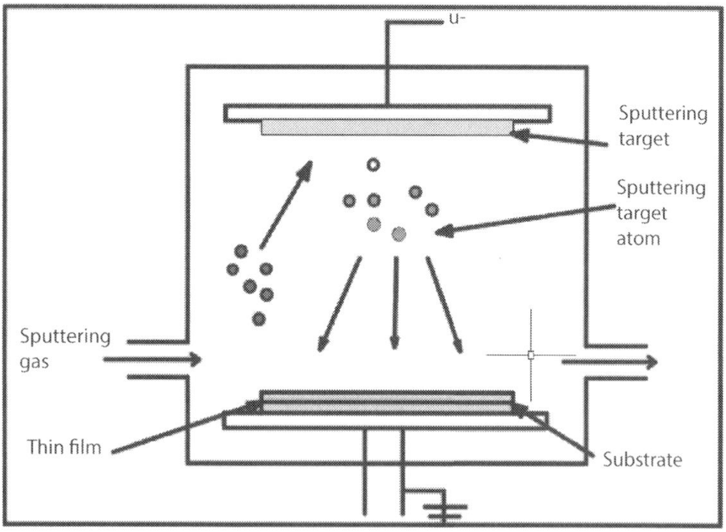

Figure 2.4 Principle of physical vapor deposition method.

The most generally utilized coating methods in physical vapor deposition methods are sputtering and evaporation methods. Sputtering is one of plasma assisted coating technique in which vapor is generated by bombardment of gaseous ions. The evaporation method is another method of generating vapor by heating materials using suitable method in vapor. The condensation mechanism has used to deposits coating material on the surface of the substrate.

2.2.5 Thermal Spray Coating

Different materials used in thermal spray coating process, namely, polymers, metallic, ceramic, and cermet. The materials have transferred to a torch or gun in the form of powders or wire or rod with above or equal to molten temperature of materials. The molten state materials are projected in a gas stream on the substrate surface. The quality of the coatings has determined by different parameters, namely, surface roughness, toughness, surface hardness, adhesive strength, coating density, and oxide contents. The most significant advantage of thermal spray coatings are wide range of coating materials and substrate materials. The coatings can be removed without any damage on the substrate and recoating can performed on the substrate surface. One of disadvantage of this method is that small or tiny and complicated shapes are difficult to coat on the substrate surface. Figure 2.5 shows general diagram of the thermal spray coating method.

2.2.6 Electrodeposition Method

The electrodeposition method can be used to fabricate nano composite coatings which include organic nanofillers, namely, PEO and PTFE [56, 57, 79] and inorganic matrices [80–84] or organic matrices [85]. Figure 2.6 shows principle of electrodeposition method.

Many papers has summarized the electrochemical method of nano composite coatings [1, 86–101].

The different organic matrices, namely, polyelectrolytes and polyethyleneimine, are electrochemically deposited on the inorganic substrate such as metal ions and ceramic nanomaterials. Compared to traditional methods of coatings, nano composite coatings displays higher hardness and heat resistive property due to presence of nano particles on the outer boundaries, it also restricts movements of ions and also recrystallization at higher temperatures.

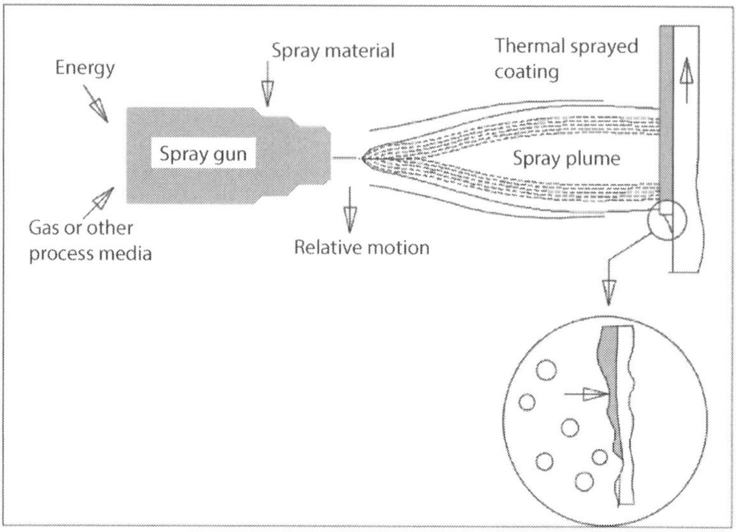

Figure 2.5 General schematic diagram of thermal spray coating processes.

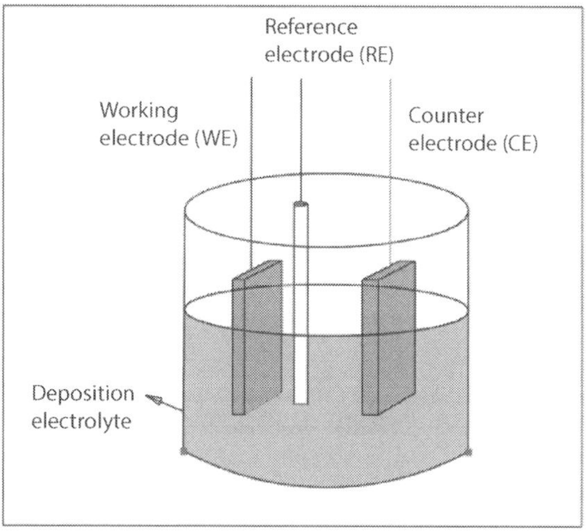

Figure 2.6 Principle of electrodeposition method.

2.2.7 Electroless Coating Method

Electroless deposition is a chemical reduction technique, in which metals like aluminium, iron and silver are coated on the fiber surface as metal ions in an aqueous solution without use of electrical energy. In this process,

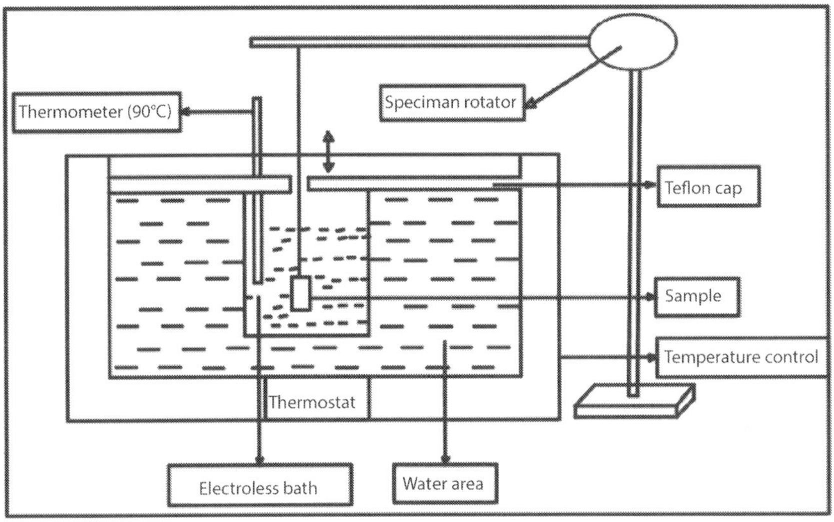

Figure 2.7 Principle of electroless coatings experimental setup.

metal ions act as electron acceptor and metal act as electron donors. Figure 2.7 shows the principle of electroless coating experimental setup. Fibers treatments by physical or chemical methods are necessary for good interaction between the electroplated layers on the surfaces. Electroless deposition process is one of most accepted methods of fabricating nano composite coatings due to its excellent corrosion [102–104], wear resistance [105, 106], and also different applications like soldering and welding applications.

2.3 Conclusions

Nowadays, nanocomposite coatings become cost effective without compromising performance. The nanocomposite coating fields are expected to cover larger application in future. The composite coatings provide enhancing properties of the materials like antibacterial, scratch resistive coatings, reflective coatings, and fire retardation properties. Classification of different coatings methods has been summarized with different coating techniques. The transparency of organic and inorganic nanocomposites has discussed. In future, smart coatings can be used as multifunctional coatings including photovoltaic cell, solar-based thermal applications, and solar-to-fuel conversion.

References

1. Nguyen-Tri, P., Nguyen, T.A., Carriere, P., Ngo Xuan, C., Nanocomposite coatings: Preparation, characterization, properties, and applications. *Int. J. Corros.*, 2018, 19, 2018.
2. Kim, H., Ra, H.N., Kim, M., Kim, H.G., Kim, S., Enhancement of barrier properties by wet coating of epoxy-ZrP nanocomposites on various inorganic layers. *Prog. Org. Coat.*, 108, 25–29, 2017.
3. Pourhashem, S., Vaezi, M.R., Rashidi, A., Bagherzadeh, M., Exploring corrosion protection properties of solvent based epoxy-graphene oxide nanocomposite coatings on mild steel. *Corros. Sci.*, 115, 78–92, 2017.
4. Shi, X., Nguyen, T.A., Suo, Z., Liu, Y., Avci, R., Effect of nanoparticles on the anticorrosion and mechanical properties of epoxy coating. *Surf. Coat. Technol.*, 204, 3, 237–245, 2009.
5. Nguyen, T. A., Protection of Steel Rebar in Salt-Contaminated Cement Mortar Using Epoxy Nanocomposite Coatings. *Int. J. Electrochemistry.*, 2018, 10, 2018.
6. Vijayan P, P., Hany El-Gawady, Y.M., Al-Maadeed, M.C., Halloysite nanotube as multifunctional component in epoxy protective coating. *Ind. Eng. Chem. Res.*, 55, 42, 11186–11192, 2016.
7. Yeh, J.-M., Huang, H.-Y., Chen, C.-L., Su, W.-F., Yu, Y.-H., Siloxane-modified epoxy resin–clay nanocomposite coatings with advanced anticorrosive properties prepared by a solution dispersion approach. *Surf. Coat. Technol.*, 200, 8, 2753–2763, 2006.
8. Schilp, S., Rosenhahn, A., Pettitt, M.E., Bowen, J., Callow, M.E., Callow, J.A., Grunze, M., Physicochemical properties of (ethylene glycol)-containing self-assembled monolayers relevant for protein and algal cell resistance. *Langmuir.*, 25, 17, 10077–10082, 2009.
9. Lowe, S., O'Brien-Simpson, N.M., Connal, L., Antibiofouling polymer interfaces: Poly (ethylene glycol) and other promising candidates. *Polym. Chem.*, 6, 2, 198–212, 2015.
10. Pelaz, B., del Pino, P., Maffre, P., Hartmann, R., Gallego, M., Rivera-Fernandez, S., Parak, W., Surface functionalization of nanoparticles with polyethylene glycol: Effects on protein adsorption and cellular uptake. *ACS Nano.*, 9, 7, 6996–7008, 2015.
11. Wang, W.-Y., Shi, J.-Y., Wang, J.-L., Li, Y.-L., Gao, N.-N., Liu, Z.-X., Lian, W.-T., Preparation and characterization of PEG-g-MWCNTs/PSf nanohybrid membranes with hydrophilicity and antifouling properties. *RSC Adv.*, 5, 103, 84746–84753, 2015.
12. Davis, A., Yeong, Y.H., Steele, A., Bayer, I.S., Loth, E., Superhydrophobic nanocomposite surface topography and ice adhesion. *ACS Appl. Mater. Interfaces.*, 6, 12, 9272–9279, 2014.
13. Sung, L.-P., Comer, J., Forster, A.M., Hu, H., Floryancic, B., Brickweg, L., Scratch behavior of nano-alumina/polyurethane coatings. *J. Coat. Technol. Res.*, 5, 4, 419–430, 2008.

14. He, S., Lu, C., Zhang, S., Facile and efficient route to polyimide-TiO2 nano-composite coating onto carbon fiber. *ACS Appl. Mater. Interfaces,* 3, 12, 4744–4750, 2011.
15. Al-Naamani, L., Dobretsov, S., Dutta, J., Burgess, J., Chitosan-zinc oxide nano-composite coatings for the prevention of marine biofouling. *Chemosphere,* 168, 408–417, 2017.
16. Tang, Y., Hu, X., Zhang, X., Guo, D., Zhang, J., Kong, F., Chitosan/titanium dioxide nanocomposite coatings: Rheological behavior and surface application to cellulosic paper. *Carbohydr. Polym.,* 151, 752–759, 2016.
17. Caldona, E.B., De Leon, A.C.C., Thomas, P.G., Naylor, D.F., III, Pajarito, B.B., Advincula, R.C., Superhydrophobic rubber-modified polybenzoxazine/SiO2 nanocomposite coating with anticorrosion, anti-ice, and superoleophilicity properties. *Ind. Eng. Chem. Res.,* 56, 6, 1485–1497, 2017.
18. Dong, H., Ye, P., Zhong, M., Pietrasik, J., Drumright, R., Matyjaszewski, K., Superhydrophilic Surfaces via Polymer–SiO2 Nanocomposites. *Langmuir.,* 26, 19, 15567–15573, 2010.
19. Toor, A., So, H., Pisano, A., Improved dielectric properties of polyvinylidene fluoride nanocomposite embedded with poly (vinylpyrrolidone)-coated gold nanoparticles. *ACS Appl. Mater. Interfaces,* 9, 7, 6369–6375, 2017.
20. Introzzi, L., Blomfeldt, T.O., Trabattoni, S., Tavazzi, S., Santo, N., Schiraldi, A., Farris, S.J.L., Ultrasound-assisted pullulan/montmorillonite bionanocom posite coating with high oxygen barrier properties. *Langmuir.,* 28, 30, 11206–11214, 2012.
21. Steele, A., Bayer, I., Loth, E., Inherently superoleophobic nanocomposite coatings by spray atomization. *Nano Lett.,* 9, 1, 501–505, 2008.
22. Asthana, A., Maitra, T., Büchel, R., Tiwari, M.K., Poulikakos, D., Multifunctional superhydrophobic polymer/carbon nanocomposites: Graphene, carbon nanotubes, or carbon black? *ACS Appl. Mater. Interfaces,* 6, 11, 8859–8867, 2014.
23. Yuan, R., Wu, S., Yu, P., Wang, B., Mu, L., Zhang, X., Superamphiphobic and electroactive nanocomposite toward self-cleaning, antiwear, and anticorrosion coatings. *ACS Appl. Mater. Interfaces,* 8, 19, 12481–12493, 2016.
24. Bogdanovic, U., Vodnik, V., Mitric, M., Dimitrijevic, S., Skapin, S.D., Zunic, V., Nanomaterial with High Antimicrobial Efficacy Copper/Polyaniline Nanocomposite. *ACS Appl. Mater. Interfaces,* 7, 3, 1955–1966, 2015.
25. Zhang, S., Sun, G., He, Y., Fu, R., Gu, Y., Chen, S., Preparation, characterization, and electrochromic properties of nanocellulose-based polyaniline nanocomposite films. *ACS Appl. Mater. Interfaces,* 9, 19, 16426–16434, 2017.
26. Shabani-Nooshabadi, M., Ghoreishi, S.M., Jafari, Y., Kashanizadeh, N.J., Electrodeposition of polyaniline-montmorrilonite nanocomposite coatings on 316L stainless steel for corrosion prevention. *J. Polym. Res.,* 21, 4, 416, 2014.
27. Fujii, S., Matsuzawa, S., Hamasaki, H., Nakamura, Y., Bouleghlimat, A., Buurma, Polypyrrole–palladium nanocomposite coating of micrometer-

sized polymer particles toward a recyclable catalyst. *Langmuir.*, 28, 5, 2436–2447, 2012.
28. Yoo, S., Kim, C., Enhancement of the meltdown temperature of a lithium ion battery separator via a nanocomposite coating. *Ind. Eng. Chem. Res.*, 48, 22, 9936–9941, 2009.
29. Fujii, S., Matsuzawa, S., Nakamura, Y., Ohtaka, A., Teratani, T., Akamatsu, K., Nawafune, H., Synthesis and characterization of polypyrrole–palladium nanocomposite-coated latex particles and their use as a catalyst for suzuki coupling reaction in aqueous media. *Langmuir.*, 26, 9, 6230–6239, 2010.
30. Farghaly, A.A. and Collinson, M., Mesoporous hybrid polypyrrole-silica nanocomposite films with a strata-like structure. *Langmuir.*, 32, 23, 5925–5936, 2016.
31. Golgoon, A., Aliofkhazraei, M., Toorani, M., Moradi, M., Rouhaghdam, A., Corrosion and wear properties of nanoclay-polyester nanocomposite coatings fabricated by electrostatic method. *Procedia Mater. Sci.*, 11, 536–541, 2015.
32. Hou, W. and Wang, Q., UV-driven reversible switching of a polystyrene/titania nanocomposite coating between superhydrophobicity and superhydrophilicity. *Langmuir.*, 25, 12, 6875–6879, 2009.
33. Cui, K.M., Tria, M.C., Pernites, R., Binag, C.A., Advincula, R., PVK/MWNT electrodeposited conjugated polymer network nanocomposite films. *ACS Appl. Mater. Interfaces.*, 3, 7, 2300–2308, 2011.
34. Ding, X., Zhou, S., Gu, G., Wu, L., A facile and large-area fabrication method of superhydrophobic self-cleaning fluorinated polysiloxane/TiO 2 nanocomposite coatings with long-term durability. *J. Mater. Chem.*, 21, 17, 6161–6164, 2011.
35. Sajjadi, S.A., Avazkonandeh-Gharavol, M.H., Zebarjad, S.M., Mohammadtaheri, M., Abbasi, M., Mossaddegh, K., A comparative study on the effect of type of reinforcement on the scratch behavior of a polyacrylic-based nanocomposite coating. *J. Coat. Technol. Res.*, 10, 2, 255–261, 2013.
36. Kaboorani, A., Auclair, N., Riedl, B., Landry, V., Mechanical properties of UV-cured cellulose nanocrystal (CNC) nanocomposite coating for wood furniture. *Prog. Org. Coat.*, 104, 91–96, 2017.
37. Behler, K.D., Stravato, A., Mochalin, V., Korneva, G., Yushin, G., Gogotsi, Y., Nanodiamond-polymer composite fibers and coatings. *ACS Nano.*, 3, 2, 363–369, 2009.
38. Facio, D.S., Mosquera, M. J., Simple strategy for producing superhydrophobic nanocomposite coatings *in situ* on a building substrate. *ACS Appl. Mater. Interfaces,* 5, 15, 7517–7526, 2013.
39. Habouti, S., Kunstmann-Olsen, C., Hoyland, J.D., Rubahn, H.-G., Es-Souni, M., *In situ* ZnO–PVA nanocomposite coated microfluidic chips for biosensing. *Appl. Phys. A.*, 115, 2, 645–649, 2014.

40. Li, C.-J., Yang, G.-J., Gao, P.-H., Ma, J., Wang, Y.-Y., Li, C.-X., Characterization of nanostructured WC-Co deposited by cold spraying. *J. Therm. Spray Technol.*, 16, 5–6, 1011–1020, 2007.
41. Kim, J., Kwon, Y., Lomovsky, O., Dudina, D., Kosarev, V., Klinkov, S., Cold spraying of *in situ* produced TiB2–Cu nanocomposite powders. *Compos. Sci. Technol.*, 67, 11–12, 2292–2296, 2007.
42. Toledano, R. and Mandler, D., Electrochemical Codeposition of Thin Gold Nanoparticles/Sol–Gel Nanocomposite Films. *Chem. Mater.*, 22, 13, 3943–3951, 2010.
43. Luo, X.-T. and Li, C.-J., Thermal stability of microstructure and hardness of cold-sprayed cBN/NiCrAl nanocomposite coating. *J. Therm. Spray Technol.*, 21, 3–4, 578–585, 2012.
44. Luo, X.-T., Yang, G.-J., Li, C.-J., Kondoh, K., High strain rate induced localized amorphization in cubic BN/NiCrAl nanocomposite through high velocity impact. *Scr. Mater.*, 65, 7, 581–584, 2011.
45. Luo, X.-T., Li, C.-J., Design, Large sized cubic BN reinforced nanocomposite with improved abrasive wear resistance deposited by cold spray. *Mater. Des.*, 83, 249–256, 2015.
46. Mišina, M., Musil, J., Kadlec, S., Composite TiN–Ni thin films deposited by reactive magnetron sputter ion-plating. *Surf. Coat. Technol.*, 110, 3, 168–172, 1998.
47. Musil, J., Karvankova, P., Kasl, J., Hard and superhard Zr–Ni–N nanocomposite films. *Surf. Coat. Technol.*, 139, 1, 101–109, 2001.
48. Musil, J., Leipner, I., Kolega, M., Nanocrystalline and nanocomposite CrCu and CrCu–N films prepared by magnetron sputtering. *Surf. Coat. Technol.*, 115, 1, 32–37, 1999.
49. Zeman, P., Čerstvý, R., Mayrhofer, P., Mitterer, C., Musil, J., Structure and properties of hard and superhard Zr–Cu–N nanocomposite coatings. *Mater. Sci. Eng. A.*, 289, 1–2, 189–197, 2000.
50. Musil, J., Zeman, P., Hrubý, H., Mayrhofer, P., ZrN/Cu nanocomposite film—A novel superhard material. Surf. *Coat. Technol.*, 120, 179–183, 1999.
51. Banovic, S., Barmak, K., Marder, A., Characterization of single and discretely-stepped electro-composite coatings of nickel-alumina. *J. Mater. Sci.*, 34, 13, 3203–3211, 1999.
52. Dong, Y., Lin, P., Wang, H., Electroplating preparation of Ni–Al2O3 graded composite coatings using a rotating cathode. *Surf. Coat. Technol.*, 200, 11, 3633–3636, 2006.
53. Woo, D., Sneed, B., Peerally, F., Heer, F., Brewer, L., Hooper, J., Osswald, S., Synthesis of nanodiamond-reinforced aluminum metal composite powders and coatings using high-energy ball milling and cold spray. *Carbon*, 63, 404–415, 2013.
54. Musil, J., Hard and superhard nanocomposite coatings. *Surf. Coat. Technol.*, 125, 1–3, 322–330, 2000.

55. Cho, S., Takagi, K., Kwon, H., Seo, D., Ogawa, K., Kikuchi, K., Multi-walled carbon nanotube-reinforced copper nanocomposite coating fabricated by low-pressure cold spray process. *Surf. Coat. Technol.*, 206, 16, 3488–3494, 2012.
56. Koleva, D., Boshkov, N., Raichevski, G., Veleva, L., Electrochemical corrosion behaviour and surface morphology of electrodeposited zinc, zinc–cobalt and their composite coatings. *Trans. Inst. Met. Finish.*, 83, 4, 188–193, 2005.
57. Wang, F., Arai, S., Endo, M., Electrochemical preparation and characterization of nickel/ultra-dispersed PTFE composite films from aqueous solution. *Mater. Trans.*, 45, 4, 1311–1316, 2004.
58. Sharma, A., Singh, A. J., Performance, Electroless Ni-P-PTFE-Al 2 O 3 Dispersion Nanocomposite Coating for Corrosion and Wear Resistance. *J. Mater. Eng. Perform.*, 23, 1, 142–151, 2014.
59. Liu, C. and Zhao, Q.J.L., Influence of surface-energy components of Ni–P–TiO2–PTFE nanocomposite coatings on bacterial adhesion. *Langmuir.*, 27, 15, 9512–9519, 2011.
60. Varela, A.I.G., Aymerich, M., García, D.N., Martín, Y.C., de Beule, P.A., Álvarez, E., Flores-Arias, M. T., Sol-Gel Glass Coating Synthesis for Different Applications: Active Gradient-Index Materials, Microlens Arrays and Biocompatible Channels. 231, 2017.
61. Sanpo, N., Lu, T.M., Cheang, P., Anti-bacterial property of cold sprayed ZnO-Al coating. *Paper presented at the 2008 International Conference on BioMedical Engineering and Informatics*, 2008.
62. Pialago, E.J.T. and Park, C. W., Cold spray deposition characteristics of mechanically alloyed Cu-CNT composite powders. *Appl. Surf. Sci.*, 308, 63–74, 2014.
63. Neerinck, D., Persoone, P., Sercu, M., Goel, A., Kester, D., Bray, D., Diamond-like nanocomposite coatings (aC: H/a-Si: O) for tribological applications. *Diam. Relat. Mater.*, 7, 2–5, 468–471, 1998.
64. Veprek, S., Haussmann, M., Reiprich, S., Shizhi, L., Dian, J., Novel thermodynamically stable and oxidation resistant superhard coating materials. *Surf. Coat. Tech.*, 86, 394–401, 1996.
65. Männling, H.-D., Patil, D., Moto, K., Jilek, M., Veprek, S., Thermal stability of superhard nanocomposite coatings consisting of immiscible nitrides. *Surf. Coat. Tech.*, 146, 263–267, 2001.
66. Wood, T., Andrews, H., Thompson, R., Badyal, J., Ion-and electron-conducting platinum-polymer nanocomposite thin films. *ACS Appl. Mater. Interfaces.*, 4, 12, 6747–6751, 2012.
67. Hou, X., Choy, K.-L., Brun, N., Serín, V., Nanocomposite coatings codeposited with nanoparticles using aerosol-assisted chemical vapour deposition. *J. Nanomater.*, 1, 2013, 2013.
68. Mattox, D.M., Deposition Processes, in: *The Foundations of Vacuum Coating Technology*, pp. 11–33, Springer, 2003.

69. Meldrum, A., Boatner, L., White, C., Atoms, Nanocomposites formed by ion implantation: Recent developments and future opportunities. *Nucl. Instrum. Methods Phys. Res. A.*, 178, 1–4, 7–16, 2001.
70. Köstenbauer, H., Fontalvo, G.A., Mitterer, C., Keckes, J., Tribological properties of TiN/Ag nanocomposite coatings. *Tribol. Lett.*, 30, 1, 53–60, 2008.
71. Willmott, P. and Huber, J., Pulsed laser vaporization and deposition. *Rev. Mod. Phys.*, 72, 1, 315, 2000.
72. Voevodin, A. and Zabinski, J., Load-adaptive crystalline–amorphous nanocomposites. *J. Mater. Sci.*, 33, 2, 319–327, 1998.
73. Liu, X., Iamvasant, C., Liu, C., Matthews, A., Leyland, A., CrCuAgN PVD nanocomposite coatings: Effects of annealing on coating morphology and nanostructure. *Appl. Surf. Sci.*, 392, 732–746, 2017.
74. Pei, Y., Chen, C., Shaha, K., De Hosson, J.T.M., Bradley, J., Voronin, S., Čada, M., Microstructural control of TiC/aC nanocomposite coatings with pulsed magnetron sputtering. *Acta. Mater.*, 56, 4, 696–709, 2008.
75. Ribeiro, E., Malczyk, A., Carvalho, S., Rebouta, L., Fernandes, J., Alves, E., Effects of ion bombardment on properties of dc sputtered superhard (Ti, Si, Al) N nanocomposite coatings. *Surf. Coat. Technol.*, 151, 515–520, 2002.
76. Yao, S.-H., Su, Y.-L., Kao, W.-H., Cheng, K.-W., Evaluation on wear behavior of Cr–Ag–N and Cr–W–N PVD nanocomposite coatings using two different types of tribometer. *Surf. Coat. Technol.*, 201, 6, 2520–2526, 2006.
77. Zhang, S., Fu, Y., Du, H., Zeng, X., Liu, Y., Magnetron sputtering of nanocomposite (Ti, Cr) CN/DLC coatings. *Surf. Coat. Technol.*, 162, 1, 42–48, 2003.
78. Aouadi, S.M., Paudel, Y., Luster, B., Stadler, S., Kohli, P., Muratore, C., Voevodin, A., Adaptive Mo 2 N/MoS 2/Ag tribological nanocomposite coatings for aerospace applications. *Tribol. Lett.*, 29, 2, 95–103, 2008.
79. Li, J., Sun, Y., Sun, X., Qiao, J., Mechanical and corrosion-resistance performance of electrodeposited titania–nickel nanocomposite coatings. *Surf. Coat. Technol.*, 192, 2–3, 331–335, 2005.
80. Dong, S.-y., Xu, B.-s., Du, L.-Z., Yang, H., Sand-wear resistance of brush electroplated nanocomposite coating in oil and its application to remanufacturing. *J. Cent. South. Univ. T.*, 12, 2, 176–180, 2005.
81. Parhizkar, N., Dolati, A., Aghababazadeh, R., Lalegani, Z., Electrochemical deposition of Ni–TiN nanocomposite coatings and the effect of sodium dodecyl sulphate surfactant on the coating properties. *Bull. Mater. Sci.*, 39, 4, 1021–1027, 2016.
82. Pradhan, A.K., Das, S.J.M., Pulse reverse electrodeposition of Cu-SiC nanocomposite coating: Effects of surfactants and deposition parameters. *Metall. Mater. Trans.*, 45, 12, 5708–5720, 2014.
83. Zhao, C., Yao, Y., He, L., Electrodeposition and characterization of Ni-W/ZrO2 nanocomposite coatings. *Bull. Mater. Sci.*, 37, 5, 1053–1058, 2014.

84. Benea, L.J.M., Electrochemical Impedance Spectroscopy and Corrosion Behavior of Co/CeO 2 Nanocomposite Coatings in Simulating Body Fluid Solution. *Metall. Mater. Trans. A.*, 44, 2, 1114–1122, 2013.
85. Weaver, C.L., LaRosa, J.M., Luo, X., Cui, X., Electrically controlled drug delivery from graphene oxide nanocomposite films. *ACS Nano.*, 8, 2, 1834–1843, 2014.
86. Chen, J.-Y., Zhou, P.-J., Li, J.-L., Li, S.-Q., Depositing Cu2O of different morphology on chitosan nanoparticles by an electrochemical method. *Carbohydr. Polym.*, 67, 4, 623–629, 2007.
87. Chen, J.-Y., Zhou, P.-J., Li, J.-L., Wang, Y., Studies on the photocatalytic performance of cuprous oxide/chitosan nanocomposites activated by visible light. *Carbohydr. Polym.*, 72, 1, 128–132, 2008.
88. Chen, Q., Zhao, L., Li, C., Shi, G.J., Electrochemical fabrication of a memory device based on conducting polymer nanocomposites. *J. Phys. Chem.*, 111, 49, 18392–18396, 2007.
89. Gajendran, P., Saraswathi, R., A., Polyaniline-carbon nanotube composites. *J. Phys. Chem. C.*, 80, 11, 2377–2395, 2008.
90. Lepiller, C., Poissonnet, S., Bonnaillie, P., Giunchi, G., Legendre, F.J., Electrochemical codeposition of copper-strontium hydroxide films from dimethylsulfoxide solutions. *J. Electrochem. Soc.*, 151, 2, D13–D23, 2004.
91. Lu, S.-Y. and Lin, I.-H.J., Characterization of polypyrrole-CdSe/CdTe nanocomposite films prepared with an all electrochemical deposition process. *J. Phys. Chem. B.*, 107, 29, 6974–6978, 2003.
92. Luo, X.-L., Xu, J.-J., Wang, J.-L., Chen, H.-Y., Electrochemically deposited nanocomposite of chitosan and carbon nanotubes for biosensor application., *Chem. Commun.*, 16 2169–2171, 2005.
93. Omastová, M. and Mičušík, M., Polypyrrole coating of inorganic and organic materials by chemical oxidative polymerisation. *Chem Pap.*, 66, 5, 392–414, 2012.
94. Pang, X., Zhitomirsky, I., Electrodeposition of composite hydroxyapatite–chitosan films. *Mater. Chem. Phys.*, 94, 2–3, 245–251, 2005.
95. Pang, X., Zhitomirsky, I., Electrodeposition of hydroxyapatite–silver–chitosan nanocomposite coatings. *Surf. Coat. Technol.*, 202, 16, 3815–3821, 2008.
96. Peipmann, R., Thomas, J., Bund, A., Electrocodeposition of nickel–alumina nanocomposite films under the influence of static magnetic fields. *Electrochim. Acta*, 52, 19, 5808–5814, 2007.
97. Tsai, Y.-C. and Hong, Y.-H., Electrochemical deposition of platinum nanoparticles in multiwalled carbon nanotube–Nafion composite for methanol electrooxidation. *J. Appl. Electrochem.*, 12, 10, 1293–1299, 2008.
98. Walcarius, A., Sibottier, E., Electrochemically-Induced Deposition of Amine-Functionalized Silica Films on Gold Electrodes and Application to Cu (II) Detection in (Hydro) Alcoholic Medium. *Electroanalysis*, 17, 19, 1716–1726, 2005.

99. Zhitomirsky, I.J., Electrophoretic deposition of organic–inorganic nanocomposites. *J. Mater. Sci.*, 41, 24, 8186–8195, 2006.
100. Baibarac, M., Gómez-Romero, P.J., Nanocomposites based on conducting polymers and carbon nanotubes: From fancy materials to functional applications. *J. Nanosci. Nanotechnol.*, 6, 2, 289–302, 2006.
101. Shao, I., Vereecken, P., Cammarata, R., Searson, P., Kinetics of particle codeposition of nanocomposites. *J. Electrochem. Soc.*, 149, 11, C610–C614, 2002.
102. Liu, J., Zhu, X., Sudagar, J., Diao, W., Yu, S.J.M.t., Increased corrosion resistance of closed-cell aluminum foams by electroless Ni-P coatings. *Mater. Trans.*, 52, 12, 2282–2284, 2011.
103. Sudagar, J., Bi, G., Jiang, Z., Li, G., Jiang, Q., Lian, J., Electrochemical polarization behaviour of electroless Ni-P deposits with different chromium-free pre-treatment on magnesium alloy. *Int. J. Electrochem. Sci.*, 6, 7, 2767–2788, 2011.
104. Sudagar, J., Venkateswarlu, K., Lian, J.J.J.o.M.E., Performance, Dry sliding wear properties of a 7075-T6 aluminum alloy coated with Ni-P (h) in different pretreatment conditions. *J. Mater. Eng. Perform.*, 19, 6, 810–818, 2010.
105. Tamilarasan, T., Rajendran, R., Sanjith, U., Rajagopal, G., Sudagar, J.J.W., Wear and scratch behaviour of electroless Ni-P-nano-TiO2: Effect of surfactants. *Wear.*, 346, 148–157, 2016.
106. Sudagar, J., Lian, J., Sha, W.J., Compounds, Electroless nickel, alloy, composite and nano coatings–A critical review. *J. Alloys Compd.*, 571, 183–204, 2013.

3

Dry Powder Coating Techniques and Role of Force Controlling Agents in Aerosol

Piyush P. Mehta[1]*, Atmaram P. Pawar[2], Kakasaheb R. Mahadik[3], Shivajirao S. Kadam[4] and Vividha Dhapte-Pawar[2]

[1]*Department of Quality Assurance, Poona College of Pharmacy, Bharati Vidyapeeth University, Pune, India*
[2]*Department of Pharmaceutics, Poona College of Pharmacy, Bharati Vidyapeeth University, Pune, India*
[3]*Department of Pharmaceutical Chemistry, Poona College of Pharmacy, Bharati Vidyapeeth University, Pune, India*
[4]*Bharati Vidyapeeth Bhavan, Bharati Vidyapeeth University, Pune, India*

Abstract
Presently, diverse methods have been well studied for the design and fabrication of versatile powder and particles. However, these techniques have few limitations on account of the need for complex manufacturing setup, the extensive volume of starting materials, and the use of toxic reagents which have a decisive impact on our ecosystem. Therefore, to surpass such issues, material scientists have explored single-step, eco-friendly dry powder coating techniques. Dry powder coating assists to develop powders and particles with miscellaneous functionalities. The present chapter underlines the importance of dry powder coating technique in the field of aerosols particularly dry powder inhaler. This chapter also contains a section detailing the process analytical techniques, the force controlling agents, advancement in medical device coating, and the use of numerical simulation in coating technology. This chapter is the first comprehensive account of dry powder coating techniques in the discipline of dry powder inhalers. Therefore, it can be of likely significance for both the industry as well as academia.

Keywords: Dry powder inhalers, aerosol, dry powder coating, mechanofusion, force controlling agents, cohesive-adhesive balance, deagglomeration, magnesium stearate

**Corresponding author*: piyu053@gmail.com

Inamuddin, Rajender Boddula, Mohd Imran Ahamed and Abdullah M. Asiri (eds.) Polymer Coatings: Technology and Applications, (41–74) © 2020 Scrivener Publishing LLC

Abbreviations

World health organization (WHO); tuberculosis (TB); lower respiratory tract infection (LRTI); chronic obstructive pulmonary disease (COPD); novel drug delivery systems (NDDS); dry powder inhalers (DPI); pressurized meter dose inhalers (pMDIs); soft mist inhalers (SMIs); cohesive-adhesive balance (CAB); spray drying (SD); dry powder coating (DPC); process analytical techniques (PAT); force controlling agents (FCAs); magnesium stearate (MgSt); Leucine (LEU); magnetically assisted impaction coating (MAIC); turbo rapid variable (TRV); X-ray photoelectron spectroscopy (XPS); rotating fluidized bed coater (RFBC); atomic force microscopy (AFM); inverse gas chromatography (IGC); quality by design (QbD); real time release (RTR); critical quality attributes (CQA); critical process parameters (CPP); glass transition temperature (Tg); hydroxypropyl methylcellulose (HPMC); polytetrafluoroethylene (PTFE); discrete element method (DEM).

3.1 Introduction

Pulmonary ailments are one of the prevailing worldwide community health crises [1]. They are described by airflow limitation, chronic inflammation, bronchoconstriction, loss of elasticity, mucus hypersecretion, and emphysema [2, 3]. Numerous factors such as air pollution (fine and ultrafine particulate matter), tobacco smoke, biomass fuels, inhalation of toxic particles, dust particles, occupational chemicals, systemic diseases, and genetics issues are responsible for the progression of pulmonary diseases [3, 4]. Recent statistics presented by the World Health Organization (WHO) showed that >235 million people suffered from respiratory diseases with >3 million mortalities reported annually due to the same [5]. A few of the most ordinary and occupational pulmonary diseases are chronic obstructive pulmonary disease (COPD), tuberculosis (TB), cystic fibrosis, pulmonary emphysema, asthma, lung cancer, pulmonary hypertension, and lower respiratory tract infection (LRTI) [6]. By knowing the severity of the diseases, various active researchers and clinicians thoroughly investigated the different drug molecules (e.g., synthetic molecules, derivatives, isolated phytoconstituents, and herbal extracts) [7–10] along with the dosage forms (e.g., tablet, capsules, and suspension) for treatment of pulmonary diseases [11]. Additionally, several novel drug delivery systems (NDDS) such as polymeric nanoparticles, lipid nanoparticles, microspheres, polymer-drug conjugates, micellar systems, and lipid vesicles were also systemically studied and clinically investigated for the management of pulmonary diseases [12]. However, by considering the pathophysiology of lung diseases and biology of the pulmonary airways, inhalation is a conceptually more suitable delivery route to augment therapeutic efficacy and patient compliance in the treatment of pulmonary diseases.

Various conventional dosage forms such as dry powder inhalers (DPI), soft mist inhalers (SMIs), pressurized meter dose inhalers (pMDIs), and nebulizers are well studied and explored for various pulmonary ailments and pathophysiological conditions [11]. DPIs have several key advantages of physiochemical stability, no need for cold chain storage, drug deposition using the patient's respiration, and the superior prospect of particle engineering as compared to other dosage forms. DPIs are the drug/device combination product comprised of three main elements, i.e., micronized drug (0.5–5 μm), right carrier (90–120 μm), and aerosol device (unit or multi-dose). From a formulation perspective, carrier molecules are more important owing to the small mass of drug (e.g., 18 mcg; Spiriva® HandiHaler®, Boehringer Ingelheim) within the final formulation [11, 13]. Cohesive-adhesive balance (CAB) between drug and carrier is fundamental for satisfactory deagglomeration of DPIs. Moreover, there are several physicochemical attributes such as carrier size, morphology, true density, and interface properties that are decisive for the satisfactory dispersion/deagglomeration of DPIs. The carrier physicochemical attributes influence the DPIs performance at different levels: molecular, particulate, and delivery. Thus, formulation scientists have taken several efforts to modify the carrier physicochemical properties in order to control the DPIs therapeutic outcome. These efforts mainly involve modification of carrier molecule using several novel processing techniques such as antisolvent crystallization [14], antisolvent crystallization using binary solvents [15], liquid crystalline phase [16], supercritical fluid crystallization [17], ultrasound-assisted crystallization [18], electrospinning assisted crystallization [19], and spray drying (SD) [20, 21]. However, these techniques have few limitations on account of the need for complex manufacturing setup, dedicated machinery, need for a large amount of starting materials which is time-consuming [22–24]. Additionally, scale-up issues and regulatory requirements make these processes more complicated [24]. To avoid such processing issues, scale-up challenges, and technical complications, material scientists explored the dry powder coating (DPC) techniques.

DPC is one-step, eco-friendly, and easy to scale-up technique wherein the physicochemical properties of carrier molecules are modified. It is a solvent-free powder processing method. DPC help to develop powders and particles with miscellaneous functionalities and properties. In DPC, equally dispersed, fine guest particles above the surface of larger host particles can be fabricated with desired properties. Acceptable blending uniformity and CAB can be achieved by modifying the carrier physicochemical properties through the DPC technique. Apart from this, various other key benefits (adequate homogeneity, flowability improvement, and dispersibility enhancement) can be achieved using DPC [25, 26].

The present chapter deals with DPC methodology and high/low shear techniques used in powder processing. This chapter also contains a dedicated section dealing with various process analytical techniques (PAT) for ensuring DPC process uniformity and consistency. Additionally, the various force controlling agents (FCAs) used in DPC is also thoroughly discussed with particular emphasis on DPIs aerodynamic performance. Furthermore, this chapter also underlines the importance of the inhaler device and capsule coating to enhance aerodynamic functioning on DPIs. In the concluding section of the chapter, significance along with applications of numerical simulation in DPC is also summarized. This chapter is the first inclusive account of dry powder coating (DPC), coating techniques, and force controlling agents (FCA) used in DPIs.

3.2 Dry Powder Coating

Dry powder coating (DPC) presents numerous key benefits such as providing a homogenous free-flowing blend, enhanced content uniformity, and low risk of contamination. Functionalized particles fabricated by DPC have tailored physicochemical properties such as improved flowability, enhanced dispersibility, good wettability, and better solubility. Moreover, it also shows improvement in catalytic, electrical, and optical properties of the final blend. Furthermore, enrichment of color, flavor, and appearance has also been well documented [25]. DPC technique is an energy-saving and eco-friendly process since it does not need any specific treatment or pre-conditioning with organic solvents or an extra drying process. In some DPC techniques, where milling is sync with coating, there is less possibility of external contamination in absence of milling aid (metal balls, glass beads) [25, 27–29].

DPC is a simple, high-shear progression wherein fine guest particles are coated to the exterior of larger host particles. Fabrication of functionalized particles using DPC technique is generally performed in three steps: de-agglomeration of the cohesive guest particles (step 1), the formation of interactive mixtures (step 2) and the redistribution of guest particles within-host particles to attain homogeneous distribution of guest particles (step 3) [30–32]. Physical or chemical types of interfacial forces are mainly present between host and guest particles during processing. DPC concept can be described as composite powder formation, micro-fabrication, mechanical treatment, hybrid mixing, mechanical dry coating, surface adsorption, engineered particles, and encapsulation phenomenon. DPC can be divided into different classes as per the extent of coating action (i) discrete (partial coating), (ii) steady film construction, as well as (iii) uninterrupted coating (encapsulation). Figure 3.1 illustrates the different types of dry powder coated particles. During DPC, absolute facade treatment can be achieved when the kinetic energy provided through DPC machine/technique is adequate to deagglomerate guest particles by offering centripetal force as well as energy for particle-particle interaction ensuring uniform and steady adsorption of fine guest molecules across the surface of host particles. However, the nature and degree of DPC is also equally dependent on the physical attributes of the guest particles

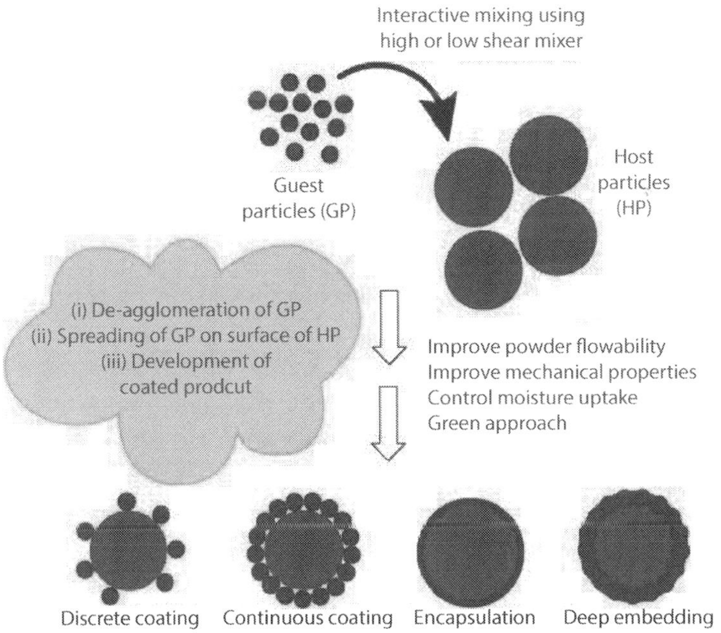

Figure 3.1 Graphical representation of dry powder coating modified after Dahmash *et al.*, 2015 [25] [Diagram represents the process of dry powder coating and types of coated particle. It is a simple and eco-friendly approach for designing multifunctional particles. Uniform coating can be achieved using high as well as low shear mixing techniques. The inner part (red-colored) is the coarse host particle while the outer part (blue colored) is the fine guest particle.]

(size, morphology, surface properties, roughness, and melting point) and mechanism involved in DPC [25, 28, 29]. In the subsequent section, various DPC processing equipments and techniques are described with particular importance on mechanism and shear rate.

3.3 Dry Powder Coating Techniques

Albeit the manufacturing steps, the formulation of an active pharmaceutical ingredient (API) or guest substance with other excipients needs to be mixed homogeneously for better pharmaceutical applications. Various tools and techniques (high shear or low shear) are available for ordered mixing or the development of interactive mixtures. High shear mixing relies on a high-intensity mixing blade while low shear mixing uses the planetary movement of the mixer arm to generate homogeneity [25]. Mechanical tumbling, cube mixer, turbula mixer, and V-blender are the commonly used techniques at laboratory scale for achieving ordered mixing. Additionally, various advanced techniques such as cyclomix, magnetically assisted impaction coating (MAIC), and/or turbo rapid variable (TRV) mixer are also explored for the development of interactive mixtures [25, 26]. These techniques can be classified on the basis of shear forces and/or mechanisms involved in DPC. Various commonly used DPC strategies are enlisted in Table 3.1.

Several high shear techniques (e.g., mechanofusion, hybridizer, TRV) are well reported in the scientific literature for DPC. Mechanofusion equipment is one of the well-known and widely applied methods for achieving uniform and reliable DPC. Mechanofusion equipment consists of a rotating cylindrical vessel and a scraper. Mechanofusion devices typically produce high mechanical shear action using a combination of sufficiently high compression and centrifugal forces causing surface interaction (or fusion) of guests and host particles. As the cylindrical assembly spin at a greater momentum, the forceful communication produces a large amount of kinetic energy with the development of a consistent guest particle layer over the host particles [30, 34]. Hybridizer is a next-generation coating technique pertinent to producing high shear mixing and milling for effective deagglomeration of guest particles followed by its uniform layering on to the host particles. It is made up of high-speed motor, a mixing vessel equipped with six blades, a powder recirculation circuit, and a coolant jacket to control the heat generated during the process. Uniform coating is achieved due to high shear action, diffusion forces, and heat produced during friction between the powder, the blades, and the vessel walls.

Table 3.1 Summary of DPC technologies with their important advantages and limitations [25, 26, 33].

Technology and manufacturer	Shear and speed	Advantages	Limitations
Mechanofusion and Hosokawa, Japan	High shear and 200 to 10,000 RPM	1. Allow fusion (physical and/or chemical) between guest and host.	1. Not suitable for thermo-labile substances and fragile host material. 2. Heat generation may lead to polymorphic transformation.
Hybridiser and Nara Machinery, Japan	High shear and 5,000 to 20,000 RPM	1. Very short processing time. 2. Powder recirculating unit guarantee homogeneity of the final blend. 3. Can work with smaller host (0.1 μm) particles.	1. Not suitable for thermo-labile substances and fragile host material. 2. Not suitable for a continuous process. 3. Heat generation may lead to polymorphic transformation.
Cyclomix	High shear and 30 m/s (Rotor speed)	1. Produces mild coating. 2. Used for coating, mixing and granulation	1. There is particle attrition or even rupture of host materials. 2. Not appropriate for unstable or brittle host materials

(Continued)

Table 3.1 Summary of DPC technologies with their important advantages and limitations [25, 26, 33]. (*Continued*)

Technology and manufacturer	Shear and speed	Advantages	Limitations
Turbo rapid variable (TRV) and GEA Group, Germany	High shear and 20 m/s (impeller tip speed)	1. Work on a single, bottom driven impeller. 2. Suitable for continuous process and scale-up. 3. Ensure homogeneity of the final blend.	1. Not suitable for thermo-labile substances and fragile host materials.
PMA™ and GEA Group, Germany	High shear	1. PMA is a bottom-drive high shear mixer. 2. Useful for blending, melt or wet pelletizing and granulation	–
Magnetically assisted impaction coating (MAIC) and Aveka, USA	High shear	1. Allow the least variation in particle shape or size. 2. Utilize a magnetic field and thus lower energy needs. 3. The minor amount of heat is generated thus suitable for thermo-labile substances	1. A significant risk of contamination due to magnetic particles. 2. Not suitable for a continuous process.

(*Continued*)

Table 3.1 Summary of DPC technologies with their important advantages and limitations [25, 26, 33]. (*Continued*)

Technology and manufacturer	Shear and speed	Advantages	Limitations
Theta composer and Tokuju, Japan	Low shear (soft coating) and 30 RPM (vessel speed); 500–3,000 RPM (rotor speed)	1. Short processing time (2 to 10 min). 2. Allow minor change in particle shape or size. 3. Low heat generation and thus no thermal deterioration. 4. Suitable for batch type process.	1. Not appropriate for a continuous coating process.
Rotating Fluidized Bed Coater (RFBC) and NJIT, USA	Low shear (soft coating)	1. Suitable for both batches as well as continues type coating process. 2. Can be utilized for very small guest and host substances.	1. Surplus aerosolization of the fine guest particles.
Turbula shaker mixer and Glen Mills Inc., USA	Low shear	1. Mixing is achieved in a closed container and thus provide dust-free blending and lower contamination risk. 2. Good powder homogeneity of the mixture owing to interaction, translation, rotation and inversion as per the geometric theory in relation to Schatz.	–

(*Continued*)

Table 3.1 Summary of DPC technologies with their important advantages and limitations [25, 26, 33]. (*Continued*)

Technology and manufacturer	Shear and speed	Advantages	Limitations
Fluid energy mill (FEM) and Sturtevant Inc., USA	Micronization with the coating	1. Can be used for nano and micron-sized guest and host substances. 2. Suitable for continuous type process.	1. Pre-mixing is required prior to FEM. 2. May lead to polymorphic transformation. 3. Need to monitor various parameters such as feeding rate, feeding and grinding pressure to ensure the formation of uniform mixture or coating.

Both these techniques support physical and chemical guest-host interactions termed as mechanochemistry [28, 29]. TRV is a high speed mixing unit composed of a single, bottom driven impeller drive. It is largely apt for the rapid mixing or coating of small quantities of API with base excipients, distinctive of inhalable products. By integrating the directional effects of the feeder blade with the high surface area of the main impeller blade, rapid uniform mixture is obtained in a reproducible way. It is available in various capacities (1, 5, 10, 20, 30, and 60 L) for batch as well as continuous manufacturing processes [35]. Various patents on TRV-based inhalation products are listed in Table 3.2.

Similar to high shear techniques, several low shear techniques were also available for DPC to achieve coating at mild conditions. Low shear techniques are mainly preferred for thermo-labile substances and fragile host material owing to controlled heat generation, compaction energy, and shear forces during coating or mixing [25]. Elliptical rotor mixer (Theta Composer) and rotating fluidized bed coater (RFBC) are the commonly used low shear techniques for DPC. In theta composer, powder coating is achieved due to compression between the elliptical vessel wall and rotor

Table 3.2 Various Turbo rapid variable (TRV) process based dry powder formulations.

Guest substances	Host substances	Key findings	Reference
Magnesium stearate	Lactose monohydrate and triphenylacetate	Achieved good blend uniformity for low dose (1 to 400 mcg) DPI formulation	[36]
Fluticasone furoate	Lactose monohydrate	Achieved uniform mixing for low dose DPI formulation	[37]
Magnesium stearate	Lactose monohydrate	Ensured uniform mixing of magnesium stearate and lactose to control chemical degradation of a drug	[38]

moving in the opposite direction [39], whereas in RFBC, uniform powder coating is produced as a result of centrifugal fluidization with the radial flow of inert gas [40]. Furthermore, the inert fluidizing gases help to enhance the collision among particles that ease de-agglomeration along with the consistent coating.

3.4 Analytical Techniques for Ensuring Coating Uniformity

The transition of particles across the powder bed during the DPC has a crucial role in the dynamics of the coating. Apt characterization of the coating quality is decisive for the superior process adaptation and optimization [41]. Comprehension of the particle dynamics can lead to a better knowledge of the overall DPC mechanism. To assess the particle dynamics, specifically high-resolution analytical tools and methods are often essential. This is mainly factual for coating analysis wherein surface energetics, surface rigidity, and surface morphology are assessed along with the calculation of the inter-particulate forces. Over the period of time, process analytical technology (PAT) like X-ray photoelectron spectroscopy (XPS), Raman analysis (in-line or on-line), atomic force microscopy (AFM), scanning electron microscopy (SEM), inverse gas chromatography (IGC) assist in understanding particle dynamics [42, 43].

US FDA defined PAT system for designing, analyzing, and monitoring manufacturing via in process measurements of critical quality and performance attributes of materials and methods with the assurance of final product quality [44]. It is a helpful and effective move toward process validation. It merges few advanced systems, tools, and procedures, for example, quality by design (QbD), root cause analysis (RCA), Pareto chart, real-time release (RTR), and total quality management (TQM) that can facilitate offline and online authentication of vital process factors. From implementation viewpoint, PAT involves three basic steps, i.e., (i) design, (ii) analysis, and (iii) control [45, 46]. PAT allows a better understanding of the critical quality attributes (CQA) and critical process parameters (CPP). In brief, PAT provides a well-structured policy for the process screening and therefore can be implemented for better process understanding as well as risk management [46]. By knowing these facts, Table 3.3 reviews the sophisticated analytical tools and techniques that have been explored in the understanding and characterization of the DPC process and particle dynamics.

3.5 Force Controlling Agents

Force controlling agents (FCAs) are commonly used in various pharmaceutical processes. They are chemically inert, odorless, and tasteless. Different types of FCAs are explored for pre-formulation and formulation development activities, i.e., (a) metallic salts of fatty acids, (b) inorganic materials, (c) polymers, (d) alkyl sulfates, (e) fatty acids, fatty alcohols, and hydrocarbons, and (f) fatty acid esters [49]. Apart from these, several natural, semi-synthetic and synthetic pulmonary surfactants or hydrophilic surface-active molecules are also utilized as FCAs (Table 3.4). When these substances are used in pulmonary formulations, they easily dispersed over the large surface of the pulmonary airways. It is assumed that this immediate diffusion and dispersion of surfactants might facilitate drug diffusion across the lung and accordingly improve clinical outcomes. Table 3.5 enlists various investigations on inhalation powders containing FCAs. Commonly used FCAs in inhalation products are thoroughly described in the subsequent section.

Table 3.3 List of sophisticated analytical techniques for understanding dry powder coating process and particle dynamics.

Technique	Advantages	Limitations	References
Atomic force Microscopy (AFM)	1. Determine surface properties at the submicron level. 2. Available in various mode, i.e., Tapping (intermittent) or contact mode (topographical analysis); Nano-indentation mode (for elastic modulus and deformation studies); Force distance measurements (for adhesion analysis) and Colloidal probe AFM (for adhesion analysis and particle surface energetics). 3. In DPIs mainly used to determine cohesive adhesive balance (CAB) between drug/guest and carrier	1. Needs specialized staff for analysis and interpretation 2. Applicable only for small particle size (nanoscale guest and submicron host) and 3. Expensive	[42, 47]
Brunauer-Emmett-Teller (BET)	1. Recognize reduction in specific surface area (SSA) and pore size before and after coating 2. Quantify the optimum level of guest material essential to achieve a uniform coating	1. Not appropriate for thermolabile substances material 2. Needs pre-conditioning for analysis	[43]

(Continued)

Table 3.3 List of sophisticated analytical techniques for understanding dry powder coating process and particle dynamics. (*Continued*)

Technique	**Advantages**	**Limitations**	**References**
Fourier-transform infrared spectroscopy (FTIR)	1. Spot the physicochemical interaction between guest and host particles	1. Needs specialized staff for analysis and interpretation	[43]
Inverse gas chromatography (IGC)	1. Determine surface energetics at a dispersive and specific level. 2. Suitable for thermolabile material 3. It can be is combined with other surface characterization methods to study surface properties	1. Needs specialized staff for analysis and interpretation 2. Expensive in nature	[48]
Scanning electron microscopy (SEM)	1. Qualitative analysis of the degree of coating 2. A very small amount of sample needed 3. Allow visual analysis of the extent of coating and qualitative in nature	1. Needs pre-conditioning for analysis	[43]

(*Continued*)

Table 3.3 List of sophisticated analytical techniques for understanding dry powder coating process and particle dynamics. (*Continued*)

Technique	Advantages	Limitations	References
Positron emission particle tracking (PEPT)	1. It allows a quantitative analysis of the motion of particles. 2. It can be is used to conclude the location of the individual particle accurately.	1. Needs specialized staff for analysis and interpretation 2. Expensive	[43]
Radioactive particle tracking (RPT)	1. It can be used to quantitative analysis of dynamics of particles within dry-coating device 2. It can be combined with different detectors to capture the response	1. Expensive	[43]
Electron Dispersion X-Ray Spectroscopy (EDX or EDS)	1. It gives the concentration of elements present in a sample. 2. In combination with SEM, can provide better visual analysis of coated guest material	1. Not appropriate for thermolabile substance 2. Small contamination might vary the outcome	[43]

(*Continued*)

Table 3.3 List of sophisticated analytical techniques for understanding dry powder coating process and particle dynamics. (*Continued*)

Technique	Advantages	Limitations	References
X-ray photoelectron spectroscopy (XPS)	1. It gives near-surface region elemental composition (2–10 nm depth) 2. High accuracy and specificity 3. More sensitive as compared to EDX	1. Needs specialized staff for analysis and interpretation 2. Not appropriate for thermolabile substances 3. Small contamination might vary the outcome 4. Expensive	[43]
Time-of-flight secondary ion mass spectrometry (TOF-SIMS)	1. It gives near-surface region elemental composition (2–5 nm depth) 2. Ultrahigh sensitivity and specificity 3. Need a very small sample for analysis	1. Needs specialized staff for analysis and interpretation 2. Small contamination might vary the outcome 3. Expensive	[43]

Table 3.4 Force controlling agents (FCA) [26].

Class	FCA
Amino acids	L-Leucine, D-Leucine, iso leucine, trileucine, methionine, lysine, valine, phenylalanine, aspartame or acesulfame K, and acesulfame potassium
Natural and synthetic lung surfactants lipids and phospholipids	Phosphatidylglycerol (PG), dipalmitoyl phosphatidylcholine (DPPC), dipalmitoyl phosphatidylethanolamine (DPPE), acyl substituted phospholipids, diacyl substituted phospholipids, dipalmitoyl phosphatidylinositol (DPPI), lecithin, lecithin-derivatives, soya lecithin, laxiric acid, magnesium lauryl sulfate (MLS), and sodium lauryl sulfate (SLS), triglycerides, i.e., hydrogenated castor oil waxy powder (Cutina® HR) and microcrystalline triglyceride (Dynasan® 118)
Saturated fatty acid	Erucic acid, stearic acid, palmitic acid, oleic acid, glyceryl behenate, lauric acid, and behenic acid
Natural and synthetic minerals	Talc, silicon dioxide, aluminum dioxide, titanium dioxide, and starch
Metal stearates and derivatives	Sodium stearyl fumarate (SSF), zinc stearate, magnesium stearate, lithium stearate, calcium stearate, sodium stearate, and sodium stearoyl lactylate (SSL)

3.5.1 Metal Stearates

Magnesium stearate (MgSt), aluminum stearate, sodium stearate, zinc stearate, and calcium stearate are commonly used metal stearates in pharmaceutical industries [60]. Among them, MgSt is the most commonly used FCA in DPIs. Various characteristic features of MgSt are displayed in Figure 3.2. MgSt is well-known and widely used functional ingredient in pharmaceutical products, cosmetics preparations, and a variety of foodstuffs. It is obtained from vegetable as well as animal resources and available as trihydrate, dihydrate, anhydrate, and amorphous forms. It is mainly used as a lubricant in conventional oral dosage forms such as powder, tablet, granules, and capsules [61]. In DPI formulations, MgSt act as a lubricant, force controlling agent, surface control agents, water barrier or stabilizer [62].

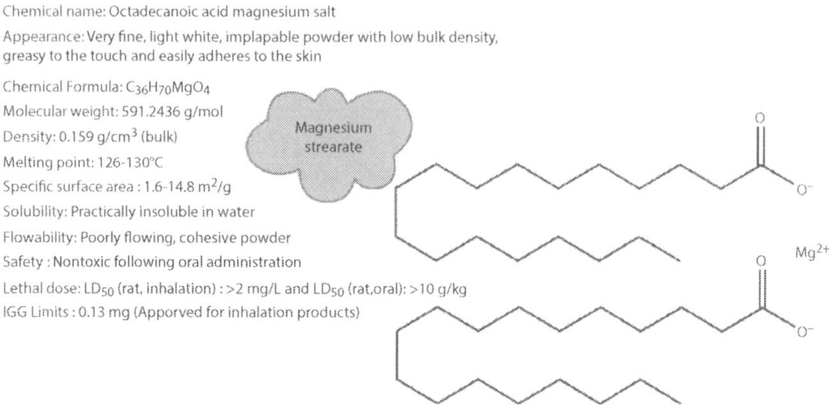

Figure 3.2 Key features of magnesium stearate for dry powder coating [Figure shows the key physical, chemical, and biological properties of magnesium stearate for successful pulmonary drug delivery].

As shown in Table 3.5, MgSt is widely used in DPI products as a tertiary ingredient. It shows promising aerosolization results by modifying the interface interaction between the micronized drug (<5 μm) particles and coarse carrier (70–150 μm) particles. Principally, MgSt regulates the inter-particulate interactions and/or CAB [62]. Another key mechanism of improved aerosolization is the dissipation of electrostatic charge on powder mixture or the formation of easily dispersible drug-excipient agglomerates [53]. It is well documented that the hydrophobization of materials with MgSt reduces the force of adhesion appreciably. More specifically, MgSt reduces the adhesion forces owing to interfacial long-range Van der Waals forces acting among the powder particles. During DPC, MgSt shapes a nanometer layering against the carrier particles which regulate the adhesive interactions between the micronized drug and sieved carrier thereby, supporting the detachment of API from the carrier upon aerosolization [53]. However, these effects are highly concentration-dependent. An optimum MgSt content is a concentration that enhances powder flow the most and results in uniform film formation to enclose the individual particle. Beyond the optimum MgSt concentration, there is a sharp drop in flowability with increased film thickness or when excess fine particles are present [26, 53]. Additionally, the specific surface area and particle size distribution (mainly, d_{50}) of MgSt are the key factors influencing its coating efficiency [60]. Moreover, MgSt is also studied to improve stability by controlling the ingress of moisture [26, 53]. By knowing these anti-adherents, anti-friction properties, and ability to modify the drug/carrier interfacial properties, MgSt is used in various commercial DPIs products (Table 3.6).

Table 3.5 Summary of inhalation powders containing FCAs.

Drug	Coating techniques	Key ingredients/ FCA	Device and flow rate (l/min)	Percentage FPF	Reference
SBS	Mechanofusion® (Rotor-type mixer)	5% MgSt	MI device and 60	69.00	[50]
SX	Mechanofusion® (Rotor-type mixer)	5% MgSt	MI device and 60	73.10	[50]
SBS	Mechanofusion®	5% MgSt	Monohaler® and 60	2.8 micron (MMAD)	[51]
Budesonide	Mechanofusion®	5% MgSt	Monohaler® and 60	2.2 micron (MMAD)	[51]
Tobramycin	Spray drying	1% Sodium Stearate	Aerolizer® and 60	84.30	[52]
SBS	Turbula® T2F mixer	10% MgSt	Rotahaler® and 60	12.6	[53]
Triamcinolone acetonide	AMS-Mini mechanofusion with Nobilta module	5% MgSt	MI device and 60	51.60	[54]
Triamcinolone acetonide	AMS-Mini mechanofusion with Nanocular module	5% MgSt	MI device and 60	48.30	[54]
SX	AMS-Mini mechanofusion with Nobilta module	5% MgSt	MI device and 60	73.10	[54]

(Continued)

Table 3.5 Summary of inhalation powders containing FCAs. (*Continued*)

Drug	Coating techniques	Key ingredients/ FCA	Device and flow rate (l/min)	Percentage FPF	Reference
SX	AMS-Mini mechanofusion with Nanocular module	5% MgSt	MI device and 60	69.30	[54]
SBS	AMS-Mini mechanofusion with Nobilta module	5% MgSt	MI and 60	68.60	[54]
SBS	AMS-Mini mechanofusion with Nanocular module	5% MgSt	MI device and 60	68.50	[54]
SBS	AMS-Mini mechanofusion with Nobilta module	2% MgSt	MI device and 60	75.4	[54]
FP	Collette MicroGral (High-shear)	Respitose® ML001 and 5% MgSt	Breezhaler® and 90	32.2	[55]
FP	Turbula® T2F mixer (Low-shear)	Respitose® ML001 and 10% MgSt	Breezhaler® and 90	26.10	[55]
SX	Collette MicroGral (High-shear)	Respitose® ML001 and 0.1% MgSt	Breezhaler® and 90	56.70	[55]
SX	Turbula® T2F mixer (Low-shear)	Respitose® ML001 and 0.5% MgSt	Breezhaler® and 90	56.30	[55]
SBS	Theta-Composer®	Pharmatose® 200M and 10% MgSt	Jethaler® and 30	~40.00	[56]

(*Continued*)

Table 3.5 Summary of inhalation powders containing FCAs. (*Continued*)

Drug	Coating techniques	Key ingredients/ FCA	Device and flow rate (l/min)	Percentage FPF	Reference
SBS	Theta-Composer®	Pharmatose® 200M and sucrose tri-stearate (J-1803F, 10%)	Jethaler® and 60	46.86	[53]
SBS	Wurster Fluidized Bed	Pharmatose® 200M (13%) aqueous solution containing HPMC (2%)	Jethaler® and 60	34.9	[57]
SBS	Vortex-Mixer	Pharmatose® 200M and 2% MgSt	Jethaler® and 60	26.0	[58]
TNRA	*Rotor-type powder mixer	Pharmatose® 325M and 1% Sucrose stearate	Jethaler® and 30	20.90	[59]
TNRA	*Rotor-type powder mixer	Pharmatose® 325M and 2% MgSt	Jethaler® and 30	34.60	[59]

SBS: salbutamol sulfate; MgSt: magnesium stearate; MI: Monodose inhaler; FPF: fine particle fraction; FP: fluticasone propionate; SX: salmeterol xinafoate; TNRA: triple neurokinin receptor antagonist and MMAD: mass median aerodynamic diameter.

Table 3.6 Commercial DPIs products containing magnesium stearate.

Market	Proprietary name	Active ingredient and dose	Company
European Union	Seebri® Breezhaler®	Glycopyrronium bromide (44 mcg)	Novartis Pharmaceuticals, UK Ltd.
European Union	Ultibro® Breezhaler®	Glycopyrronium bromide (85 mcg) and indacaterol maleate (43 mcg)	Novartis Pharmaceuticals, UK Ltd.
European Union	*Fostair NEXThaler®	Beclometasone dipropionate (100 mcg) and formoterol fumarate dihydrate (6 mcg)	Chiesi Ltd.
European Union	Incruse Ellipta®	Umeclidinium bromide (55 mcg)	Glaxo Smith Kline, UK
European Union	Anoro Ellipta®	Vilanterol trifenatate (55 mcg) and umeclidinium bromide (22 mcg)	Glaxo Smith Kline, UK
European Union	Relvar Ellipta®	Fluticasone furoate (184 mcg) and vilanterol trifenatate (22 mcg)	Glaxo Smith Kline, UK
European Union	Trelegy Ellipta®	Fluticasone furoate (92 mcg), vilanterol trifenatate (55 mcg) and umeclidinium bromide (22 mcg)	Glaxo Smith Kline, UK
United States	Utibron®	Glycopyrrolate (15.6 mcg) and indacaterol maleate (27.5 mcg)	Sunovion Pharmaceuticals Inc.

*Available in 200/6 mcg strength.

3.5.2 Amino Acids

Amino acids (AA) are the endogenous, natural compounds which may have fewer safety problems for drug targeting or delivery to the pulmonary airways. AA and/or derivatives have a suitable molecular weight (0.25 to 1,000 kDa) and therapeutic properties for pulmonary therapy. Moreover, AA's isomers and racemic mixtures may be used for formulation development and drug delivery [26, 56]. A superior amount of AA can modify the aerosolization pattern of potent antibiotics exclusive of apparent cytotoxicity [26]. Moreover, AA and derivatives are used as stabilizers for various bio-molecular formulations employing potent biomolecules and bacteriophages [56]. In the case of DPIs, leucine (LEU) is the most frequently explored ingredient.

The aerodynamic outcome of particles and powders can be enhanced by applying a surface layer of hydrophobic amino acid, i.e., LEU [63, 64]. During co-SD with a drug, LEU is known to supersaturate quickly and provide a surface coating to drug particles [65]. LEU shows good lubrication properties for inhalable powders and particles. LEU (5%) often leads to the fabrication of irregular (wrinkled/corrugated) particles [66]. Such particle structure with reduced inter-particulate interaction provides better CAB as compared to smooth particle [66]. Together, corrugated particles showed a mean particle size suitable for pulmonary delivery. Hence, particles surface covered by LEU generally has superior flow properties, dispersion, as well as aerosolizing behavior [56]. Table 3.7 enlists various investigations on inhalation powders containing LEU. As shown in Table 3.7, LEU was also utilized in mixture among additional ingredients. Within a binary or ternary powder system with a suitable sugar carrier (lactose/mannitol/trehalose), LEU helps to stabilize biomacromolecules and enhance the aerosolization performance [67]. It also acts as a drug dispersion enhancer for chitosan containing customized drug release preparation including various molecules such as curcumin [68], diltiazem hydrochloride [69], and levofloxacin [70]. Recently, spray-dried capreomycin: LEU (80:20) inhalation powder was explored in a clinical study to manage TB [81]. In brief, on the basis of the final concentration, method of preparation and chemical nature of the drug, LEU can act as an aerosol enhancer, stabilizer, and flow enhancer.

Trileucine is a tri-peptide produced from three L-LEU residues. It is another essential amino acid used in DPI formulations. As compared to leucine, it has a high glass transition temperature (T_g) (104°C), which demonstrate better physicochemical sturdiness of trileucine. Tg of SD trileucine is highly dependent on the pH and concentration of anions,

Table 3.7 Summary of inhalation powders containing amino acids as FCAs.

Formulation system	Method/technique	Key ingredients/ amino acids	Device and flow rate (l/min)	Percentage FPF	Reference
Mannitol	SD	Leucine	Monodose inhaler and 60	80.10	[71]
Mannitol	SD	Leucine (10%)	Monodose inhaler and 60	80.46	[71]
Trehalose	SD	Leucine (10%)	Monodose inhaler and 60	66.51	[71]
Curcumin-nanoparticle	SFD	Leucine, chitosan and mannitol	Powder entrainment tube and 85	49.10	[72]
Diltiazem hydrochloride nanoparticle	W/O emulsification	Chitosan, leucine	Rotahaler® and 60	21.00	[73]
Levofloxacin microparticles	SD	Chitosan, leucine (4:1)	Cyclohaler® and 60	52.24	[74]
Isoniazid nanoparticles	SD	Chitosan, leucine (10%), tripolyphosphate	Cyclohaler® and 60	45.00	[75]
Immune globulin G	SD	Cysteine (50%)	Cyclohaler® and 60	70.51	[76]

(*Continued*)

Table 3.7 Summary of inhalation powders containing amino acids as FCAs. (*Continued*)

Formulation system	Method/technique	Key ingredients/ amino acids	Device and flow rate (l/min)	Percentage FPF	Reference
Immune globulin G	SD	Phenylalanine (50%)	Cyclohaler® and 60	62.43	[76]
Immune globulin G	SD	Leucine (50%)	Cyclohaler® and 60	58.68	[76]
Gentamicin	SD	Trileucine (2%)	PDS and 28.3	46.80	[77]
Gentamicin	SD	Trileucine (15%)	PDS and 28.3	49.30	[77]
Netilmicin	SD	Trileucine (2%)	PDS and 28.3	48.80	[77]
Netilmicin	SD	Trileucine (15%)	PDS and 28.3	62.40	[77]
Budesonide	SD	D-leucine (1%)	RS01 model 7 and 60	30.07	[78]
Disodium cromoglycate	SD	Leucine (2%)	Osmohaler® and 100	71.90	[79]
Disodium cromoglycate	SD	Methionine (40%)	Osmohaler® and 100	62.42	[80]

FPF: fine particle fraction; SD: Spray drying; SFD: Spray freeze-drying and PDS: pulmonary delivery system (Nektar Therapeutics, CA).

cations, and zwitterions of the system. The dispersibility of trileucine is mainly governed by the pH of a solution and is poor at pH 7. In comparison to leucine, it has better surface activity and poor aqueous (approx. 3-fold) solubility and is amorphous following the SD. Its aqueous solubility and pH-dependent behavior promote the formation of low density highly corrugated particles with reduced inter-particulate interaction and cohesiveness; which eventually can control the solid-state aggregation with improvement in the flowability and dispersibility of powder [26, 77].

3.6 Inhaler Device and Capsule Coating

Various active researchers explored the DPC techniques, FCAs for tailoring the performance of powder and particles. Few formulation scientists initiated the idea of coating inhaler device and capsules with suitable material to improve the dispersion and de-agglomeration performance of DPIs. These interesting cases related to the inhaler device and capsules coating are described here.

Behara *et al.* 2014, studied the effect of CC_1-3Dm inhaler and hydroxypropyl methylcellulose (HPMC) capsule surfaces coating effect on the aerodynamic performance of albuterol sulfate. Device (CC_1-3Dm) and HPMC capsule were carefully coated using lower surface energy material, i.e., polytetrafluoroethylene (PTFE) suspension (LU™ 708, Sprayon Products, OH). Additionally, the HPMC capsule was layered using low surface energy powders, i.e., leucine and MgSt. PTFE coating of the CC_1-3Dm inhaler and HPMC capsule has a considerable impact on capsule shell drug retention, mutually with reducing mouthpiece (MP) and capsule chamber retention related to uncoated inhaler device. PTFE coated CC_1-3Dm inhaler displayed 1.27 and 1.25-fold improvement in emitted dose as related to uncoated resin and metal array device, respectively. PTFE coated CC_1-3Dm inhaler showed 1.45 and 1.12-fold reduction in MP retention than the uncoated resin and metal array device, respectively. Moreover, PTFE coated CC_1-3Dm inhaler also showed a marked 4-fold reduction in capsule chamber retention as related to both uncoated inhalers. Furthermore, PTFE coated HPMC capsules showed a 1.93 and 2.21-fold reduction in capsule retention as compared to capsules actuated from uncoated resin and metal array device, respectively. Coating the capsule shell with the ingredient of lower surface energy also showed a significant impact on the aerosolization performance of albuterol sulfate. Capsule coating with MgSt showed a decrease (1.54-fold) in the capsule shell drug retention as related to the uncoated capsule. However, there was no considerable variation among capsule shell drug retention for the MgSt and LEU capsule coatings. This shows that the chemical nature of coating material may not perform a crucial function in lessening capsule shell drug retention. In summary, the coating of the inhaler and/or capsule has a major impact on the aerosolization functioning of the drug substance. However, extra efforts are needed to explore the effect of coating materials size and capsule aperture orientation on powder dispersion [82].

In a study by Heng *et al.* (2013), inhalers (i.e., Rotahaler® and Aerolizer®) and capsules (size 3 HPMC) were coated carefully by ethanol-based MgSt suspensions (0.05, 0.1, and 0.3 g/mL) to improve the overall aerodynamic fate of DPI. Spherical shape, smooth surface spray-dried lysozyme (1.91 μm) particles were used as a model drug. The aerodynamic assessment was performed using a multi-stage liquid impinger (MSLI, Copley, UK) at a flow rate of 60 L/min. Aerolizer® device containing coated capsule (0.3 g/mL) showed a maximum reduction in capsule shell retention with a marked 89% powder reduction as compared to the uncoated capsule. In a similar manner, the coated Aerolizer® device (0.3 g/mL) exhibited a maximum reduction in device retention and displayed a marked 52% powder reduction as compared to the uncoated device. Coated (0.3 g/mL) Aerolizer® device, as well as a capsule, demonstrated a marked 79% and 69% reduction in powder retention as compared to the uncoated counterparts, respectively. Similar to the Aerolizer® device, coating (0.3 g/mL) of Rotahaler® device and capsules showed a maximum reduction in device and capsule powder retention. Coating (0.3 g/mL) of both the Rotahaler® device and capsule showed a marked 73% and 53% reduction in powder retention as compared to their uncoated counterparts, respectively. In both cases, a significant reduction in powder retention was mainly attributed to the ability of MgSt to shape superior Langmuir-type films which impart anti-adherent features to the coated face. MgSt coating is known to reduce the adhesion between the spray-dried lysozyme particles and the interior facade of the capsule shell and device, which sequentially elevates the aerosolization performance [83]. However, a lot of active efforts need to be invested for commercial translation of such coating technologies.

3.7 Numerical Simulation

The primary theory of DPC includes the mixing of guest (drug) particles with host (carrier) particles using a suitable mixing technique [26]. Although this idea is simple, the mechanism involved in DPC is multifaceted and interrelated [84]. The blending of API and carrier particles is a decisive key process that regulates overall DPI functioning [85]. While interactions between API and carrier particles, for instance, molecular van der Waals forces, Coulombic electrostatic charges, and moisture uptake (capillary action), are prime parameters that conclude how the drug particles adhere and detach from the carrier molecules [26, 86]. Ultimately, drug-carrier interactions and other physiochemical properties (particle size, morphology, interface attributes) of powder blend have an influence on the overall performance of DPIs [86]. Thus, to understand these complex and multifaceted systems, various researchers explored in-silico modeling, i.e., numerical modeling, simulation, and computational fluid dynamics (CFD) experiments. Few interesting case studies related to particle-particle interaction and particle-mixing equipment interactions are described in the succeeding section.

Yang *et al.* (2013) studied the particle adhesion mechanism during powder blending using discrete element method (DEM) analysis. The author showed that the critical sticking velocity and particle radius had a crucial influence on the work of adhesion. While, during mixing, optimal vibration velocity frequency and amplitude are fundamental to maximize the contact number. DEM revealed that the particle sticking potential reduces as the frequency or amplitude is augmented [86]. The same research group also studied the electrostatic interface during particle mixing in DPIs by DEM analysis. The DEM analysis is explored to probe the outcome of electrostatic energy on a DPI wherein API particles and pre-charged carrier particles are blended in a vibrating container. During mixing the pre-charged carriers were found to control the electrostatic interactions between the API and carrier particles while the size (capacity) of the mixing container had a major impact on mixing performance. The mixing efficacy in the electrostatic state is superior to the van der Waals state when the vibrational frequency and amplitude are lower [86]. Similar to these investigations, various flow simulations, fine particle tracking, and multi-scale modeling analysis were performed to understand particle-particle interaction during mixing and ultimately to optimize the DPIs [87]. Modeling-based DPI development is a good policy for exploring its usefulness in decision-making during the optimization of carried-based formulations.

3.8 Conclusion

Dry powder coating is a potential strategy for improving the aerodynamic performance of dry powder inhalers. As summarized in this chapter, dry particle coating, as well as force controlling agents, have shown to be a feasible approach for the development of versatile particles with improved aerosolization characteristics. A variety of processing techniques and a range of force controlling agents are available for materialize functionalized inhalable powders and particles. Additionally, the dry powder coating practice is a single step as well as easy to scale-up. However, it is well evident that, in the upcoming era, extensive experimental efforts are needed to optimize and understand the in-depth mechanism of eco-friendly dry powder coating technology.

References

1. Salvi, S. *et al.*, The burden of chronic respiratory diseases and their heterogeneity across the states of India: The Global Burden of Disease Study 1990–2016. *Lancet Glob. Health*, 6, 12, e1363–e1374, 2018.
2. Weber, S., Zimmer, A., Pardeike, J., Solid lipid nanoparticles (SLN) and nanostructured lipid carriers (NLC) for pulmonary application: A review of the state of the art. *Eur. J. Pharm. Biopharm.*, 86, 1, 7–22, 2014.
3. Rogers, L.K. and Cismowski, M.J., Oxidative stress in the lung–the essential paradox. *Curr. Opin. Toxicol.*, 7, 37–43, 2018.
4. Mehta, P. *et al.*, Phytoconstituent based dry powder inhalers as biomedicine for the management of pulmonary diseases. *Biomed. Pharmacother.*, 108, 828–8375, 2018.
5. World Health Organization (WHO), Chronic respiratory diseases. https://www.who.int/respiratory/en/.
6. Mehta, P. *et al.*, Dendrimers for pulmonary delivery: Current perspectives and future challenges. *New J. Chem.*, 43, 22, 8396–8409, 2019.
7. Mehta, P. *et al.*, Pharmacokinetic profile of phytoconstituent (s) isolated from medicinal plants-a comprehensive review. *J. Tradit. Complement. Med.*, 5, 4, 207–227, 2015.
8. Mehta, P. and Dhapte, V., A comprehensive review on pharmacokinetic profile of some traditional chinese medicines. *New J. Sci.*, 2016, 2016.
9. Heinrich, M. *et al.*, *Fundamentals of Pharmacognosy and Phytotherapy E-Book*, Elsevier Health Sciences, 2012.
10. Mehta, P., Lohidasan, S., Mahadik, K.R., Pharmacokinetic behaviour of clinically important TCM prescriptions. *Orient. Pharm. Exp. Med.*, 17, 3, 171–188, 2017.
11. Mehta, P., Dry powder inhalers: A focus on advancements in novel drug delivery systems. *J. Drug Deliv.*, 2016.
12. Kuzmov, A. and Minko, T., Nanotechnology approaches for inhalation treatment of lung diseases. *J. Controlled Release*, 219, 500–518, 2015.
13. Hoppentocht, M. *et al.*, Technological and practical challenges of dry powder inhalers and formulations. *Adv. Drug Deliv. Rev.*, 75, 18–315, 2014.

14. Kaialy, W. and Nokhodchi, A., Antisolvent crystallisation is a potential technique to prepare engineered lactose with promising aerosolisation properties: Effect of saturation degree. *Int. J. Pharm.*, 437, 1–2, 57–69, 2012.
15. Kaialy, W. *et al.*, Characterisation and deposition studies of recrystallised lactose from binary mixtures of ethanol/butanol for improved drug delivery from dry powder inhalers. *AAPS J.*, 13, 1, 30–43, 2011.
16. Patil, S.S., Mahadik, K.R., Paradkar, A.R., Liquid crystalline phase as a probe for crystal engineering of lactose: Carrier for pulmonary drug delivery. *Eur. J. Pharm. Sci.*, 68, 43–50, 2015.
17. Muhammad, S.A.F.S. *et al.*, The effect of lactose micro-spherical crystals prepared by conditioning with supercritical fluid on salbutamol sulphate inhalation performance. *J. Supercrit. Fluids*, 71, 92–1015, 2012.
18. Dhumal, R.S. *et al.*, Ultrasound assisted engineering of lactose crystals. *Pharm. Res.*, 25, 12, 2835–2844, 2008.
19. Mehta, P.P. and Pawar, V.S., Electrospun nanofiber scaffolds: Technology and applications, in: *Applications of nanocomposite materials in drug delivery*, pp. 509–573, Woodhead Publishing, 2018.
20. Mehta, P. *et al.*, Effect of USP Induction Ports, Glass Sampling Apparatus, and Inhaler Device Resistance on Aerodynamic Patterns of Fluticasone Propionate-Loaded Engineered Mannitol Microparticles. *AAPS PharmSciTech*, 20, 5, 197, 2019.
21. Kamble, R., Palve, P., Mehta, P., Preparation and evaluation of amorphous olmesartan medoxomil with porous silica microparticles using spray-drying technique. *J. Adv. Pharm. Educ. Res.*, Oct–Dec 4, 4, 2014.
22. Kamble, R.N. *et al.*, Synthesis, solid state characterization and antifungal activity of ketoconazole cocrystals. *J. Pharm. Investig.*, 48, 5, 541–549, 2018.
23. Mohite, R. *et al.*, Synthesis of fisetin co-crystals with caffeine and nicotinamide using the cooling crystallization technique: Biopharmaceutical studies. *New J. Chem.*, 43, 34, 13471–13479, 2019.
24. Kuminek, G. *et al.*, Cocrystals to facilitate delivery of poorly soluble compounds beyond-rule-of-5. *Adv. Drug Deliv. Rev.*, 101, 143–1665, 2016.
25. Dahmash, E.Z. and Mohammed, A.R., Functionalised particles using dry powder coating in pharmaceutical drug delivery: Promises and challenges. *Expert Opin. Drug Deliv.*, 12, 12, 1867–1879, 2015.
26. Mehta, P., Imagine the superiority of dry powder inhalers from carrier engineering. *J. Drug Deliv.*, 2018, 2018.
27. Honda, H. *et al.*, Preparation of monolayer particle coated powder by the dry impact blending process utilizing mechanochemical treatment. *Colloids Surf., A: Physicochemical and Engineering Aspects*, 82, 2, 117–128, 1994.
28. Ishizaka, T. *et al.*, Preparation of drug-diluent hybrid powders by dry processing. *J. Pharm. Pharmacol.*, 41, 6, 361–368, 1989.
29. Pfeffer, R. *et al.*, Synthesis of engineered particulates with tailored properties using dry particle coating. *Powder Technol.*, 117, 1–2, 40–67, 2001.

30. Gera, M. *et al.*, Mechanical methods for dry particle coating processes and their applications in drug delivery and development. *Recent Pat. Drug Deliv. Formul.*, 4, 1, 58–81, 2010.
31. Alonso, M. and Alguacil, F.J., Dry mixing and coating of powders. *Rev. Metal.*, 35, 5, 315–328, 1999.
32. Alonso, M., Satoh, M., Miyanami, K., The effect of random positioning on the packing of particles adhering to the surface of a central particle. *Powder Technol.*, 62, 1, 35–40, 1990.
33. PMA™ High Shear Processor, https://www.gea.com/en/products/high-shear-mixer-granulators.jsp.
34. Beach, L.E., Effect of dry particle coating on the properties of cohesive pharmaceutical powders, A Dissertation Submitted to the Faculty of New Jersey Institute of Technology, in Chemical Engineering, Department of Chemical, Biological and Pharmaceutical Engineering, Page No. 1-144, 2011.
35. GEA Pharma Systems, https://www.gea.com/en/applications/pharma/solid-dosage/solid-dosage_dry-bonding.jsp.
36. Pharmaceutical formulations comprising 4-hexyl)amino]-1-hydroxyethyl}-2-(hydroxymethyl)phenol. US 2011/0319371 A1.
37. Dry powder compositions of glycopyrrolate and magnesium stearate. EP 2 611 423 B1.
38. Pharmaceutical formulations comprising magnesium stearate. WO2005004 845A1.
39. Endoh, S. *et al.*, Experimental and theoretical analysis of mechanical coating process of particles with the theta composer. *Chem. Eng. Commun.*, 191, 10, 1259–1270, 2004.
40. Quevedo, J. *et al.*, Fluidization of nanoagglomerates in a rotating fluidized bed. *AIChE J.*, 52, 7, 2401–2412, 2006.
41. Zhou, Q.T. and Morton, D.A.V., Drug–lactose binding aspects in adhesive mixtures: Controlling performance in dry powder inhaler formulations by altering lactose carrier surfaces. *Adv. Drug Deliv. Rev.*, 64, 3, 275–284, 2012.
42. Weiss, C., McLoughlin, P., Cathcart, H., Characterization of dry powder inhaler formulations using atomic force microscopy. *Int. J. Pharm.*, 494, 1, 393–407, 2015.
43. Dahmash, E.Z. and Mohammed, A.R., Characterization and surface-profiling techniques for composite particles produced by dry powder coating in pharmaceutical drug delivery. *Drug Discov. Today*, 21, 4, 550–561, 2016.
44. FDA (Food Drug Administration), *Guidance for Industry: PAT – A Frame work for Innovative Pharmaceutical Development, Manufacture, and Quality Assurance*, 2004, http://www.fda.gov/downloads/Drugs/Guidance ComplianceRegulatoryInformation/Guidances/UCM070305.pdf.
45. Knop, K. and Kleinebudde, P., PAT-tools for process control in pharmaceutical film coating applications. *Int. J. Pharm.*, 457, 2, 527–536, 2013.

46. Mehta, P.P. and Dhapte, V.V., Propulsive PAT paradigm: Optimization of freeze drying process. *Int. J. Pharm. Sci. Rev. Res.*, 28, 240–246, 2014.
47. Bunker, M., Davies, M., Roberts, C., Towards screening of inhalation formulations: Measuring interactions with atomic force microscopy. *Expert Opin. Drug Deliv.*, 2, 4, 613–624, 2005.
48. Jones, M.D., Young, P., Traini, D., The use of inverse gas chromatography for the study of lactose and pharmaceutical materials used in dry powder inhalers. *Adv. Drug Deliv. Rev.*, 64, 3, 285–293, 2012.
49. Morin, G. and Briens, L., The effect of lubricants on powder flowability for pharmaceutical application. *AAPS PharmSciTech*, 14, 3, 1158–1168, 2013.
50. Morton, D. *et al.*, Surface modification of micronized drug powders to improve aerosolization via mechanical dry powder coating. *Proceedings of the The electronic conference on pharmaceutical sciences*, 2011.
51. Begat, P. *et al.*, The role of force control agents in high-dose dry powder inhaler formulations. *J. Pharm. Sci.*, 98, 8, 2770–2783, 2009.
52. Parlati, C. *et al.*, Pulmonary spray dried powders of tobramycin containing sodium stearate to improve aerosolization efficiency. *Pharm. Res.*, 26, 5, 1084–1092, 2009.
53. Tay, T., Das, S., Stewart, P., Magnesium stearate increases salbutamol sulphate dispersion: What is the mechanism? *Int. J. Pharm.*, 383, 1–2, 62–69, 2010.
54. Zhou, Q.T. *et al.*, Improving aerosolization of drug powders by reducing powder intrinsic cohesion via a mechanical dry coating approach. *Int. J. Pharm.*, 394, 1–2, 50–59, 2010.
55. Jetzer, M.W. *et al.*, Investigations on the mechanism of magnesium stearate to modify aerosol performance in dry powder inhaled formulations. *J. Pharm. Sci.*, 107, 4, 984–998, 2018.
56. Chen, L. *et al.*, Amorphous powders for inhalation drug delivery. *Adv. Drug Deliv. Rev.*, 100, 102–1155, 2016.
57. Tan, B.M.J. *et al.*, Particle surface roughness–Its characterisation and impact on dry powder inhaler performance. *Pulm. Drug Deliv.*, 199–222, 2015.
58. Minne, A. *et al.*, Optimization of the aerosolization properties of an inhalation dry powder based on selection of excipients. *Eur. J. Pharm. Biopharm.*, 70, 3, 839–844, 2008.
59. Shott, M., Morton, D., Davies, R., *A method for preparing a dry powder formulation suitable for inhalation*, Vectura Limited, 2012.
60. Wang, J., Wen, H., Desai, D., Lubrication in tablet formulations. *Eur. J. Pharm. Biopharm.*, 75, 1, 1–15, 2010.
61. Rowe, R.C., Sheskey, P., Quinn, M., *Handbook of pharmaceutical excipients*, Libros Digitales-Pharmaceutical Press, 2009.
62. Shur, J. *et al.*, From single excipients to dual excipient platforms in dry powder inhaler products. *Int. J. Pharm.*, 514, 2, 374–383, 2016.
63. Wang, L., Zhang, Y., Tang, X., Characterization of a new inhalable thymopentin formulation. *Int. J. Pharm.*, 375, 1–2, 1–7, 2009.

64. Pham, D.-D. *et al.*, Pulmonary delivery of pyrazinamide-loaded large porous particles. *Eur. J. Pharm. Biopharm.*, 94, 241–2505, 2015.
65. Boraey, M.A. *et al.*, Improvement of the dispersibility of spray-dried budesonide powders using leucine in an ethanol-water cosolvent system. *Powder Technol.*, 236, 171–1785, 2013.
66. Hoppentocht, M. *et al.*, The Cyclops for pulmonary delivery of aminoglycosides; a new member of the Twincer™ family. *Eur. J. Pharm. Biopharm.*, 90, 8–155, 2015.
67. Sou, T. *et al.*, The effect of amino acid excipients on morphology and solid-state properties of multi-component spray-dried formulations for pulmonary delivery of biomacromolecules. *Eur. J. Pharm. Biopharm.*, 83, 2, 234–243, 2013.
68. Yu, H. *et al.*, Dry powder aerosols of curcumin-chitosan nanoparticle complex prepared by spray freeze drying and their antimicrobial efficacy against common respiratory bacterial pathogens. *Colloids Surf., A: Physicochemical and Engineering Aspects*, 504, 34–425, 2016.
69. Muhsin, M.D.A. *et al.*, Effects of chemical conjugation of L-leucine to chitosan on dispersibility and controlled release of drug from a nanoparticulate dry powder inhaler formulation. *Mol. Pharm.*, 13, 5, 1455–1466, 2016.
70. Merchant, Z. *et al.*, Engineering hydrophobically modified chitosan for enhancing the dispersion of respirable microparticles of levofloxacin. *Eur. J. Pharm. Biopharm.*, 88, 3, 816–829, 2014.
71. Sou, T. *et al.*, Investigating the interactions of amino acid components on a mannitol-based spray-dried powder formulation for pulmonary delivery: A design of experiment approach. *Int. J. Pharm.*, 421, 2, 220–229, 2011.
72. Yu, H. *et al.*, Dry powder aerosols of curcumin-chitosan nanoparticle complex prepared by spray freeze drying and their antimicrobial efficacy against common respiratory bacterial pathogens. *Colloids Surf., A: Physicochemical and Engineering Aspects*, 504, 34–425, 2016.
73. Muhsin, M.D., George, G., Beagley, K., Ferro, V., Wang, H., Islam, N., Effects of Chemical Conjugation of l-Leucine to Chitosan on Dispersibility and Controlled Release of Drug from a Nanoparticulate Dry Powder Inhaler Formulation. *Mol. Pharm.*, 13, 1455–66, 2016.
74. Merchant, Z. *et al.*, Engineering hydrophobically modified chitosan for enhancing the dispersion of respirable microparticles of levofloxacin. *Eur. J. Pharm. Biopharm.*, 88, 3, 816–829, 2014.
75. Pourshahab, P.S. *et al.*, Preparation and characterization of spray dried inhalable powders containing chitosan nanoparticles for pulmonary delivery of isoniazid. *J. Microencapsul.*, 28, 7, 605–613, 2011.
76. Faghihi, H. *et al.*, The use of amino acids to prepare physically and conformationally stable spray-dried IgG with enhanced aerosol performance. *Int. J. Pharm.*, 466, 1–2, 163–171, 2014.

77. Lechuga-Ballesteros, D. *et al.*, Trileucine improves aerosol performance and stability of spray-dried powders for inhalation. *J. Pharm. Sci.*, 97, 1, 287–302, 2008.
78. Schoubben, A. *et al.*, D-leucine microparticles as an excipient to improve the aerosolization performances of dry powders for inhalation. *Eur. J. Pharm. Sci.*, 130, 54–645, 2019.
79. Li, L. *et al.*, l-Leucine as an excipient against moisture on in vitro aerosolization performances of highly hygroscopic spray-dried powders. *Eur. J. Pharm. Biopharm.*, 102, 132–1415, 2016.
80. Yu, J. *et al.*, Protection of hydrophobic amino acids against moisture-induced deterioration in the aerosolization performance of highly hygroscopic spray-dried powders. *Eur. J. Pharm. Biopharm.*, 119, 224–2345, 2017.
81. Dharmadhikari, A.S. *et al.*, Phase I, single-dose, dose-escalating study of inhaled dry powder capreomycin: A new approach to therapy of drug-resistant tuberculosis. *Antimicrob. Agents Chemother.*, 57, 6, 2613–2619, 2013.
82. Behara, S.R.B. *et al.*, Development of a high efficiency dry powder inhaler: Effects of capsule chamber design and inhaler surface modifications. *Pharm. Res.*, 31, 2, 360–372, 2014.
83. Heng, D. *et al.*, Novel alternatives to reduce powder retention in the dry powder inhaler during aerosolization. *Int. J. Pharm.*, 452, 1–2, 194–200, 2013.
84. Torotwa, I. and Ji, C., A Study of the Mixing Performance of Different Impeller Designs in Stirred Vessels Using Computational Fluid Dynamics. *Designs*, 2, 1, 10, 2018.
85. Selvam, P. *et al.*, Micronized drug adhesion and detachment from surfaces: Effect of loading conditions. *Aerosol Sci. Technol.*, 45, 1, 81–87, 2011.
86. Yang, J., Wu, C.-Y., Adams, M., DEM analysis of the effect of electrostatic interaction on particle mixing for carrier-based dry powder inhaler formulations. *Particuology*, 23, 25–30, 2015.
87. Sommerfeld, M., Cui, Y., Schmalfuß, S., Potential and constraints for the application of CFD combined with Lagrangian particle tracking to dry powder inhalers. *Eur. J. Pharm. Sci.*, 128, 299–3245, 2018.

4

Superhydrophobic Polymer Coatings

Amir Ershad Langroudi

Color and Surface Coatings Group, Polymer Processing Department, Iran Polymer and Petrochemical Institute, Tehran, Iran

Abstract

A surface is considered as superhydrophobic in the case of having more than about 150° the apparent contact angle and hysteresis and sliding contact angles less than 10° and 5°, respectively. In addition, it must have a good physical and mechanical stability. Many of artificial hydrophobic polymer coatings are prepared by nature inspiration like lotus leaf. Furthermore, the superhydrophobic coatings with additional functions have recently been developed in various fields of applications such as anti-icing, self-cleaning, water repelling, and anticorrosion coatings. The aim of this chapter is reviewed briefly the important parameters for preparation of superhydrophobic coatings, working mechanisms, as well as fundamental models of superhydrophobicity.

Keywords: Superhydrophobic, models, surface tension, self-cleaning, coatings application

Abbreviations

r	Parameter related to Surface roughness
φ_n	Surface fraction related to material of n
θ and θ_x	Contact angle is usually used for water and x is a specific index from the following list:
	$_Y$ index of Young equation
	$_w$ index of Wenzel model
	$_c$ index of Cassie–Baxter Model
	$_h$ Hysteresis

Email: A.ershad@ippi.ac.ir; Ershadamir@yahoo.com

Inamuddin, Rajender Boddula, Mohd Imran Ahamed and Abdullah M. Asiri (eds.) Polymer Coatings: Technology and Applications, (75–94) © 2020 Scrivener Publishing LLC

	$_a$ Advancing contact angle on a sloping surface (at the lower side)
	$_r$ Receding contact angle on a sloping surface (at the upper Side)
γ_z	Surface tension and z is a specific index from the following list:
	$_{SG}$ Properties at the solid-gas interface
	$_{SL}$ Properties at the solid-liquid interface
	$_{LG}$ Properties at the liquid-gas interface
	$_{LV}$ Properties at the liquid-vapor interface
g	the gravitational constant
$\Delta\rho$	the mass density difference of the fluids
λ_c	capillary length

4.1 Introduction

Many scientific technologies are bioinspired, and among them, superhydrophobic coatings in the field of surface technologies led to a series of innovations and new applied researches [1–3]. Hydrophobic coatings are used for different applications to conserve of historical monuments, self-cleaning for window glasses and clothing articles, anti-icing for outdoor applications such as automotive and airplane, etc. The innovation of superhydrophobic coating and research papers are progressively improved. Figure 4.1 indicates the number of articles in superhydrophobic topic from Google Scholar Web. As it seen from the presented data in this figure, the research articles indicate about 10 times growth in hydrophobic coatings topic during two decades.

Nature has always been a human inspiration in the development of science and technology. Many interdisciplinary researches are based on this background (see Table 4.1). Superhydrophobic materials are among the most prominent examples from nature in surface research and physical chemistry of the surface interface. The development of superhydrophobic materials has led to a number of new research innovations applied across a wide range of fields.

4.2 Theoretical Background

A shape of suspended liquid drop is specified by resultant of two forces, i.e., the surface tension ($\gamma_{lv}\times$ d) and gravity ($\rho g d^3$). The first one creates a spherical form to minimize surface area. However, the latter attempts to

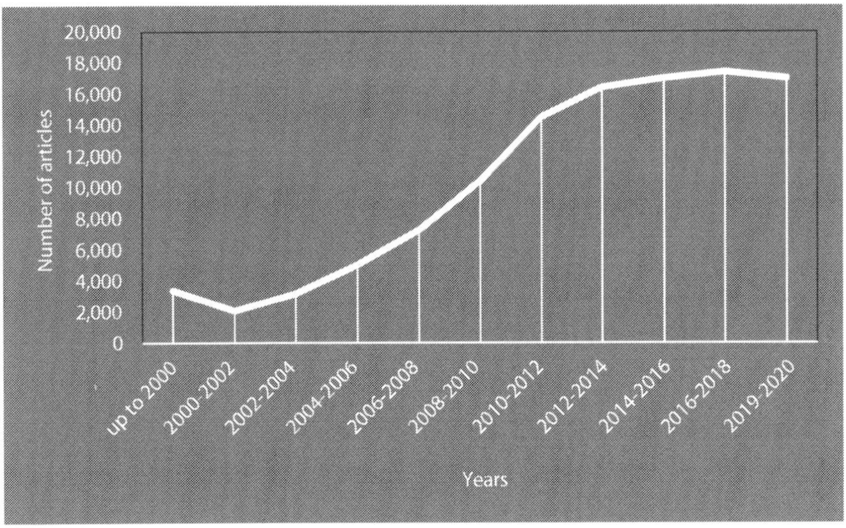

Figure 4.1 An increase trend of the number of research papers with title of superhydrophoic coating in the past two decades.

Table 4.1 Some characteristics of natural examples of hydrophobic surface [4].

Lotus leaves	Low drag and adhesion
Rice leaves	Anisotropic wettability and low adhesion
Butterfly wings	Low and directional adhesion
Gecko feet	High adhesion
Red rose petals	High adhesion
Snail shell	Oleophilic

flatten the drop. The effect of the size of drop on its shape is determined by comparing it with the capillary length.

The capillary length or capillary constant is a fundamental physical property that balances gravity (body forces) and surface tension (Laplace pressure). The capillary length depends on type of liquids and ambient conditions [5, 6].

In the case of a liquid drop immersed in another fluid (e.g., air), the capillary length denoted λ_c is given by.

$$\lambda_C = \sqrt{\frac{\gamma_{LV}}{\Delta\rho . g}}$$

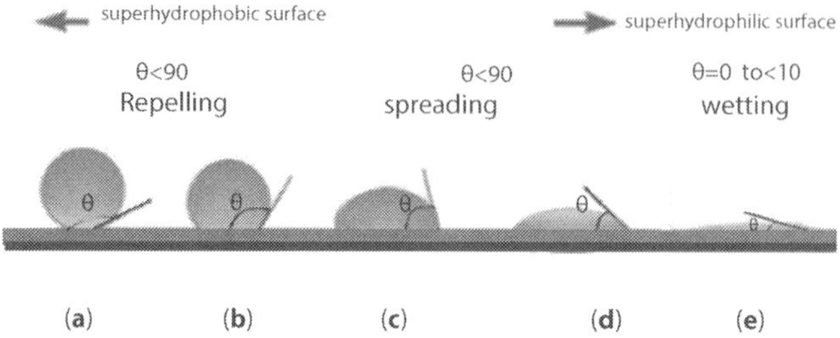

Figure 4.2 Schematic presentation of water droplet on the rigid surface: (a) super hydrophobic, (b) hydrophobic, (c) and (d) hydrophilic, (e) super hydrophilic.

The liquid drop is considered small when its diameter is smaller than the capillary constant. The capillary length is about 3 mm (2.71 mm) for drop water in air at room temperature (20°C) on Earth [7, 8].

The surface hydrophobicity can be measured by placing of a small liquid droplet (e.g., water droplet) on a rigid surface in an equilibrium state. Figure 4.2 shows schematically four categories of the rigid surfaces based on the water contact angle measurement: (1) superhydrophobic, the contact angle (θ) is more than 150°, (2) hydrophobic when θ is in the range of less than 150° to 90°, (3) hydrophilic when θ is less than 90° up to 10°, and (4) superhydrophilic when the contact angle (θ) is less than 10° and near zero degree [9].

The contact angle measurement is based on the surface substrate condition like as smooth or roughness, homogenous, and the fluid surrounding conditions such as temperature.

The Leidenfrost effect describes the hot solid surface that makes a water vapor film when the water drops deposit on it. In this case, the surface indicated the apparent water contact angle at about 180° with a hysteresis contact angle in the range of 0° [3, 10].

4.2.1 Young's Equation

Thomas Young formulated an equation, called "Young's equation" based on the shape of a liquid drop on the solid surface for quantification of

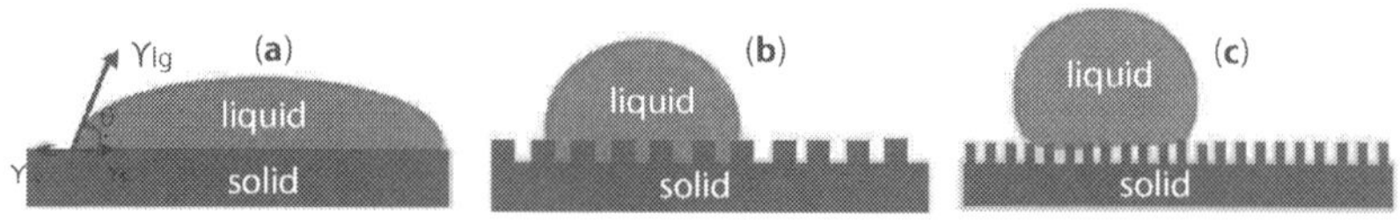

Figure 4.3 Schematic image of a liquid drop on the surface by different models: (a) Young, (b) Wenzel, and (c) Cassie-Baxter.

wettability [11]. When the small liquid drop is set on a flat homogenous rigid and horizontal surface where θ represents the liquid contact angle as seen in Figure 4.3a, three surface tension interfaces become a predominant factor: surface, liquid, and vapor. "Young's equation" establish a relationship between three interfacial surface tensions (i.e., solid-vapor, solid-liquid, and liquid-gas or vapor) as

$$\cos\theta_Y = \frac{\gamma_{SV} - \gamma_{SL}}{\gamma_{Lg}}, \tag{4.1}$$

where θ_Y, γ_{Lg} γ_{SL}, and γ_{sv} represent the contact angle (by Young model), the surface tensions in the following interfaces: liquid-gas, solid-liquid, and solid-vapor, as seen in Figure 4.3a.

The level of surface hydrophobicity can be determined from Young's contact angle (θ_Y) and it can be considered as a characteristic of the surface substance [12]. However, the surface roughness is not seen in calculating the young's contact angle by this equation [13].

4.2.2 Wenzel Model

The Young's equation is modified by Wenzel model [14]. In this model, the roughness of solid surface is considered while the liquid droplets fill the rough channels on the surface as seen in Figure 4.3b.

Wenzel equation relates the apparent contact angle to the surface roughness as given in Equation (4.2):

$$\cos\theta_w = r \times \frac{\gamma_{Sv} - \gamma_{SL}}{\gamma_{Lg}} = rcos\theta_Y, \tag{4.2}$$

where r is surface roughness as the ratio of actual to geometric surface area and θ_w is the apparent contact angle in this model, respectively.

According to this expression, the roughness of solid surface slightly increases in two interfacial surface tensions (i.e., γ_{SV} and γ_{SL}). Therefore, if the apparent contact angle calculates by Wenzel's model, it is usually more than the contact angle obtained by Young's model. Therefore, a hydrophobic solid surface may be considered as super-hydrophobic if roughness is considered [3, 15].

However, the calculated contact angle in the Wenzel model will be more than it from Young's equation, the droplet movement is more difficult.

It can be attributed to high surface contact of liquid droplet without considering the air (vapor) in the voids (i.e., high adhesion).

4.2.3 Cassie-Baxter Model

Cassie and Baxter expanded the Wenzel's equation by considering the shape of the droplet with increasing surface roughness so that the air (vapor) fills the surface cavities. In this case, droplet cannot enter the cavities and it is in contact with the peak of surface roughness as seen in Figure 4.3c [16].

In this case, a part of liquid droplet connects with the peaks and creates an interface in solid surface and liquid droplet while the rest of liquid droplet touches with the vapor or gas in the cavities. This part of liquid droplet makes an interface between the liquid and vapor or gas in the cavities. Therefore, the apparent contact angle, θ_c in Cassie–Baxter's model can be calculated as following:

$$cos\theta_c = \varphi_1.cos\theta_Y - \varphi_2 = \varphi_1.cos\theta_Y - 1 + \varphi_1, \tag{4.3}$$

where φ_1 represents fraction of air (i.e., gas or vapor) in the voids and φ_2 is fraction of solid surface [17, 18].

In this model, the liquid droplets only touch the tip of surface peaks and rolling off from the substrate's surface in a low sliding contact angle (i.e., it is the minimum angle at which a liquid droplet begins to roll down in a sloping surface). From this point of view, the sliding angle is also an essential factor besides the apparent contact angle, for evaluation of surface hydrophobicity [19].

In this regard, the surfaces have a high contact angle and may be don't have a low sliding angle. A phenomenon is seen in the red rose petals by a high surface roughness in nano size and also a surface with high hydrophobicity. The water droplets cannot roll off from the inclined surface because of the high adhesion forces [20].

In addition, the hysteresis contact angle (θ_h) is another important parameter in surface hydrophobicity. The droplet makes an asymmetrical shape on a sloping surface with two contact angles: advancing (θ_a) at the lower side and receding (θ_r) at the upper Side. The difference between these two contact angles is called as the hysteresis contact angle [21–23].

Therefore, it is necessary for creating a super hydrophobic surface, the coating possess a high apparent contact angle, and the low hysteresis contact angle.

Another important parameter in the preparation of super hydrophobic coating is surface morphology, which may have different microstructures containing various combinations of nano and/or mico levels on a

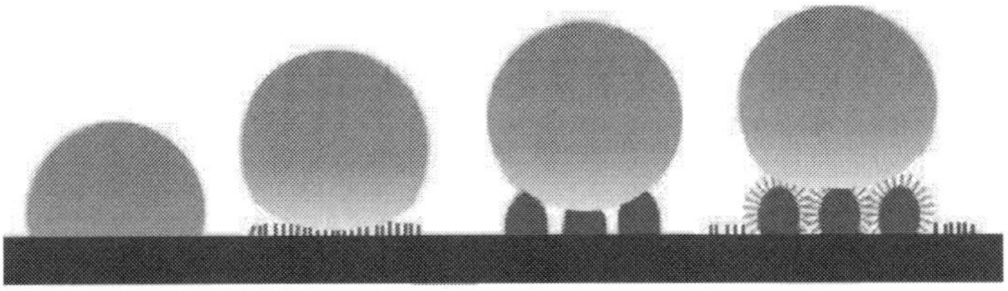

Figure 4.4 The effect of surface microstructure on changing hydrophobic to superhydrophobic behavior.

rough surface (see Figure 4.4). One study found that the effect of surface roughness was not as effective as its chemical nature on hysteresis [24]. In another study, it has been attempted to distinguish between the advancing contact angle (θ_a) and the droplet mobility [25].

4.3 Physical and Chemical Texturing

Physical texturing is an essential process for eliminating surface imperfections, impurities, and surface oxides especially from metallic surfaces. This process may require chemical texturing with coating techniques to enhance surface topography. Some coating methods for metallic surfaces are mentioned in the following.

4.3.1 Cleaning Process

The first step of any coating process is the surface cleaning. It is essential for eliminating of contaminants and getting durability, uniformity, and reproducibility [26]. In addition, it is advisable in most cases to coat immediately after cleaning the surface. The cleaning processes established in literature can be classified as follows: (1) wet chemical procedures use different mixes of organic solvents, acidic and/or bases solutions at various conditional variables such as times, strengths and temperatures. (2) Dry methods; this category can be pointed to, for example, (Ar or O_2) plasma, laser, UV-ozone, pyrolysis treatments. Furthermore, mechanical process can be considered as a pre-cleaning step.

4.3.2 Wet Chemical Reaction

One of the popular methods to increase hydrophobicity of surface is wet chemical etching in the presence of suitable chemical agents and ultrasonic waves. It forms a superhydrophobic surface by creating nano and/or micro structures morphology [27].

4.3.3 Sol-Gel Process

Modified and functionalized silanes and alkoxysilanes can be applied on a wide range of substrates such as ceramics, metals, glass, by using sol-gel processes in different techniques like dip coating, spin coating, to create a superhydrophobic surface with silica nanostructured lattices [28].

4.3.4 Immersion Coated

The reagents are used effectively and the test time is shorter in this method. By etching and using hydrophobic precursors such as 1H, 1H, 2H 2H per fluro decyl triethoxysilane, it is possible to create uniform micro-nanostructures with superhydrophobic properties [29].

Furthermore, it is reported that superhydrophobic and self-cleaning coatings were prepared on aluminum surface by an acidic etching (HCl and HNO_3) then treated with lauric acid [30].

4.3.5 Electrochemical Deposition

The superhydrophobic surfaces were prepared on copper metal surface by electrochemical deposition in dilute ethanolic stearic acid solution and by direct current. The surfaces had water contact angles of 153° and hysteresis of 2° [31]. This superhydrophobic property is attributed to the small-scale nanoporous microstructure of surface and also a hydrophobic copolymer coated on this morphology by CVD electrochemical deposition.

4.3.6 Ion Irradiation or Implantation

High-energy ions can play a key role in surface modification, depending on the ions and its types. Surface hydrophobic modification can be accomplished by ion implants with low-energy ions up to several hundred kV. In this method, the interaction of ions with substances and ions at the material level is the determining factor in hydrophobicity and superhydrophobicity. The type of ion beams used, for example, argon or nitrogen, affects the composition of the surface and changes its roughness. The effects of ion beam irradiation on ultra-hydrophilic films showed that the material morphology could change to hydrophobic in nature [32, 33].

4.3.7 Plasma Treatment

Plasma treatment generally uses ionized gas, which is created by the use of excess energy on the gas. This energy can be used to etch and clean the surface, to fabricate surface nanoparticles and also create nano and micro roughness [30, 31]. In addition, these operations can be applied at low temperature for surface modification of polymers, for example, polystyrene (PS) substrates. Plasma treatment has been performed on copper joints to prevent corrosion in these materials. By plasma etching on copper surfaces with functionalized fluorinated compound, a hydrophobic coating with micro/nano roughness was created which showed WCA as high as 163° and a low θ_h at about 1° (see Figure 4.5) [34, 35].

Recently, all types of hydrophobic and anti-reflex coatings on glass have been reviewed. According to this paper, hydrophobic coatings comprise the main categories:

1. Plasma surface treatment,
2. Sol-gel process,
3. DC sputtering.

Though among them, the sol-gel method, including effective parameters, solvent and precursor selection, and limitations, has been emphasized. On the other hand, this type of coating is also applied to anti-reflection (AR) coatings through the sol-gel method. Furthermore, the challenges of this method, including delicate cleaning methods and water-based hydrophobic coatings, which are suitable for biological and eye-contact applications, have also been investigated [26].

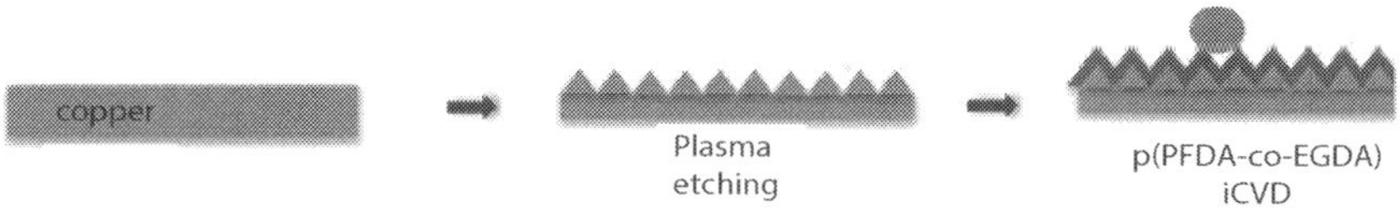

Figure 4.5 The micro and nano roughness is created on the metallic substrate by O_2-Ar-plasma etchings for fabrication of low surface energy. In the following step, a hydrophobic copolymer, containing 1H, 1H, 2H, 2H perfluoro decyl acrylate (PFDA) - ethylene glycol diacrylate (EGDTA) was coated on the etched metallic surface via initiated CVD [35].

4.4 Development of Superhydrophobic Coatings With Nanoparticles

The use of nanoparticles is one of the common ways for the preparation of superhydrophobic coatings. In this section, some of recent progress will be introduced in this field as well as the fabrication methods [36].

4.4.1 CNT Nanoparticles

A simple and effective way for constructing superhydrophobic hybrid membranes is proposed through the covalent bonding of the hydrophobic polymers such as polystyrene to the modified carbon nanotube. The membranes have fire-retardant properties and high performance in oil/water separation.

The methodology for manufacturing superhydrophobic polymer/CNTs hybrid membranes is surface modification of CNTs by creating some photocatalytic modifications, -COOH and -OH groups on carbon nano particle surfaces which can be followed by grafting a typical hydrophobic polymer brushes, for example, PS in a photo-polymerization process. The superhydrophobic hybrid membranes can separate water from oil emulsions with a wide range of absorption and good recycling [37].

Recently, thermomechanical properties of super hydrophobic coatings have been investigated. The thermal cycling was applied on the multi-walled carbon nanotube and silicone nanocomposite coatings. Stress relaxation and wetting properties of silicone and multi-walled carbon nanotube (MWNT) nanocomposite were studied. Super hydrophobic coating for anti-frosting applications is used for energy conservation and safety reasons. Safety factor is a critical parameter which may be concerned in cold weathering regions by wide fluctuation of daily temperature. However, the thermomechanical cycling which can be supported by the superhydrophobic and anti-icing coatings should be quantitatively studied under stress relaxation and wetting behavior.

The silicone - MWNT nanocomposite coatings indicated a microstructure with a lot of open nanopores between the nanoparticles explained as the Cassie-Baxter model. In addition, a superhydrophobic coating should be an adjustment of thermal expansion coefficient by the substrate to avoid of thermal stress.

It has been reported that the amount of stress relaxation for the open microstructure of silicone –30% MWNT nanocomposite reaches about 80% of the value for its same composite with a closed and non-porous microstructure. In addition, the structure and composition of the superhydrophobic silicone - MWNT coating are important fabrication parameters that should be controlled to withstand a lot of thermal cycles. Furthermore, the nucleation and growth of ice crystal can be delayed by size and shape of nanoparticles in anti-icing and hydrophobic coatings [38].

4.4.2 Carbon-Based Fillers

Nano graphene and graphene oxide have been considered for its good electrical properties, thermal conductivity, simple, and inexpensive synthesis methods just as probability to apply them for synthesizing of various types of nanoscale materials, especially super-hydrophobic, anti-corrosion, and anti-microbial coatings. The preparation of stable dispersed nanoparticles in various interfaces such as liquid-liquid and air-liquid interfaces and dual functional agents is the important factors in the preparation of this kind of coatings [39].

Graphene and carbon nanotubes can be used for superhydrophobic coatings due to their nonpolar structure and inherently high surface area. The study exhibited that modification of graphene aerogels with a silane coupling agent induces superhydrophobicity. Furthermore, the amount of modified graphene aerogels was investigated on the receding water contact angle. The results indicated that graphene aerogels are hydrophobic. In addition, a fluorinated treatment can be induced an increment in superhydrophobic property. The result also indicated the modified graphene aerogels compare with the carbon nanomaterials generate higher superhydrophobic property. Moreover, the synthesize procedure is capable to produce in scale up production. It was shown that the hydrophobicity of coating can be improved by modification of graphene's surface with 1H, 1H, 2H, 2H per fluoro decyl tri chlorosilane (FDTS). The results indicated the contact angle can increase from about 125° for unmodified graphene to 160° where it is modified [40].

4.4.3 Silica-Based Superhydrophobic Nanocoatings

Anti-icing coating is an important application of ultra-hydrophobic material. This kind of coating can be used in aircrafts, ships, refrigerators, wind turbines, etc [41, 42].

All hydrophobic coatings are not anti-icing. However, the formation of ice can be delayed at the presence of hydrophobic surfaces. Furthermore, the microstructure of surface and consequently their superhydrophobic properties can be severely reduced by repeating the freeze-thawing cycle [43–45].

A series of nanocomposites were fabricated as anti-icing coating based on organic modification of silanes as a coupling agent and silica nano particles in a range of 20 nm up to 20 μm with cross-linked the acrylic polymers. The result indicates that the surface-modified silica particles have an inherent water contact angle about 110°. In addition, all of the nanocomposite coatings except it containing 20μm silica particles possessed the superhydrophobic properties with the water contact angles more than 150° and the hysteresis contact angle less than 2° by a maximum in superhydrophobicity for sample containing 50 nm silica particles. It can be emphasized that the critical size of silica nano particles on the coated surface are among of important parameters in the anti-icing properties as well as their superhydrophobicity but in two various length ranges. The probability of icing indicates an increasing curve as a function of silica particle size. It can be described by a heterogeneous nucleation model. Therefore, any superhydrophobic coating is not necessarily an anti-icing surface and it is important to achieve these dual properties by designing the surface morphology and tuning surface textures in various length scales [46].

4.5 Transparent Superhydrophobic Coatings for Self-Cleaning Applications

The self-cleaning surface can be formulated in two main ways: in the first process by combining a super-hydrophilic and photocatalyst coating or in the second way by a combination of super-hydrophobic and photocatalyst coating [47–49].

The presence of photocatalytic nanoparticles such as titanium dioxide nanoparticles with anatase structure on the superhydrophobic surface can decompose contaminants and impurities of organic origin and the water flow or raindrops washed away the decomposed particles and dusts by formation of a film of water in a superhydrophilic surface or the rolling effect in a superhydrophobic coating (see Figure 4.6) [50–53].

Therefore, the use of transparent self-cleaning coatings can prevent dirt build-up on the surface and increase surface efficiency for various applications such as solar panels, tiles in the bathroom, glass industries for various applications like building window and automotive glass, architectural protection, and monument conservation, etc [36, 54, 55].

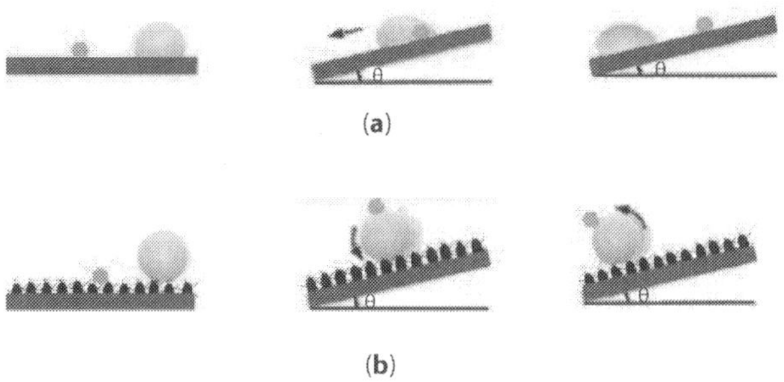

Figure 4.6 Schematic presentation of self-cleaning surface: (a) normal and (b) superhydrophobic (the water drops roll over the slipped surface and clean it from the decomposed organic dirt).

4.6 Superhydrophobic Coatings With Additional Self-Cleaning Function

Superhydrophobic coatings with additional self-cleaning function may be divided into four general categories.

4.6.1 Nanoparticles in Coating

Titanium dioxide nanoparticles in anatase or rutile structures are common materials for preparation of self-cleaning coatings due to its unique physical stability and photocatalytic activity for self-degradation of organic containments [56, 57].

4.6.2 Plant Leaves

Many plant leaves and the most prominent of them, the lotus leaf, are hydrophobic and self-cleaning with a small sliding angle, which is the source of much scientific and industrial research. Lotus effect is known as the intrinsic hydrophobic effect from epicuticular wax as well as isotropic dual roughness of surface in micro and nano ranges [58, 59].

4.6.3 Animal (Gecko Setae)-Inspired

A few animals in nature, such as Gecko and crawlies, have an extraordinary ability to attach and detach from the surface. The extraordinary adhesion of Gecko foots is ascribed to millions microscopic arrays of well-aligned leg hair having about 110 μm length and 5 μm diameter that come apart many of smaller nano range (200–500 nm) ends named spatula [60, 61]. However, the toes are highly hydrophobic but they can adhere strongly to both hydrophilic and hydrophobic surface whatever the polarity of surface chemistry. The high adhesion is attributed to Van der Waals force instead of surface polarity [62–65].

4.6.4 Marine Organisms–Inspired Antifouling Self-Cleaning

The recent researches on polymer-based marine biofouling coatings, for example, skins of shark and whale are reviewed in literature. Based on this study, the adhesion behavior is an important parameter in synthesis of polymeric coating for anti biofouling in marine applications [58, 66, 67].

Recently, the role of the surface topography of the particle contact area and its adhesion has been investigated. The geometry of the surface perturbation affects the contact area and the particle anisotropy on this surface. The local stiffness can be affected by surface topography which determines the strength of adhesion. In this study, topographic features of the surface have been studied systematically and each of them is discussed on the desirable surface and nature of adhesion in the design of corrugated topographic surfaces [64].

4.7 Summary and Outlook

Superhydrophobic surfaces with special features and additional functionalities like self-cleaning, antibacterial, corrosion-resistant, anti-stain, anti-pollution, snow-paste, dust-proof, and many more functionalities are widely used in different industries.

Extensive research is underway to build hydrophobic surfaces and coatings inspired by nature-based materials. Examples of the nature and understanding of their mechanisms and theoretical models provide scientists with an interest and challenge to develop durable superhydrophobic surfaces with nanosynthetic processes by surface modification by chemical and physical methods. However, despite the numerous methods proposed for the manufacture of superhydrophobic surfaces, the mechanical strength and durability of such surfaces for various applications remains a major challenge. Because the superhydrophobic property of these surfaces depends on the exterior area and if it is damaged or wiped out, the surface hydrophobicity disappears, so the challenge is to build volumetric hydrophobic materials that their bulk has superhydrophobic property. In this case, it can exhibit durable superhydrophobicity without considering eventual surface damage. Hydrophobic surfaces in the next generation should be multipurpose. The hydrophobicity is combine with additional functions, like transparency, superelophobic, self-cleaning, de-icing, disinfection, and low-cost coating. Further studies should be conducted towards scale up manufacturing processes using cheaper, environmentally friendly raw materials to prepare them for wider industrial applications.

References

1. Sharma, D.K., Baghel, V., Kumar, R., Avasthi, D.K., Sikarwar, B.S., Recent Developments in Fabrication of Super-Hydrophobic Surfaces: A Review, in: *Advances in Industrial and Production Engineering*, K. Shanker, R. Shankar, R. Sindhwani (Eds.), pp. 127–140, Springer, Singapore, 2019.
2. Si, Y., Dong, Z., Jiang, L., Bioinspired designs of superhydrophobic and superhydrophilic materials. *ACS Cent. Sci.*, 4, 1102, 2018.
3. Jeevahan, J., Chandrasekaran, M., Joseph, G.B., Durairaj, R.B., Mageshwaran, G.J.J.O.C.T., Superhydrophobic surfaces: A review on fundamentals, applications, and challenges. *J. Coat. Tech. Res.*, 15, 231, 2018.
4. Sethi, S.K. and Manik, G., Recent progress in super hydrophobic/hydrophilic self-cleaning surfaces for various industrial applications: A review. *Polym. Plast. Technol. Eng.*, 57, 1932, 2018.
5. Kim, S.H., Fabrication of Superhydrophobic Surfaces. *J. Adhes. Sci. Technol.*, 22, 235, 2008.
6. Shirtcliffe, N.J., McHale, G., Atherton, S., Newton, M.I., An Introduction to Superhydrophobicity. *Adv. Coll. Interface Sci.*, 161, 124, 2010.
7. Aarts, D.G.A.L., Capillary Length in a Fluid–Fluid Demixed Colloid–Polymer Mixture. *J. Phys. Chem. B.*, 109, 7407, 2005.
8. https://en.wikipedia.org/wiki/Capillary_length.
9. Koch, K. and Barthlott, W., Superhydrophobic and Superhydrophilic Plant Surfaces: An Inspiration for Biomimetic Materials. *Philos. Trans. R. Soc. A*, 367, 1487, 2009.
10. Quere, D. and Reyssat, M., Non-adhesive Lotus and Other Hydrophobic Materials. *Philos. Trans. R. Soc. A*, 366, 1539, 2008.
11. Young, T., An Essay on the Cohesion of Fluids. *Philos. Trans. R. Soc. London*, 95, 65, 1805.
12. Crick, C.R. and Parkin, I.P., Preparation and Characterisation of Super-Hydrophobic Surfaces. *Chem. Eur. J.*, 16, 3568, 2010.
13. Wang, X.S., Cui, S.W., Zhou, L., Xu, S.H., Sun, Z.W., Zhu, R.Z., A Generalized Young's Equation for Contact Angles of Droplets on Homogeneous and Rough Substrates. *J. Adhes. Sci. Technol.*, 28, 161, 2014.

14. Wenzel, R.N., Resistance of Solid Surfaces to Wetting by Water. *Ind. Eng. Chem.*, 28, 988, 1936.
15. Dorrer, C. and Ruhe, J., Some Thoughts on Superhydrophobic Wetting. *Soft Matter*, 5, 51, 2009.
16. Cassie, A.B.D. and Baxter, S., Wettability of Porous Surfaces. *Trans. Faraday Soc.*, 40, 546, 1944.
17. McHale, G., Cassie and Wenzel: Were They Really So Wrong? *Langmuir*, 23, 8200, 2007.
18. Bhushan, B. and Jung, Y.C., Natural and Biomimetic Artificial Surfaces for Superhydrophobicity, Self-Cleaning, Low Adhesion, and Drag Reduction. *Prog. Mater. Sci.*, 56, 1, 2011.
19. Miwa, M., Nakajima, A., Fujishima, A., Hashimoto, K., Watanabe, T., Effects of the Surface Roughness on Sliding Angles of Water Droplets on Superhydrophobic Surfaces. *Langmuir*, 16, 5754, 2000.
20. Feng, L., Zhang, Y., Xi, J., Zhu, Y., Wang, N., Xia, F., Jiang, L., Petal Effect: A Superhydrophobic State with High Adhesive Force. *Langmuir*, 24, 4114, 2008.
21. Lam, C.N.C., Wu, R., Lia, D., Hair, M.L., Neumann, A.W., Study of the Advancing and Receding Contact Angles: Liquid Sorption as a Cause of Contact Angle Hysteresis. *Adv. Coll. Interface. Sci.*, 96, 169, 2002.
22. Eral, H.B., Mannetje, D.J.C.M., Oh, J.M., Contact Angle Hysteresis: A Review of Fundamentals and Applications. *Colloid Polym. Sci.*, 291, 247, 2013.
23. Gao, L. and McCarthy, T.J., Contact Angle Hysteresis Explained. *Langmuir*, 22, 6234, 2006.
24. Extrand, C.W. and Kumagai, Y., Liquid Drops on an Inclined Plane: The Relation between Contact Angles, Drop Shape and Retentive Force. *J. Colloid Interface Sci.*, 170, 515, 1994.
25. Pierce, E., Carmona, F.J., Amirfazli, A., Understanding of Sliding and Contact Angle Results in Tilted Plate Experiments. *Colloids Surf. A Physicochem. Eng. Asp.*, 323, 73, 2008.
26. Dey, T. and Naughton, D., Cleaning and anti-reflective (AR) hydrophobic coating of glass surface: A review from materials science perspective. *J. Sol.-Gel. Sci. Technol.*, 77, 1, 2016.
27. Pan, L., Dong, H., Bi, P., Facile preparation of super- hydrophobic copper surface by HNO3 etching technique with the assistance of CTAB and ultrasonication. *Appl. Surf. Sci.*, 257, 1707, 2010.
28. Parale, G.V., Mahadik, B.D., Phadtare, D.V., Paisal, A.A., Park, H.H., Wategaonkar, S.B., Dip coated superhydrophobic and anticorrosive silica coatings. *Int. J. Mater. Sci. Eng.*, 4, 60, 2016.
29. Kong, X.C., Yang, G., Yu, L., Zhang, P., Preparation and characterization of slice-like $Cu_2(OH)_3NO_3$ superhydrophobic structure on copper foil. *Appl. Surf. Sci.*, 254, 7255, 2008.
30. Varshney, P., Mohapatra, S.S., Kumar, A., Fabrication of mechanically stable superhydrophobic aluminium surface with excellent self-cleaning and anti-fogging properties. *Biomimetics*, 2, 2, 1, 2017.

31. Huang, Y.H., Wu, J.T., Yang, S.Y., Direct fabricating patterns using stamping transfer process with PDMS mold of hydrophobic nanostructures on surface of micro-cavity. *Microelectron. Eng.*, 88, 849, 2011.
32. Lakshmi, S.V.B.G. and Avasthi, K.D., Tuning of wettability of PANI-GNP composites using keV energy ions. *Nucl. Instr. Methods Phys. B.*, 379, 152, 2016.
33. Singh, B.U., Agarwal, C.D., Khan, A.S., Kumar, M., Tripathi, A., Singhal, R., Panigrahi, K.B., Avasthi, K.D., Engineering of hydrophilic and plasmonic properties of Ag thin film by atom beam irradiation. *Appl. Surf. Sci.*, 258, 1464, 2011.
34. Jafari, R., Asadollahi, S., Farzaneh, M., Applications of plasma technology in development of superhydrophobic surfaces: A review. *Plasma Chem Plasma Process*, 33, 177, 2013.
35. Vilaro, I., Yague, L.J., Borros, S., Superhydrophobic copper surfaces with anti-corrosion properties fabricated by solventless CVD methods. *Appl. Mater. Interfaces*, 9, 1057, 2017.
36. Das, S., Kumar, S., Samal, S.K., Mohanty, S., Nayak, S.K., A review on superhydrophobic polymer nanocoatings: Recent development and applications. *Ind. Eng. Chem. Res.*, 57, 2727, 2018.
37. Gu, J., Xiao, P., Chen, J., Liu, F., Huang, Y., Li, G., Zhang, J., Chen, T., Robust preparation of superhydrophobic polymer/carbon nanotube hybrid membranes for highly effective removal of oils and separation of water-in-oil emulsions. *J. Mat. Chem. A*, 2, 15268, 2014.
38. Sohn, Y., Kim, D., Lee, S., Song, J.Y., Park, S., Kim, H., Ko, Y., Chu, K., Han, I., Correlation between Thermo-mechanical Reliability and Superhydrophobic Nature of CNT Composite Coatings. *Int. Symp. Microelect.*, 1, 000086–000091, 2014. *Inter. Microelect. Assem. Packag. Soc.*, 2014.
39. Advincula, R.C., Graphene Based Nanomaterials and Coatings. *Corrosion 2019, NACE Int.*, 2019.
40. Lin, Y., Ehlert, G.J., Bukowsky, C., Sodano, H.A., Superhydrophobic functionalized graphene aerogels. *ACS Appl. Mater. Interfaces*, 3, 2200, 2011.
41. Zhang, P. and Lv, F.Y., A review of the recent advances in superhydrophobic surfaces and the emerging energy-related applications. *Energy*, 82, 1068, 2015.
42. Kreder, M.J., Alvarenga, J., Kim, P., Aizenberg, J., Design of anti-icing surfaces: Smooth, textured or slippery? *Nat. Rev. Mater.*, 1, 15003, 2016.
43. Boinovich, L.B., Emelyanenko, A.M., Ivanov, V.K., Pashinin, A.S., Durable Icephobic Coating for Stainless Steel. *ACS Appl. Mater. Interfaces*, 5, 2549, 2013.
44. Hejazi, V., Sobolev, K., Nosonovsky, M., From Superhydrophobicity to Icephobicity: Forces and Interaction Analysis. *Sci. Rep.*, 3, 1, 2013.
45. Sarkar, D.K. and Farzaneh, M., Superhydrophobic Coatings with Reduced Ice Adhesion. *J. Adhes. Sci. Technol.*, 23, 1215, 2009.

46. Cao, L., Jones, A.K., Sikka, V.K., Wu, J., Gao, D., Anti-icing superhydrophobic coatings. *Langmuir*, 25, 12444, 2009.
47. Ganesh, V.A., Raut, H.K., Nair, A.S., Ramakrishna, S., A review on self-cleaning coatings. *J. Mater. Chem.*, 21, 16304, 2011.
48. Tadanaga, K., Morinaga, J., Matsuda, A., Minami, T., Superhydrophobic–superhydrophilic micropatterning on flowerlike alumina coating film by the sol–gel method. *Chem. Mater.*, 12, 590, 2000.
49. Roach, P., Shirtcliffe, N.J., Newton, M.I., Progress in superhydrophobic surface development. *Soft Matter*, 4, 224, 2008.
50. Blossey, R., Self-cleaning surfaces—Virtual realities. *Nat. Mater.*, 2, 301, 2003.
51. Sun, R.D., Nakajima, A., Fujishima, A., Watanabe, T., Hashimoto, K., Photoinduced surface wettability conversion of ZnO and TiO_2 thin films. *J. Phys. Chem. B*, 105, 1984, 2001.
52. Hashimoto, K., Irie, H., Fujishima, A., TiO_2 photocatalysis: A historical overview and future prospects. *Jap. J. Appl. Phys.*, 44, 8269, 2005.
53. Liu, H., Gao, S.W., Cai, J.S., He, C.L., Mao, J.J., Zhu, T.X., Chen, Z., Huang, J.Y., Meng, K., Zhang, K.Q., Al-Deyab, S., Recent progress in fabrication and applications of superhydrophobic coating on cellulose-based substrates. *Materials*, 9, 124, 2016.
54. Somasundaram, S. and Kumaravel, V., Application of Nanoparticles for Self-Cleaning Surfaces, in: *Emerging Nanostructured Materials for Energy and Environmental Science. Environmental Chemistry for a Sustainable World*, vol. 23, S. Rajendran, M. Naushad, K. Raju, R. Boukherroub (Eds.), pp. 471–498, Springer, Cham, 2019.
55. Nishimoto, S. and Bhushan, B., Bioinspired self-cleaning surfaces with superhydrophobicity, superoleophobicity, and superhydrophilicity. *Rsc Advances*, 3, 671, 2013.
56. Lai, Y., Tang, Y., Gong, J., Gong, D., Chi, L., Lin, C., Chen, Z., Transparent superhydrophobic/superhydrophilic TiO_2-based coatings for self-cleaning and anti-fogging. *J. Mater. Chem.*, 22, 7420, 2012.
57. Kamegawa, T., Shimizu, Y., Yamashita, H., Superhydrophobic surfaces with photocatalytic self-cleaning properties by nanocomposite coating of TiO_2 and polytetrafluoroethylene. *Adv. Mater.*, 24, 3697, 2012.
58. Liu, K. and Jiang, L., Bio-inspired self-cleaning surfaces. *Annu. Rev. Mater. Res.*, 42, 231, 2012.
59. Liu, K., Yao, X., Jiang, L., Recent developments in bio-inspired special wettability. *Chem. Soc. Rev.*, 39, 3240, 2010.
60. Hansen, W.R. and Autumn, K., Evidence for Self-Cleaning in Gecko Setae. *Proc. Natl. Acad. Sci.*, 102, 385, 2005.
61. Geim, A.K., Dubonos, S.V., Grigorieva, I.V., Novoselov, K.S., Zhukov, A.A., Shapoval, S.Y., Microfabricated Adhesive Mimicking Gecko Foot-Hair. *Nat. Mater.*, 2, 461, 2003.

62. Wen, G., Guo, Z., Liu, W., Biomimetic polymeric superhydrophobic surfaces and nanostructures: From fabrication to applications. *Nanoscale*, 9, 3338, 2017, 2017.
63. Cui, M., Wang, B., Wang, Z., Nature-Inspired Strategy for Anticorrosion. *Adv. Eng. Mater.*, 2019, https://doi.org/10.1002/adem.201801379
64. Erramilli, S. and Genzer, J., Influence of surface topography attributes on settlement and adhesion of natural and synthetic species. *Soft Matter*, 15, 4045, 2019.
65. Zhu, J., Tian, Y., Liu, X., Rapid fabrication of super-hydrophobic surfaces of silicon wafers with excellent anisotropic wetting. *Microsyst. Technol.*, 25, 237, 2019.
66. Kirschner, C.M. and Brennan, A.B., Bio-inspired antifouling strategies. *Annu. Rev. Mater. Res.*, 42, 211, 2012.
67. Nir, S. and Reches, M., Bio-inspired antifouling approaches: The quest towards non-toxic and non-biocidal materials. *Curr. Opin. Biotechnol.*, 39, 48, 2016.

5

Superhydrophobic Coatings Applications

Hamidreza Parsimehr and Amir Ershad Langroudi*

Color and Surface Coating Group, Polymer Processing Department, Iran Polymer and Petrochemical Institute, Tehran, Iran

Abstract

The superhydrophobic coatings have various applications that determined by the substrate, the molecular structure of the coating, and the environmental conditions. The final properties of the coatings depend on the physicochemical characteristics of the substrate and coating. According to the substrate and coating properties, the application of the superhydrophobic coating were determined. Therefore, in first step, substrate and molecular structure of the superhydrophobic coatings were investigated. Afterwards, in the second step, application of these coatings is divided into the three major group and examined.

Keywords: Superhydrophobic, repellent, surface tension, self-cleaning, coatings application

5.1 Introduction

In order to investigate the application of superhydrophobic coatings, two steps were examined. The first step is the study about the nature of the superhydrophobic coatings and substrates. The second step focused on the study about superhydrophobic coatings applications which originate from physicochemical characteristics of the coatings and substrates. In first step, physicochemical characteristics of the substrates and coatings as two significant issues were examined. This step has significant role to understand the nature of the superhydrophobic coatings which determine the properties and application of the superhydrophobic coatings in different situations. The first part investigates the nature of the substrate and coating divided

**Corresponding author*: A.ershad@ippi.ac.ir; Ershadamir@yahoo.com

Inamuddin, Rajender Boddula, Mohd Imran Ahamed and Abdullah M. Asiri (eds.) Polymer Coatings: Technology and Applications, (95–120) © 2020 Scrivener Publishing LLC

into two subsets which include study about substrates and superhydrophobic coatings, respectively. The final application of the superhydrophobic coatings was determined according to the physical and chemical properties of the substrate and coatings which are showed in Figure 5.1.

The first significant issue in the first step is the substrate characteristics. Different types of substances are used as substrates including building materials [1–4] (especially brick-like substances) [5–7], Coating [8–10], glass [11–15], leather [16], metal [17] (especially steel) [18–21], paper [22–25], plastics [26–28], solar cell [29–32], textiles [33–37] (especially cotton) [38–41], and wood [8, 42–44] were investigated. The substrates physicochemical characteristics have significant influence on the choice of the coatings type. The substrate roughness [45] and the polar bonds content [46] in substrate are the two major subject in physicochemical characteristics of substrates. The second issue in the first step is about nature of Superhydrophobic coatings and their chemical structure. Superhydrophobic attribute exists in some organic and inorganic substances which originates from very low surface tension of these substances [47, 48]. The two significant issues in

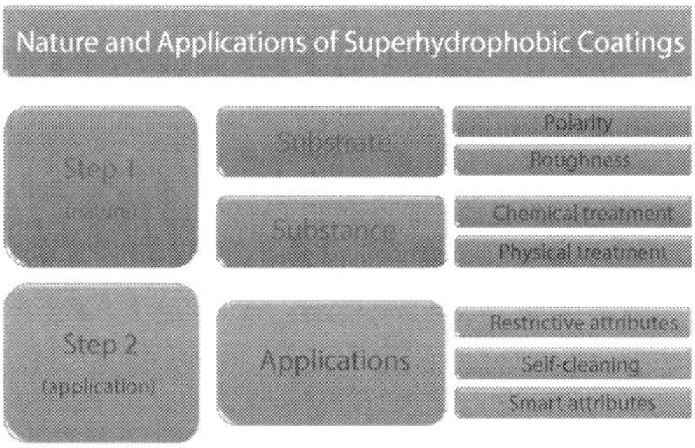

Figure 5.1 Investigation of the superhydrophobic coatings applications in two steps.

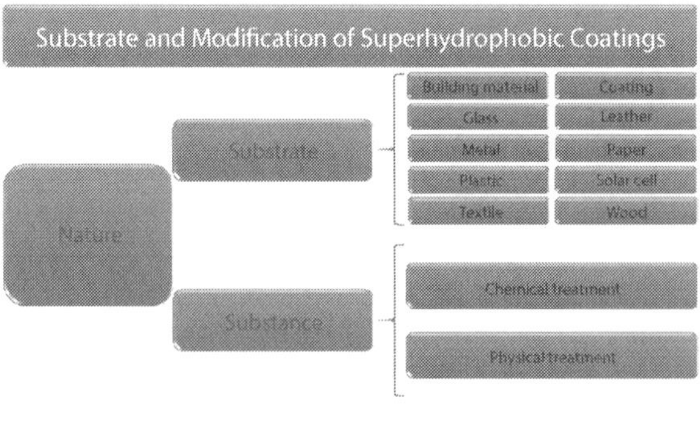

Figure 5.2 Superhydrophobic coatings substrates and substances.

the first step were showed in the Figure 5.2. After reviewing these two significant issues in the first step (nature of materials), the coatings could be classified according to their application into the three main group.

5.2 Step I

The coating and substrate characterization have significant influence in different applications like superhydrophobic coatings applications. Therefore, physical and chemical properties of the substrate and coating must be investigated in the first step.

5.2.1 Substrate

Roughness is the most important factor in the physical properties of the substrate. Roughness growth in the nanoscale could lead to the form superhydrophobic in the surfaces [18]. Also, chemical structure of substrate determined interactions between substrate and coating. Substrate roughness and substrate chemical bonds are two important properties have influence on the coating adhesion and durability which determines the type of coating [49, 50]. Not only the substrate roughness is affected on the coatings adhesion but also the roughness of the coating effects on the properties of the superhydrophobic coating [51, 52].

5.2.2 Substance

Superhydrophobic attribute could produce by chemical and physical techniques. However, the chemical methods have more variety than the physical methods. The chemical methods include organic and inorganic substances. Various organic and inorganic substances have the ability to produce superhydrophobic property on the surface which are showed in Figure 5.3. Organic substances in four group and inorganic substances in three group could produce superhydrophobic property on the surface. Organic materials are long chain hydrocarbons, organofluorine, organosilicon, and organic aerogel. The inorganic materials are inorganic aerogels, inorganic nanoparticles and nanowires. The long chain hydrocarbons are lauric acid [9, 53], stearic acid [54], glycerol stearoyl ester [42], octadecyl isocyanate [55], Octadecyl mercaptan [56], n-Octadecylamine [57], and myristic acid [58]. Cellulose aerogels [57, 59] and carbonaceous aerogel [60] as organic aerogels are the last group of superhydrophobic organic substances. The organofluorine compounds including water-dispersed perfluorinated

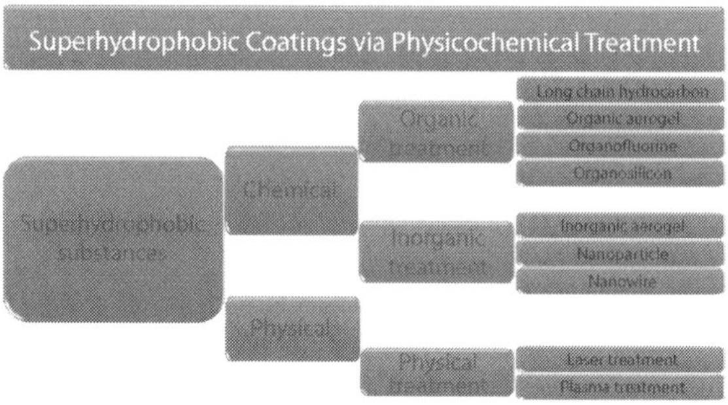

Figure 5.3 Superhydrophobic approach via chemical and physical techniques.

acrylic copolymer [61], 1H,1H,2H,2H-perfluorodecanethiol [36, 41], 1H,1H,2H,2H-Perfluorooctyltriethoxysilane [20, 21, 60], Fluorinated ethylene propylene [31], 1H,1H,2H,2H-perfluorodecyltrichlorosilane [62], 1H,1H,1H,2H-perfluorodecyl trimethoxysilane [63], Polytetrafluoroethylene [35], Perfluoroalkyl methacrylic copolymer [44], poly(2-perfluorooctylethyl methacrylate) [64], and Heptacedafluoro-1,1,2,2-tetra-hydrodecyltrichlorosilane [17, 65].

Also, the organosilicon substances are methyltrichlorosilane [37], Polydimethylsiloxane [15, 66, 67], Tetraethyl orthosilicate [15], Hexadecyltrimethoxysilane [68], Hexamethyldisilazane [29, 69], (3-Glycidyloxypropyl) trimethoxysilane [29], chlorotrimethylsilane [70], polyhedral oligomeric silsesquioxane-graphene oxide [71], Hexadecyltrimethoxysilane [72], and Heptaisobutyl-Polyhedral oligomeric silsesquioxane-propyl methacrylate [73]. Superhydrophobic inorganic materials were divided into the three major group. Silica aerogels [74] as inorganic aerogels were used to produce superhydrophobic property in different studies. Also, nanosilica [10, 75–77], Gold (III) chloride tetrahydrate ($HAuCl_4$) [78], and silver(I) oxide (Ag_2O) [79] nanosilver [36] were used as inorganic nanoparticles. Finally, Copper hydroxide nanowires [80], Copper Nanowire [80], ZnO nanowires [58, 77], carbon nano fiber silicone nanofilament [27, 81], and γ-Al_2O_3 [82] as nanowires. The physical techniques through roughness surface alteration were produced superhydrophobic surfaces including laser [18, 83, 84] and plasma [85, 86] treatment.

5.3 Step II

The coatings application categorized into the three main parts: self-cleaning, restrictive attributes, and smart attributes that are showed in

Figures 5.4 and 5.5. Self-cleaning is referred to the spontaneous decontamination of the surfaces against different liquid and solid pollutants. In other words, self-cleaning properties prevent sticking various types of materials on the surface. In order to approach self-cleaning superhydrophobic attribute, the contact angle between surface and pollutant must be increased that lead to decrease adhesion between pollutant and the surface. The restrictive attributes refer to some properties which prevent to occur some unfavorable events on the surface like fungal growth or corrosion beginning. Preventing such undesirable issues have great importance because these unfavorable events cause significant financial loss to the industries around the world. The smart attributes refer to some properties which could change in different situations like self-healing, self-assembly, and stimuli responsive properties. All superhydrophobic applications were examined in this section.

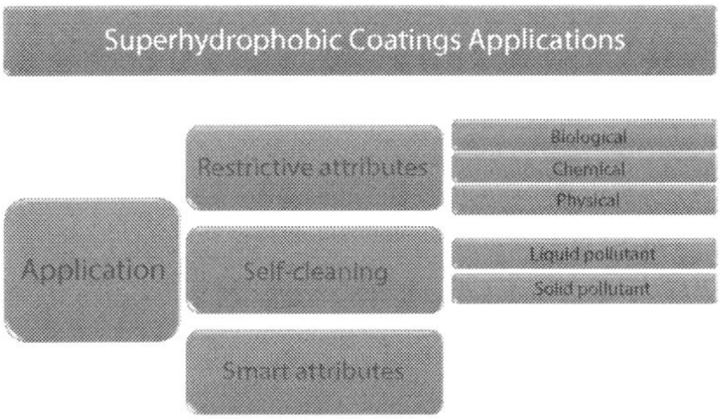

Figure 5.4 Superhydrophobic coatings applications.

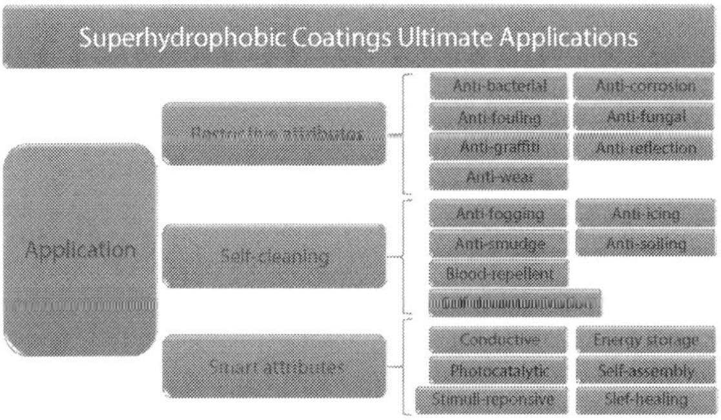

Figure 5.5 Superhydrophobic coatings application in different situations.

5.3.1 Restrictive Attributes

The adhesion of various materials on the surface could make other negative secondary phenomena in addition to the dirt and unpleasant appearance which showed in Figure 5.6. One of the most important of these undesirable phenomena is the corrosion that annually causes immense financial losses to the industry. The surface superhydrophobic attribute could prevent the materials adhering on the surface and consequently prevent these undesirable properties.

5.3.1.1 Biological Agents

Living organisms can stick to the surface and lead to surface contamination. Also, it can lead to other inappropriate issues such as surface corrosion. Therefore, it is important to prevent the living organism's adhesion to the surface. Mainly, organisms that can adhere to the surface are divided into macro and nano scales. The macro scale includes moss, and also the nano scale includes algae, bacteria, and fungi.

5.3.1.1.1 Anti-Bacterial

The bacteria attachment on the surface in addition to reduce surface efficiency can lead to some unpleasant events such as corrosion [87, 88]. Also, bacteria presence on the surface causes various infections and diseases. Therefore, preparation of surfaces which can repel or destroy bacteria has significant importance. The superhydrophobic anti-bacterial surfaces have been prepared via different chemical treatment including tannic acid treatment by silver nanoparticles [36], modification of silver nanoparticles through chitosan and 1H,1H,2H,2H-perfluorodecanethiol [41],

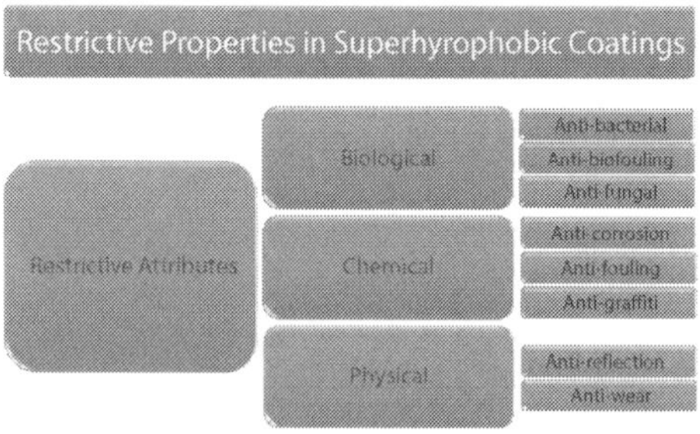

Figure 5.6 Superhydrophobic coatings applications to prevent inappropriate phenomenon.

polydimethylsiloxane treatment by copper nanowires [80], and also zinc oxide nanoparticles treatment by (3-Aminopropyl)triethoxysilane and hexadecyltrimethoxysilane [68]. Applying these coatings in addition to have self-cleaning superhydrophobic feature will also reduce disease and infection probability.

5.3.1.1.2 Anti-Biofouling

Fouling is usually has been originated from other surface phenomena which occurs on the surface. The fouling phenomenon lead to reduce surface efficiency and even corrosion and degradation of surfaces. The fouling materials include living organisms (biofouling like algae [89]) and non-living substance (organic and inorganic) [90] that biofouling is discussed in this section. The non-living fouling will be discussed in the chemical agents. The anti-biofouling feature through superhydrophobic surfaces was obtained. Different substances such as P25 TiO_2 nanoparticles coated with 1H,1H,2H,2H-perfluorooctyltriethoxysilane [91], Cu-electrodeposited Stainless Steel surface via electrodeposition oxidation and vapor deposition [92], elastomeric silicone/graphene oxide γ-Al_2O_3 hybrid nanorods [82], poly(dimethylsiloxane)-poly(tert-butylacrylate)-poly(methacrylolsobutyl POSS) [73], and Sepiolite coated with cinnamic acid and myristic acid have been used to prepare superhydrophobic anti-biofouling coatings [93].

5.3.1.1.3 Anti-Fungal

The fungi attachment on the surface like bacteria organism lead to reduce surface efficiency and some unpleasant events such as corrosion [94, 95]. The chemical treatment of wood by cellulose stearoyl esters and glycerol stearoyl esters [42] and also surface treatment of braze aluminium by heptadecafluoro-1,1,2,2-tetrahydrodecyl trichlorosilane were obtained the superhydrophobic anti-fungal surfaces.

5.3.1.2 Chemical Agents

Chemical pollutants cause various types of pollution, reduce performance, and even destroy in different surfaces. In general, the corrosion phenomenon is recognized as the most important issue in surface-limiting substances. There are many parameters contributing to the corrosion phenomenon. Also, the graffiti and adhesion of various non-living substances to the surface are classified as chemical pollutants.

5.3.1.2.1 Anti-Corrosion

The corrosion phenomenon is one of the most important challenges that human beings face. This phenomenon has many different types and various parameters affect in corrosion. Corrosion damages billions of dollars every year to industries of all countries. Therefore, the corrosion phenomenon investigation has been always interested to scientists and industries. The non-adhering to the surface is an ideal idea to prevent the initiation and propagation of corrosion [96]. Hence, superhydrophobic coatings and surfaces are special choice between corrosion inhibitors. There are widespread researches that have been done to prepare corrosion immunity through superhydrophobicity. The anti-corrosion feature through superhydrophobicity has obtained using various chemical treatment such as polyhedral oligomeric silsesquioxane-graphene oxide [71], Zn-Ni-Co coating modification with myristic acid [58], high density polyethylene solution [14], steel modified by 1H,1H,2H,2H-perfluoroalkyltriethoxysilane [20], Polysulfone treated by carbon nanotubes [97], Polydimethylsiloxane [98], zinc treated by NH_4F [99], rubber modified with 1H,1H,2H,2H-perfluorodecanethiol [100], discotic columnar liquid crystals [101], paper treated by hydrophobic fumed silica and carbon nanotube [102], and also covering nylon mesh [103] as physical treatment have been investigated.

5.3.1.2.2 Anti-Fouling

The adhesion and interaction of various organic and inorganic compounds with the surface lead to widespread problems. Therefore, the superhydrophobicity feature can prevent these phenomena [90]. Zinc nanoparticles modified by stearic acid [104] and SiO_2 modified by 1H,1H,2H,2H-perfluorodecyltriethoxysilane lead to prepare superhydrophobic anti-fouling coatings [105].

5.3.1.2.3 Anti-Graffiti

The graffiti especially as vandalism is an important issue that has received serious attention in recent years. This type of contamination has considerable removing cost and also scientists investigate different methods to prevent graffiti contamination. The aluminium coated by Polysilazane-based coatings have hydrophobicity and anti-graffiti feature [106]. Also, the hydroxyl-terminated hyper-branched polyester coating precursor treated by mono-hydroxyl-terminated polydimethylsiloxane obtained hydrophobic anti-graffiti coating. Although obtaining anti-graffiti via superhydrophobic coatings has not been investigated completely but the contact angle analyses of these two researches were determine more than 100° which means hydrophobicity increase anti-graffiti feature.

5.3.1.3 *Physical Agents*

The anti-reflective and anti-wear properties recognized as physical features which have been obtained through surfaces superhydrophobicity. The variability of physical features is not as widespread as the biological and chemical properties. Also, the effect of hydrophobicity on these properties is less than the biological and chemical features. However, the hydrophobicity phenomenon can improve these features. The impact of superhydrophobicity on other physical features has not yet been fully investigated and needs further research in this field at the future.

5.3.1.3.1 Anti-Reflection

The superhydrophobic anti-reflective coatings have been obtained via various chemical treatment such as replication process of polydimethylsiloxane treated by heptacedafluoro-1,1,2,2-tetra-hydrodecyl-trichlorosilane [65], SiO_2 treated by hexamethyldisilazane [69], polydimethylsiloxane nanocone cluster microstructures with different aspect ratios from anodized aluminum oxide template [107]. Also, using tetraethyl orthosilicate and hexamethyldisilazane as superhydrophobic sol and antireflective layer through (3-Glycidyloxypropyl)trimethoxysilane and octadecylamine led to prepare superhydrophobic anti-reflective coating [29].

5.3.1.3.2 Anti-Wear

It has been needed inorganic and metallic substances with hard nature to prepare the anti-wear feature. The common superhydrophobic substances such as organofluorine and organosilicon which based on organic compounds have not significant abrasion resistance. Therefore, the anti-wear superhydrophobic surfaces have been prepared using inorganic or metallic surfaces which have good abrasion resistance through superhydrophobicity. The polyhedral oligomeric silsesquioxane grafted by graphene oxide [71] as inorganic substance and also copper coated by $CuZn_5$ and $AuZn_3$ [78] as metallic to prepare anti-wear superhydrophobic surfaces were investigated.

5.3.2 Self-Cleaning

Typically, separating different types of contamination from the surfaces require considerable time and energy. Therefore, self-cleaning coatings

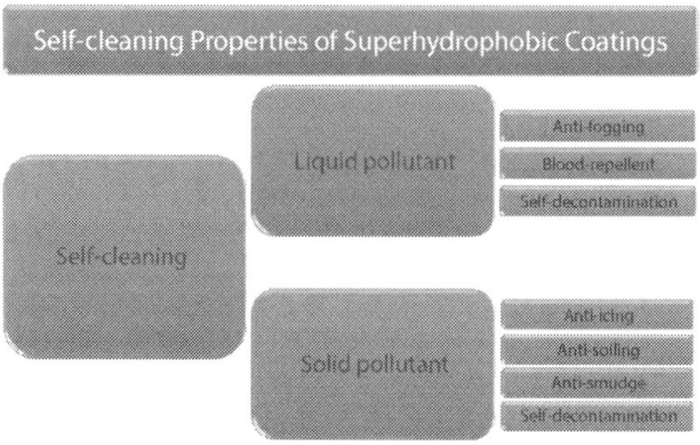

Figure 5.7 Superhydrophobic coatings application to clean up surfaces in different situations.

that were capable of spontaneously pollutant decontamination from the surface were rapidly expanded. The hydrophobicity attribute is one of the most important self-cleaning mechanisms in the coatings. The superhydrophobicity phenomenon which oriented from very low surface tension prevents contaminants adhesion on the surface. According to their physical form, the pollutants were divided into solid and liquid forms which showed in Figure 5.7. Typically, solid contaminants adhere to surfaces through physical interactions and static electricity while the adhesion forces of liquid contaminants are polar interactions in most cases.

5.3.2.1 Liquid Pollutants

The liquid phase pollutants typically have significant surface adhesion that lead to adhere to different surfaces. Liquid pollutants include polar and nonpolar compounds. The chemical structure influences on the type of interaction between the pollutant and the surface. Water is the most widespread liquid phase compound in the environment and also polar interactions are stronger than nonpolar interactions. It can be concluded that the self-cleaning feature via superhydrophobicity have significant importance. There are several researches about sticking liquid phase pollutants that have been done which are discussed in this section.

5.3.2.1.1 Anti-Fogging

The temperature reduction causes water vapor condensation and then formed water thin layer on the surfaces. Also, further lowering of temperature leads to ice formation on the surface. The fog has been formed easily on different surfaces especially metals and glass. It could originate from considerable surface adhesion of water and polar structure of these substances. Therefore, superhydrophobic attribute can prevent fog formation on surfaces. The anti-fogging superhydrophobic properties on the aluminium surface was obtained via chemical etching and surface treatment with lauric acid [9, 53].

5.3.2.1.2 Blood-Repellent

The blood cleansing from the surfaces has considerable importance because the blood has high adhesion tendency which originates from polar bonds in blood chemical structure. Also, the blood remaining on the different surfaces leads to infection spread and transmission of pathogens. Therefore, it is important to prepare coatings or surfaces that blood does not stick to them. The blood repellency feature has been obtained through different chemical and physical treatment methods including silica nanoparticles modification by chlorotrimethylsilane [70], titania nanoflower surfaces via hydrothermal synthesis in presence of HF [108], roll-to-roll (R2R) manufacturing by shrinking a prestressed thermoplastic with a thin layer of silver and calcium [28], and laser ablation of titanium [109].

5.3.2.1.3 Self-Decontamination

Various types of liquid phase materials lead to contamination of different surfaces. The most important liquid phase pollutants are water-based mixtures and various solvents. The liquid compounds can cause surface contamination either in pure form and mixture with other substances. The liquid pollutants not only cause dirt and surface contamination but can also cause other destructive phenomena such as corrosion on the surfaces. Prevention of liquid substances remaining on the surface can prevent the occurrence of negative phenomena on the surfaces.

5.3.2.2 Solid Pollutants

The solid phase pollutants especially airborne particles cause significant pollution at various surfaces particularly in public places. Solid contaminants can be attached on the surface by static electricity, van der Waals, and hydrogen interaction. Therefore, extensive research has been done on this subject.

5.3.2.2.1 Anti-Icing

Ice formation on the surfaces in glacial conditions could lead to various types of problems on the surfaces such as corrosion and reduced surface performance. Therefore, preventing ice formation on the surfaces in glacial conditions is very important. There are various chemical [75] and physical [18] methods to create these types of coatings. Unlike the anti-soiling attribute which typically associated with physical interactions, the coatings anti-icing attribute is related to the ability of surfaces to prevent the ice formation. The ice formation prevention is originated from the prevention of hydrogen bonds between ice and surface. The superhydrophobic surfaces could have produce anti-icing feature through different methods such as the treatment of silica nanoparticles by 1H,1H,2H,2H-perfluorodecyltriethoxysilane [75] and polyzwitterion brushes [62] as chemical treatment, and also laser treatment of stainless steel surface [18] and plasma etching [85] as physical treatment.

5.3.2.2.2 Anti-Soiling

Dust particles on the surface are solid contaminant samples which can easily remove by the help of water-repellent coatings because water carries dust particles with it [10]. Also, the combination of anti-dust and anti-reflective properties will significantly reduce the absorption of dust and airborne particles on the surface [76]. The superhydrophobic surfaces with anti-soiling feature have been produced through chemical treatment functionalization of SiO_2 nanoparticles by different organofluorine substances including heptadecafluoro- 1,1,2,2-tetrahydrodecyl trichlorosilane [76], Tridecafluoro-1,1,2,2-tetrahydrooctyl trichlorosilane [110], and fluorinated acrylic emulsion [10] have been obtained.

5.3.2.2.3 Anti-Smudge

The superhydrophobic coatings can prevent adhering soot and other airborne particles to the surface. It have considerable important to keep surfaces clean against airborne particles and vehicle smoke pollution especially in crowded public places [111, 112]. The surfaces which have superhydrophobic and anti-smudge properties were obtained through chemical treatment by hydroxyl terminated polydimethylsiloxane and mono-di-hydroxy-terminated polydimethylsiloxane which are safer than fluorinated compounds [111, 112]. Also, surface treatment of SiO_2 nanoparticles by poly diallyldimethylammonium chloride and fluorosurfactant layer leads to produce superhydrophobic anti-smudge surfaces [113, 114].

5.3.2.2.4 Self-Decontamination

Various types of solid pollutants can be placed on the surfaces which ice and airborne dust are the most common solid pollutants. Also, other contaminants can absorb onto the surface and contaminate the surface. It is also possible to absorb and transport other hazardous substances such as microbial and chemical contaminants through solid contaminants. Harmful and hazardous substances can be absorbed onto the surface of airborne solid pollutants and transported by wind. Therefore, it is important to remove solid contaminants from the surfaces. The repellent attribute of the superhydrophobic coatings causes easily separate different types of solid contamination. Other important aspect is the prevention of sticking moss [104, 105], fungi [17], chemical deposits [20], and other substances on the surface.

5.3.3 Smart Attributes

The preparation of polymers or coatings that can response to the external stimulant or, in other words, stimuli responsive has expanded significantly in recent years. In addition to the superhydrophobic attribute, these coatings also have an important ability to enhance their performance.

5.3.3.1 Conductivity

The most of polymer coatings and surfaces are electrically insulated. Therefore, conductive coating preparation had significant advantages. It is possible to prepare superhydrophobic and conductive properties simultaneously through chemical modification. This issue has been investigated in various recent studies. One-step vapor-phase polymerization of poly(3,4-ethylenedioxythiophene) in the presence of fluorinated decyl polyhedral oligomeric silsesquioxane and fluorinated alkyl silane [115], polypropylene treatment by vapor-grown carbon fiber and ketjenblack [116], textile coated by silver nanowire as conductive part, and also using carbon nanotube, polytetrafluoroethylene nanoparticles and fluoroacrylic polymer superhydrophobic part [35], polyurethane binding multi-walled carbon nanotube [117], superhydrophobic conductive sponge via WS_2-Ag nanoparticles modified by polydimethylsiloxane [66] and glass treated by silicone nanofilaments and carbon nanofibers [81] are some of researches which examine about superhydrophobic conductive coatings.

5.3.3.2 *Energy Storage*

The global warming, environmental and air pollution, fossil fuel reserves reduction, and the ozone layer depletion led to increased attention to renewable energies. The superhydrophobic coatings have important role to spreading renewable energies and energy storage. There has been extensive research on this topic in recent years which expected that the amount of research will expand significantly in the coming years. The superhydrophobic coatings cause to remain clean different surfaces against various contaminants. Therefore, it prevents the efficiency reduction of renewable energies generation or storage devices. However, this field requires much more researches. Beeswax coated by perfluorinated acrylic copolymer [61] disposable cotton balls treatment by surface-coating of Fe_3O_4 particles and post hydrophobic-modified treatment with long-chain silanes [60] and also microencapsulation of n-docosane with nanostructured ZnO/SiO_2 shell [77] are the examples of researches about this modern issue.

5.3.3.3 *Photocatalytic*

The simultaneous formation of superhydrophobic and photocatalytic properties due to increase surface responsiveness to environmental events. Hence, in addition to the contact angle, the presence of ultra violet ray can influence in the surface self-cleaning properties. This new issue has been found in recent researches such as one-step reaction of octadecyl isocyanate with TiO_2 nanoparticles [55], Ag_2O particles [79], and also glass beads modification by TiO_2 nanoparticles and silicone nanofilaments [27].

5.3.3.4 *Self-Assembly*

The self-assembly process is internal organization of mixtures which proceed in certain orientation without external directions. The self-assembled substances have special characteristics because of their special morphology. The self-assembled superhydrophobic coatings have special properties which examined in different researches that include pillar array formation in silk nanowires [118], triblock copolymer based on poly(dimethylsiloxane)-poly(tert-butyl acrylate)-poly(methacrylolsobutyl POSS) [73], and the pig fat coated by biopolymer and Fe_3O_4 [119].

5.3.3.5 *Self-Healing*

Self-healing as a superior feature in coatings leads to make coatings more durable. The mechanical stresses can damage the surface of superhydrophobic coatings. Therefore, the simultaneous development of superhydrophobicity and self-healing properties can increase the efficiency and durability of coatings. The surface treatment by hydrolyzed fluoroalkyl silane and fluorinated-decyl polyhedral oligomeric silsesquioxane [120], one-step vapor-phase polymerization of poly(3,4-ethylenedioxythiophene) in the presence of fluorinated decyl polyhedral oligomeric silsesquioxane and fluorinated alkyl silane [115], polydimethylsiloxane with flat lotus leaf [98], wood coated by polydimethylsiloxane as the base coat and waterborne perfluoroalkyl methacrylic copolymer-TiO_2 as the top coat [44], fabrics coated by polydopamine and stearic acid emulsion [54], cement modified by 1H,1H,1H,2H-perfluorodecyl trimethoxysilane [63], and also biomimetic polydimethylsiloxane via replication of shark skin and surface-initiated atom transfer radical polymerization 2-perfluorooctylethyl methacrylate [64].

5.3.3.6 *Stimuli-Responsive*

The ability of the polymers to respond to external stimuli is recognized as an advanced and novel feature. These smart polymers are capable to external stimuli responding. Therefore, research on these polymers is expanding at astonishing rate [121, 122]. Also, smart superhydrophobic coatings have been expanded in recent years. Some of the most important of these studies include poly-3-methylthiophene-carbon membranes as electro-responsive carbon membranes with reversible superhydrophobicity/superhydrophilicity switch for efficient oil/water separation [123], magnet-responsive fabrics treated by hexadecyltrimethoxysilane, polydopamine, and Fe_3O_4 nanoparticles [72], magnetically-responsive using polydimethylsiloxane and iron particles [67], and also fire-retarding magnetic-responsive superhydrophobic sponges through polydopamine treated by Fe_3O_4 nanoparticles and n-octadecyl mercaptan.

5.3.3.7 *Multifunctional Superhydrophobic Coatings*

The coatings which have at least three substantial and effective properties are called multifunctional coatings. Multiple appropriate properties of the multifunctional coatings lead to increase their applications [124]. The superhydrophobic multifactorial coatings can be prepared via chemical and physical modification methods. However, chemical modification methods have more variety and versatility. The multifunctional superhydrophobic coatings which have been prepared via chemical treatment such as one-step vapor-phase polymerization of poly(3,4-ethylenedioxythiophene) in the presence of fluorinated decyl polyhedral oligomeric silsesquioxane and fluorinated alkyl silane [115], silica treatment by polymethylhydrosiloxane [125], polydimethylsiloxane with flat lotus leaf [98], copper coated by $CuZn_5$ and $AuZn_3$ [78], polyhedral oligomeric silsesquioxane-graphene oxide [71], TiO_2 treatment by octadecyl isocyanate [55], rubber treated by polydopamine and Ag nanoparticles [100], zinc sheet modified by NH_4F [99], cement modified by 1H,1H,1H,2H-perfluorodecyl trimethoxysilane [63], glass treated by silicone nanofilaments and carbon nanofibers [81], and SiO_2 nanoparticles treated by fluorinated acrylic resin emulsion [10].

Also, the superhydrophobic multifactorial coatings have been obtained through physical methods such as nylon mesh cover peeling from the surface of the concrete [103] and laser-induced graphene [83].

5.4 Conclusions and Summary

Useful properties of superhydrophobic coatings lead to increase attention and research about this issue. Non-stick surfaces cause to increase health and significant reduction in cleaning and protection costs. Various methods of chemical and physical modification to prepare superhydrophobic surfaces have been investigated. The superhydrophobic coatings not only lead to self-cleaning feature but also have some other valuable features. The superhydrophobic coatings are classified through substrate type, coating type, and also the characteristics which expected of the coating. Nowadays, the most attentions are paid to prepare smart coatings as well as multifunctional coatings. It originates from the ability to respond to external stimuli as well as to the ability to function in different environmental conditions. Preparation of superhydrophobic surfaces or coatings that are simultaneously both responsive to external stimuli and multifunctional have extraordinary importance. This significant challenge is being widely pursued by scientists that appropriate progress has been made in recent years.

References

1. Zhu, C., Lv, J., Chen, L., Lin, W., Zhang, J., Yang, J., Feng, J., Dark, heat-reflective, anti-ice rain and superhydrophobic cement concrete surfaces. *Constr. Build. Mater.*, 220, 21–28, 2019.

2. Gao, Y., He, B., Xiao, M., Fang, Z., Dai, K., Study on properties and mechanisms of luminescent cement-based pavement materials with superhydrophobic function. *Constr. Build. Mater.*, 165, 548–559, 2018.
3. She, W., Wang, X., Miao, C., Zhang, Q., Zhang, Y., Yang, J., Hong, J., Biomimetic superhydrophobic surface of concrete: Topographic and chemical modification assembly by direct spray. *Constr. Build. Mater.*, 181, 347–357, 2018.
4. Wang, F., Lei, S., Ou, J., Xue, M., Li, C., Li, W., Superhydrophobic calcium aluminate cement with super mechanical stability. *Ind. Eng. Chem. Res.*, 58, 10373–10382, 2019.
5. Zhang, X., Zhi, D., Sun, L., Zhao, Y., Tiwari, M.K., Carmalt, C.J., Parkin, I.P., Lu, Y., Super-durable, non-fluorinated superhydrophobic free-standing items. *J. Mater. Chem. A*, 6, 357–362, 2018.
6. Xu, H., Clarke, A., Rothstein, J.P., Poole, R.J., Viscoelastic drops moving on hydrophilic and superhydrophobic surfaces. *J. Colloid Interface Sci.*, 513, 53–61, 2018.
7. Wu, Y., Shen, Y., Tao, J., He, Z., Xie, Y., Chen, H., Jin, M., Hou, W., Facile spraying fabrication of highly flexible and mechanically robust superhydrophobic F-SiO 2 @PDMS coatings for self-cleaning and drag-reduction applications. *New J. Chem.*, 42, 18208–18216, 2018.
8. Xing, Y., Xue, Y., Song, J., Sun, Y., Huang, L., Liu, X., Sun, J., Superhydrophobic coatings on wood substrate for self-cleaning and EMI shielding. *Appl. Surf. Sci.*, 436, 865–872, 2018.
9. Varshney, P. and Mohapatra, S.S., Durable and regenerable superhydrophobic coatings for brass surfaces with excellent self-cleaning and anti-fogging properties prepared by immersion technique. *Tribol. Int.*, 123, 17–25, 2018.
10. Zhu, C., Lin, W., Chen, L., Lv, J., Zhang, J., Feng, J., Deep color, heat-reflective, superhydrophobic and anti-soiling coatings with waterborne silicone emulsion. *Sol. Energy Mater. Sol. Cells*, 199, 129–135, 2019.
11. Lin, Y., Han, J., Cai, M., Liu, W., Luo, X., Zhang, H., Zhong, M., Durable and robust transparent superhydrophobic glass surfaces fabricated by a femtosecond laser with exceptional water repellency and thermostability. *J. Mater. Chem. A*, 6, 9049–9056, 2018.
12. Zhao, S., Zhao, J., Wen, M., Yao, M., Wang, F., Huang, F., Zhang, Q., Cheng, Y.B., Zhong, J., Sequentially Reinforced Additive Coating for Transparent and Durable Superhydrophobic Glass. *Langmuir*, 34, 11316–11324, 2018.
13. Huang, Q.Z., Fang, Y.Y., Liu, P.Y., Zhu, Y.Q., Shi, J.F., Xu, G., A novel strategy for durable superhydrophobic coating on glass substrate via using silica chains to fix silica particles. *Chem. Phys. Lett.*, 692, 33–37, 2018.
14. Cheng, Y., Wu, B., Ma, X., Lu, S., Xu, W., Szunerits, S., Boukherroub, R., Facile preparation of high density polyethylene superhydrophobic/superoleophilic coatings on glass, copper and polyurethane sponge for self-cleaning, corrosion resistance and efficient oil/water separation. *J. Colloid Interface Sci.*, 525, 76–85, 2018.

15. Li, S., Page, K., Sathasivam, S., Heale, F., He, G., Lu, Y., Lai, Y., Chen, G., Carmalt, C.J., Parkin, I.P., Efficiently texturing hierarchical superhydrophobic fluoride-free translucent films by AACVD with excellent durability and self-cleaning ability. *J. Mater. Chem. A*, 6, 17633–17641, 2018.
16. Gurera, D. and Bhushan, B., Fabrication of bioinspired superliquiphobic synthetic leather with self-cleaning and low adhesion. *Colloids Surf. A Physicochem. Eng. Asp.*, 545, 130–137, 2018.
17. Lee, J.W. and Hwang, W., Fabrication of a superhydrophobic surface with fungus-cleaning properties on brazed aluminum for industrial application in heat exchangers. *Appl. Surf. Sci.*, 442, 461–466, 2018.
18. Wang, H., He, M., Liu, H., Guan, Y., One-Step Fabrication of Robust Superhydrophobic Steel Surfaces with Mechanical Durability, Thermal Stability, and Anti-icing Function. *ACS Appl. Mater. Interfaces*, 11, 28, 25586–25594, 2019.
19. Tudu, B.K. and Kumar, A., Robust and durable superhydrophobic steel and copper meshes for separation of oil-water emulsions. *Prog. Org. Coat.*, 133, 316–324, 2019.
20. Gao, X. and Guo, Z., Mechanical stability, corrosion resistance of superhydrophobic steel and repairable durability of its slippery surface. *J. Colloid Interface Sci.*, 512, 239–248, 2018.
21. Kim, J.H., Mirzaei, A., Kim, H.W., Kim, S.S., Facile fabrication of superhydrophobic surfaces from austenitic stainless steel (AISI 304) by chemical etching. *Appl. Surf. Sci.*, 439, 598–604, 2018.
22. Khanjani, P., King, A.W.T., Partl, G.J., Johansson, L.S., Kostiainen, M.A., Ras, R.H.A., Superhydrophobic Paper from Nanostructured Fluorinated Cellulose Esters. *ACS Appl. Mater. Interfaces*, 10, 11280–11288, 2018.
23. Li, H., He, Y., Yang, J., Wang, X., Lan, T., Peng, L., Fabrication of food-safe superhydrophobic cellulose paper with improved moisture and air barrier properties. *Carbohydr. Polym.*, 211, 22–30, 2019.
24. Wen, G. and Guo, Z., Nonflammable superhydrophobic paper with biomimetic layered structure exhibiting boiling-water resistance and repairable properties for emulsion separation. *J. Mater. Chem. A*, 6, 7042–7052, 2018.
25. Wen, G., Gao, X., Tian, P., Zhong, L., Wang, Z., Guo, Z., Modifier-free fabrication of durable and multifunctional superhydrophobic paper with thermostability and anti-microbial property. *Chem. Eng. J.*, 346, 94–103, 2018.
26. Latthe, S.S., Sutar, R.S., Kodag, V.S., Bhosale, A.K., Kumar, A.M., Kumar Sadasivuni, K., Xing, R., Liu, S., Self – cleaning superhydrophobic coatings: Potential industrial applications. *Prog. Org. Coat.*, 128, 52–58, 2019.
27. Zhang, X., Liu, S., Salim, A., Seeger, S., Hierarchical Structured Multifunctional Self-Cleaning Material with Durable Superhydrophobicity and Photocatalytic Functionalities. *Small*, 1–19, 15, 1901822, 2019.
28. Nokes, J.M., Liedert, R., Kim, M.Y., Siddiqui, A., Chu, M., Lee, E.K., Khine, M., Reduced Blood Coagulation on Roll-to-Roll, Shrink-Induced Superhydrophobic Plastics. *Adv. Healthc. Mater.*, 5, 593–601, 2016.

29. Zhi, J. and Zhang, L.Z., Durable superhydrophobic surface with highly anti-reflective and self-cleaning properties for the glass covers of solar cells. *Appl. Surf. Sci.*, 454, 239–248, 2018.
30. zhi Zhang, L., jian Pan, A., rong Cai, R., Lu, H. Indoor experiments of dust deposition reduction on solar cell covering glass by transparent super-hydrophobic coating with different tilt angles. *Sol. Energy*, 188, 1146–1155, 2019.
31. Roslizar, A., Dottermusch, S., Vüllers, F., Kavalenka, M.N., Guttmann, M., Schneider, M., Paetzold, U.W., Hölscher, H., Richards, B.S., Klampaftis, E., Self-cleaning performance of superhydrophobic hot-embossed fluoropolymer films for photovoltaic modules. *Sol. Energy Mater. Sol. Cells*, 189, 188–196, 2019.
32. Jung, M.H., Hydrophobic perovskites based on an alkylamine compound for high efficiency solar cells with improved environmental stability. *J. Mater. Chem. A*, 7, 14689–14704, 2019.
33. Yeerken, T., Yu, W., Feng, J., Xia, Q., Liu, H., Durable superamphiphobic aramid fabrics modified by PTFE and FAS for chemical protective clothing. *Prog. Org. Coat.*, 135, 41–50, 2019.
34. Su, X., Li, H., Lai, X., Zhang, L., Liao, X., Wang, J., Chen, Z., He, J., Zeng, X., Dual-Functional Superhydrophobic Textiles with Asymmetric Roll-Down/Pinned States for Water Droplet Transportation and Oil-Water Separation. *ACS Appl. Mater. Interfaces*, 10, 4213–4221, 2018.
35. Jia, L.C., Zhang, G., Xu, L., Sun, W.J., Zhong, G.J., Lei, J., Yan, D.X., Li, Z.M., Robustly Superhydrophobic Conductive Textile for Efficient Electromagnetic Interference Shielding. *ACS Appl. Mater. Interfaces*, 11, 1680–1688, 2019.
36. Bu, Y., Zhang, S., Cai, Y., Yang, Y., Ma, S., Huang, J., Yang, H., Ye, D., Zhou, Y., Xu, W., Gu, S., Fabrication of durable antibacterial and superhydrophobic textiles via *in situ* synthesis of silver nanoparticle on tannic acid-coated viscose textiles. *Cellulose*, 26, 2109–2122, 2019.
37. Cheng, Y., Zhu, T., Li, S., Huang, J., Mao, J., Yang, H., Gao, S., Chen, Z., Lai, Y., A novel strategy for fabricating robust superhydrophobic fabrics by environmentally-friendly enzyme etching. *Chem. Eng. J.*, 355, 290–298, 2019.
38. Chauhan, P., Kumar, A., Bhushan, B., Self-cleaning, stain-resistant and anti-bacterial superhydrophobic cotton fabric prepared by simple immersion technique. *J. Colloid Interface Sci.*, 66–74, 535, 2019.
39. Xue, C.H., Fan, Q.Q., Guo, X.J., An, Q.F., Jia, S.T., Fabrication of superhydrophobic cotton fabrics by grafting of POSS-based polymers on fibers. *Appl. Surf. Sci.*, 465, 241–248, 2019.
40. Yang, M., Liu, W., Jiang, C., He, S., Xie, Y., Wang, Z., Fabrication of superhydrophobic cotton fabric with fluorinated TiO_2 sol by a green and one-step sol-gel process. *Carbohydr. Polym.*, 197, 75–82, 2018.
41. Liu, G., Xiang, J., Xia, Q., Li, K., Lan, T., Yu, L., Superhydrophobic cotton gauze with durably antibacterial activity as skin wound dressing. *Cellulose*, 26, 1383–1397, 2019.

42. Yao, Y., Gellerich, A., Zauner, M., Wang, X., Zhang, K., Differential antifungal effects from hydrophobic and superhydrophobic wood based on cellulose and glycerol stearoyl esters. *Cellulose*, 25, 1329–1338, 2018.
43. Wang, Y., Vitas, S., Burgert, I., Cabane, E., Beech wood cross sections as natural templates to fabricate superhydrophobic surfaces. *Wood Sci. Technol.*, 53, 985–999, 2019.
44. Tu, K., Wang, X., Kong, L., Guan, H., Facile preparation of mechanically durable, self-healing and multifunctional superhydrophobic surfaces on solid wood. *Mater. Des.*, 140, 30–36, 2018.
45. Deng, W. and Kesari, H., Depth-dependent hysteresis in adhesive elastic contacts at large surface roughness. *Sci. Rep.*, 9, 1–12, 2019.
46. Ramezanzadeh, M., Sanaei, Z., Ramezanzadeh, B., The influence of steel surface treatment by a novel eco-friendly praseodymium oxide nanofilm on the adhesion and corrosion protection properties of a fusion-bonded epoxy powder coating. *J. Ind. Eng. Chem.*, 62, 427–435, 2018.
47. Pan, S., Guo, R., Björnmalm, M., Richardson, J.J., Li, L., Peng, C., Bertleff-Zieschang, N., Xu, W., Jiang, J., Caruso, F., Coatings super-repellent to ultralow surface tension liquids. *Nat. Mater.*, 17, 1040–1047, 2018.
48. Kim, A., Lee, C., Kim, J., Durable, scalable, and tunable omniphobicity on stainless steel mesh for separation of low surface tension liquid mixtures. *Surf. Coat. Technol.*, 344, 394–401, 2018.
49. Rezzoug, A., Abdi, S., Kaci, A., Yandouzi, M., Thermal spray metallisation of carbon fibre reinforced polymer composites: Effect of top surface modification on coating adhesion and mechanical properties. *Surf. Coat. Technol.*, 333, 13–23, 2018.
50. Brely, L., Bosia, F., Pugno, N., The influence of substrate roughness, patterning, curvature, and compliance in peeling problems. *Bioinspir. Biomim.*, 13, 2, 2018.
51. Gao, S., Dong, X., Huang, J., Dong, J., Cheng, Y., Chen, Z., Lai, Y., Co-solvent induced self-roughness superhydrophobic coatings with self-healing property for versatile oil-water separation. *Appl. Surf. Sci.*, 459, 512–519, 2018.
52. Huang, X., Kong, X., Cui, Y., Ye, X., Wang, X., Shi, B., Durable superhydrophobic materials enabled by abrasion-triggered roughness regeneration. *Chem. Eng. J.*, 336, 633–639, 2018.
53. Varshney, P., Lomga, J., Gupta, P.K., Mohapatra, S.S., Kumar, A., Durable and regenerable superhydrophobic coatings for aluminium surfaces with excellent self-cleaning and anti-fogging properties. *Tribol. Int.*, 119, 38–44, 2018.
54. Dong, X., Gao, S., Huang, J., Li, S., Zhu, T., Cheng, Y., Zhao, Y., Chen, Z., Lai, Y., A self-roughened and biodegradable superhydrophobic coating with UV shielding, solar-induced self-healing and versatile oil-water separation ability. *J. Mater. Chem. A*, 7, 2122–2128, 2019.
55. Huang, Z., Gurney, R.S., Wang, T., Liu, D., Environmentally durable superhydrophobic surfaces with robust photocatalytic self-cleaning and self-healing

properties prepared via versatile film deposition methods. *J. Colloid Interface Sci.*, 527, 107–116, 2018.
56. Liu, P., Zhang, Y., Liu, S., Zhang, Y., Du, Z., Qu, L., Bio-inspired fabrication of fire-retarding, magnetic-responsive, superhydrophobic sponges for oil and organics collection. *Appl. Clay Sci.*, 172, 19–27, 2019.
57. Gao, R., Xiao, S., Gan, W., Liu, Q., Amer, H., Rosenau, T., Li, J., Lu, Y., Mussel Adhesive-Inspired Design of Superhydrophobic Nanofibrillated Cellulose Aerogels for Oil/Water Separation. *ACS Sustain. Chem. Eng.*, 6, 9047–9055, 2018.
58. Xiang, T., Han, Y., Guo, Z., Wang, R., Zheng, S., Li, S., Li, C., Dai, X., Fabrication of Inherent Anticorrosion Superhydrophobic Surfaces on Metals. *ACS Sustain. Chem. Eng.*, 6, 5598–5606, 2018.
59. He, J., Zhao, H., Li, X., Su, D., Zhang, F., Ji, H., Liu, R., Superelastic and superhydrophobic bacterial cellulose/silica aerogels with hierarchical cellular structure for oil absorption and recovery. *J. Hazard. Mater.*, 346, 199–207, 2018.
60. Lu, Y., Niu, Z., Yuan, W., Multifunctional magnetic superhydrophobic carbonaceous aerogel with micro/nano-scale hierarchical structures for environmental remediation and energy storage. *Appl. Surf. Sci.*, 480, 851–860, 2019.
61. Naderizadeh, S., Heredia-Guerrero, J.A., Caputo, G., Grasselli, S., Malchiodi, A., Athanassiou, A., Bayer, I.S., Superhydrophobic Coatings from Beeswax-in-Water Emulsions with Latent Heat Storage Capability. *Adv. Mater. Interfaces*, 6, 1–11, 2019.
62. Liang, B., Zhang, G., Zhong, Z., Huang, Y., Su, Z., Superhydrophilic Anti-Icing Coatings Based on Polyzwitterion Brushes. *Langmuir*, 35, 1294–1301, 2019.
63. Pang, B., Qian, J., Zhang, Y., Jia, Y., Ni, H., Pang, S.D., Liu, G., Qian, R., She, W., Yang, L., Liu, Z., 5S multifunctional intelligent coating with super-durable, superhydrophobic, self-monitoring, self-heating, and self-healing properties for existing construction application. *ACS Appl. Mater. Interfaces*, 11, 29242–29254, 2019.
64. Liu, Y., Gu, H., Jia, Y., Liu, J., Zhang, H., Wang, R., Zhang, B., Zhang, H., Zhang, Q., Design and preparation of biomimetic polydimethylsiloxane (PDMS) films with superhydrophobic, self-healing and drag reduction properties via replication of shark skin and SI-ATRP. *Chem. Eng. J.*, 356, 318–328, 2019.
65. Kim, S., Hwang, H.J., Cho, H., Choi, D., Hwang, W., Repeatable replication method with liquid infiltration to fabricate robust, flexible, and transparent, anti-reflective superhydrophobic polymer films on a large scale. *Chem. Eng. J.*, 350, 225–232, 2018.
66. Jing, X. and Guo, Z., Multifunctional WS_2&M-AgNPs superhydrophobic conductive sponges for application in various sensors. *New J. Chem.*, 43, 5287–5296, 2019.

67. Yang, C., Wu, L., Li, G., Magnetically Responsive Superhydrophobic Surface: *In Situ* Reversible Switching of Water Droplet Wettability and Adhesion for Droplet Manipulation. *ACS Appl. Mater. Interfaces*, 10, 20150–20158, 2018.
68. Agrawal, N., Tan, J.S.J., Low, P.S., Fong, E.W.M., Lai, Y., Chen, Z., Green Synthesis of Robust Superhydrophobic Antibacterial and UV-Blocking Cotton Fabrics by a Dual-Stage Silanization Approach. *Adv. Mater. Interfaces*, 6, 1–10, 2019.
69. Liang, Z., Li, W., Dong, B., Sun, Y., Tang, H., Zhao, L., Wang, S., Double-function SiO 2 -DMS coating with antireflection and superhydrophobic surface. *Chem. Phys. Lett.*, 716, 211–214, 2019.
70. Li, Z., Nguyen, B.L., Cheng, Y.C., Xue, J., MacLaren, G., Yap, C.H., Durable, flexible, superhydrophobic and blood-repelling surfaces for use in medical blood pumps. *J. Mater. Chem. B*, 6, 6225–6233, 2018.
71. Ye, Y., Zhang, D., Li, J., Liu, T., Pu, J., Zhao, H., Wang, L., One-step synthesis of superhydrophobic polyhedral oligomeric silsesquioxane-graphene oxide and its application in anti-corrosion and anti-wear fields. *Corros. Sci.*, 147, 9–21, 2019.
72. Fu, S., Zhou, H., Wang, H., Ding, J., Liu, S., Zhao, Y., Niu, H., Rutledge, G.C., Lin, T., Magnet-responsive, superhydrophobic fabrics from waterborne, fluoride-free coatings. *RSC Adv.*, 8, 717–723, 2018.
73. Zhou, X., Lee, Y.Y., Chong, K.S.L., He, C., Superhydrophobic and slippery liquid-infused porous surfaces formed by the self-assembly of a hybrid ABC triblock copolymer and their antifouling performance. *J. Mater. Chem. B*, 6, 440–448, 2018.
74. Rao, A.V., Elastic superhydrophobic and water glass-based silica aerogels and applications. *J. Sol-Gel Sci. Technol.*, 90, 28–54, 2019.
75. Wang, N., Tang, L., Tong, W., Xiong, D., Fabrication of robust and scalable superhydrophobic surfaces and investigation of their anti-icing properties. *Mater. Des.*, 156, 320–328, 2018.
76. Jang, G.G., Smith, D.B., List, F.A., Lee, D.F., Ievlev, A.V., Collins, L., Park, J., Polizos, G., The anti-soiling performance of highly reflective superhydrophobic nanoparticle-textured mirrors. *Nanoscale*, 10, 14600–14612, 2018.
77. Sun, K., Liu, H., Wang, X., Wu, D., Innovative design of superhydrophobic thermal energy-storage materials by microencapsulation of n-docosane with nanostructured ZnO/SiO_2 shell. *Appl. Energy*, 237, 549–565, 2019.
78. Cheng, Y., Lu, S., Xu, W., Cao, K., Li, J., Zheng, Y., Controllable fabrication of superhydrophobic alloys surface on copper substrate for self-cleaning, anti-icing, anti-corrosion and anti-wear performance. *Surf. Coat. Technol.*, 333, 61–70, 2018.
79. Jiang, W., Fu, H., Zhu, Y., Yue, H., Yuan, S., Liang, B., Floatable superhydrophobic Ag_2O photocatalyst without a modifier and its controllable wettability by particle size adjustment. *Nanoscale*, 10, 13661–13672, 2018.

80. Tripathy, A., Kumar, A., Sreedharan, S., Muralidharan, G., Pramanik, A., Nandi, D., Sen, P., Fabrication of Low-Cost Flexible Superhydrophobic Antibacterial Surface with Dual-Scale Roughness. *ACS Biomater. Sci. Eng.*, 4, 2213–2223, 2018.
81. Saddiqi, N.U.H. and Seeger, S., Chemically Resistant, Electric Conductive, and Superhydrophobic Coatings. *Adv. Mater. Interfaces*, 6, 1–9, 2019.
82. Selim, M.S., El-Safty, S.A., Fatthallah, N.A., Shenashen, M.A., Silicone/graphene oxide sheet-alumina nanorod ternary composite for superhydrophobic antifouling coating. *Prog. Org. Coat.*, 121, 160–172, 2018.
83. Tittle, C.M., Yilman, D., Pope, M.A., Backhouse, C.J., Robust Superhydrophobic Laser-Induced Graphene for Desalination Applications. *Adv. Mater. Technol.*, 3, 1–10, 2018.
84. Ye, R., James, D.K., Tour, J.M., Laser-Induced Graphene. *Acc. Chem. Res.*, 51, 1609–1620, 2018.
85. Nguyen, T.B., Park, S., Lim, H., Effects of morphology parameters on anti-icing performance in superhydrophobic surfaces. *Appl. Surf. Sci.*, 435, 585–591, 2018.
86. Barz, J., Haupt, M., Oehr, C., Hirth, T., Grimmer, P., Stability and water wetting behavior of superhydrophobic polyurethane films created by hot embossing and plasma etching and coating. *Plasma Process. Polym.*, 16, 8–10, 2019.
87. Liu, H., Gu, T., Zhang, G., Cheng, Y., Wang, H., Liu, H., The effect of magneticfield on biomineralization and corrosion behavior of carbon steel induced by iron-oxidizing bacteria. *Corros. Sci.*, 102, 93–102, 2016.
88. Liu, H., Gu, T., Asif, M., Zhang, G., Liu, H., The corrosion behavior and mechanism of carbon steel induced by extracellular polymeric substances of iron-oxidizing bacteria. *Corros. Sci.*, 114, 102–111, 2017.
89. Asmone, A.S. and Chew, M.Y.L., An investigation of superhydrophobic self-cleaning applications on external building façade systems in the tropics. *J. Build. Eng.*, 17, 167–173, 2018.
90. Genzer, J. and Efimenko, K., Recent developments in superhydrophobic surfaces and their relevance to marine fouling: A review. *Biofouling*, 22, 339–360, 2006.
91. Hwang, G.B., Page, K., Patir, A., Nair, S.P., Allan, E., Parkin, I.P., The Anti-Biofouling Properties of Superhydrophobic Surfaces are Short-Lived. *ACS Nano*, 12, 6050–6058, 2018.
92. Ouyang, Y., Zhao, J., Qiu, R., Hu, S., Zhang, Y., Wang, P., Bioinspired superhydrophobic and oil-infused surface: Which is the better choice to prevent marine biofouling? *Colloids Surf. A Physicochem. Eng. Asp.*, 559, 297–304, 2018.
93. Razavi, S.M.R., Oh, J., Haasch, R.T., Kim, K., Masoomi, M., Bagheri, R., Slauch, J.M., Miljkovic, N., Environment-Friendly Anti-Biofouling Superhydrophobic Coatings. *ACS Sustain. Chem. Eng.*, 7, 14509–14520, 2019.

94. Qu, Q., Wang, L., Li, L., He, Y., Yang, M., Ding, Z., Effect of the fungus, Aspergillus niger, on the corrosion behaviour of AZ31B magnesium alloy in artificial seawater. *Corros. Sci.*, 98, 249–259, 2015.
95. Qu, Q., Li, S., Li, L., Zuo, L., Ran, X., Qu, Y., Zhu, B., Adsorption and corrosion behaviour of Trichoderma harzianum for AZ31B magnesium alloy in artificial seawater. *Corros. Sci.*, 118, 12–23, 2017.
96. Vazirinasab, E., Jafari, R., Momen, G., Application of superhydrophobic coatings as a corrosion barrier: A review. *Surf. Coat. Technol.*, 341, 40–56, 2018.
97. Zhu, Y., Sun, F., Qian, H., Wang, H., Mu, L., Zhu, J., A biomimetic spherical cactus superhydrophobic coating with durable and multiple anti-corrosion effects. *Chem. Eng. J.*, 338, 670–679, 2018.
98. Qian, H., Xu, D., Du, C., Zhang, D., Li, X., Huang, L., Deng, L., Tu, Y., Mol, J.M.C., Terryn, H.A., Dual-action smart coatings with a self-healing superhydrophobic surface and anti-corrosion properties. *J. Mater. Chem. A*, 5, 2355–2364, 2017.
99. Guo, J., Qian, H., Liu, P., Ma, J., Fabrication of durability superhydrophobic LDH coating on zinc sheet surface via NH4F-assisted *in-situ* growth and post-modification for enhancing anti-corrosion and anti-icing. *Appl. Clay Sci.*, 180, 105182, 2019.
100. Wang, L., Wang, H., Huang, X.W., Song, X., Hu, M., Tang, L., Xue, H., Gao, J., Superhydrophobic and superelastic conductive rubber composite for wearable strain sensors with ultrahigh sensitivity and excellent anti-corrosion property. *J. Mater. Chem. A*, 6, 24523–24533, 2018.
101. Liu, Y.C., Huang, W.J., Wu, S.H., Lee, M., Yeh, J.M., Chen, H.H., Excellent superhydrophobic surface and anti-corrosion performance by nanostructure of discotic columnar liquid crystals. *Corros. Sci.*, 138, 1–7, 2018.
102. Li, Q., Liu, H., Zhang, S., Zhang, D., Liu, X., He, Y., Mi, L., Zhang, J., Liu, C., Shen, C., Guo, Z., Superhydrophobic Electrically Conductive Paper for Ultrasensitive Strain Sensor with Excellent Anticorrosion and Self-Cleaning Property. *ACS Appl. Mater. Interfaces*, 11, 21904–21914, 2019.
103. Song, J., Li, Y., Xu, W., Liu, H., Lu, Y., Inexpensive and non-fluorinated superhydrophobic concrete coating for anti-icing and anti-corrosion. *J. Colloid Interface Sci.*, 541, 86–92, 2019.
104. Feng, Y., Chen, S., Frank Cheng, Y., Stearic acid modified zinc nano-coatings with superhydrophobicity and enhanced antifouling performance. *Surf. Coat. Technol.*, 340, 55–65, 2018.
105. Shao, Y., Han, M., Wang, Y., Li, G., Xiao, W., Li, X., Wu, X., Ruan, X., Yan, X., He, G., Jiang, X., Superhydrophobic polypropylene membrane with fabricated antifouling interface for vacuum membrane distillation treating high concentration sodium/magnesium saline water. *J. Membr. Sci.*, 579, 240–252, 2019.
106. Rossi, S., Deflorian, F., Fedel, M., Polysilazane-based coatings: Corrosion protection and anti-graffiti properties. *Surf. Eng.*, 35, 343–350, 2019.

107. Ma, J., Ai, Y., Kang, L., Liu, W., Ma, Z., Song, P., Zhao, Y., Yang, F., Wang, X., A Novel Nanocone Cluster Microstructure with Anti-reflection and Superhydrophobic Properties for Photovoltaic Devices. *Nanoscale Res. Lett.*, 13, 0–5, 2018.
108. Movafaghi, S., Leszczak, V., Wang, W., Sorkin, J.A., Dasi, L.P., Popat, K.C., Kota, A.K., Hemocompatibility of superhemophobic titania surfaces. *Adv. Health. Mater.* 6, 1600717, 2017.
109. Moradi, S., Hadjesfandiari, N., Toosi, S.F., Kizhakkedathu, J.N., Hatzikiriakos, S.G., Effect of Extreme Wettability on Platelet Adhesion on Metallic Implants: From Superhydrophilicity to Superhydrophobicity. *ACS Appl. Mater., Interfaces*, 8, 17631–17641, 2016.
110. Polizos, G., Sharma, J.K., Smith, D.B., Tuncer, E., Park, J., Voylov, D., Sokolov, A.P., Meyer, H.M., Aman, M., Anti-soiling and highly transparent coatings with multi-scale features. *Sol. Energy Mater. Sol. Cells*, 188, 255–262, 2018.
111. Zhong, X., Hu, H., Yang, L., Sheng, J., Fu, H., Robust Hyperbranched Polyester-Based Anti-Smudge Coatings for Self-Cleaning, Anti-Graffiti, and Chemical Shielding. *ACS Appl. Mater. Interfaces*, 11, 14305–14312, 2019.
112. Zhong, X., Sheng, J., Fu, H., A novel UV/sunlight-curable anti-smudge coating system for various substrates. *Chem. Eng. J.*, 345, 659–668, 2018.
113. Brown, P.S. and Bhushan, B., Mechanically durable, superomniphobic coatings prepared by layer-by-layer technique for self-cleaning and anti-smudge. *J. Colloid Interface Sci.*, 456, 210–218, 2015.
114. Brown, P.S. and Bhushan, B., Mechanically durable, superoleophobic coatings prepared by layer-by-layer technique for anti-smudge and oil-water separation. *Sci. Rep.*, 5, 1–9, 2015.
115. Wang, H., Zhou, H., Gestos, A., Fang, J., Niu, H., Ding, J., Lin, T., Robust, electro-conductive, self-healing superamphiphobic fabric prepared by one-step vapour-phase polymerisation of poly(3,4-ethylenedioxythiophene) in the presence of fluorinated decyl polyhedral oligomeric silsesquioxane and fluorinated alkyl silane. *Soft Matter*, 9, 277–282, 2013.
116. Cai, Z., Shen, L., Wang, X., Guo, Q., Fabrication of a bulk superhydrophobic conductive material by mechanical abrasion. *Compos. Sci. Technol.*, 164, 238–247, 2018.
117. Luo, X., Wei, M., Cao, M., Ren, H., Feng, J., Wear-resistant and conductive superhydrophobic coatings with nest-like structure prepared by a one-step spray-drying method. *Chem. Eng. Process. – Process. Intensif.*, 131, 27–33, 2018.
118. Gustafsson, L., Jansson, R., Hedhammar, M., van der Wijngaart, W., Structuring of Functional Spider Silk Wires, Coatings, and Sheets by Self-Assembly on Superhydrophobic Pillar Surfaces. *Adv. Mater.*, 30, 1–8, 2018.
119. Xie, J., Yang, Y., Gao, B., Wan, Y., Li, Y.C., Cheng, D., Xiao, T., Li, K., Fu, Y., Xu, J., Zhao, Q., Zhang, Y., Tang, Y., Yao, Y., Wang, Z., Liu, L., Magnetic-Sensitive Nanoparticle Self-Assembled Superhydrophobic Biopolymer-Coated

Slow-Release Fertilizer: Fabrication, Enhanced Performance, and Mechanism. *ACS Nano*, 13, 3320–3333, 2019.

120. Wang, H., Zhou, H., Gestos, A., Fang, J., Lin, T., Robust, superamphiphobic fabric with multiple self-healing ability against both physical and chemical damages. *ACS Appl. Mater. Interfaces*, 5, 10221–10226, 2013.
121. Roth, P.J. and Lowe, A.B., Stimulus-responsive polymers. *Polym. Chem.*, 8, 10–11, 2017.
122. Wei, M., Gao, Y., Li, X., Serpe, M.J., Stimuli-responsive polymers and their applications. *Polym. Chem.*, 8, 127–143, 2017.
123. Du, L., Quan, X., Fan, X., Chen, S., Yu, H., Electro-responsive carbon membranes with reversible superhydrophobicity/superhydrophilicity switch for efficient oil/water separation. *Sep. Purif. Technol.*, 210, 891–899, 2019.
124. Ragesh, P., Anand Ganesh, V., Nair, S.V., Nair, A.S., A review on "self-cleaning and multifunctional materials". *J. Mater. Chem. A*, 2, 14773–14797, 2014.
125. Cho, E.C., Chang-Jian, C.W., Chen, H.C., Chuang, K.S., Zheng, J.H., Hsiao, Y.S., Lee, K.C., Huang, J.H., Robust multifunctional superhydrophobic coatings with enhanced water/oil separation, self-cleaning, anti-corrosion, and anti-biological adhesion. *Chem. Eng. J.*, 314, 347–357, 2017.

6

Adsorptive Polymer Coatings

Aneela Sabir[1]*, Muhammad Hamad Zeeshan[2], Muhammad Shafiq[1], Rafi Ullah Khan[1,2] and Karl I. Jacob[3]

[1]*Department of Polymer Engineering and Technology, University of the Punjab, Lahore, Pakistan*
[2]*Institute of Chemical Engineering and Technology (ICET), University of the Punjab, Lahore, Pakistan*
[3]*School of Materials Science and Engineering, Georgia Institute of Technology, Atlanta, Georgia, USA*

Abstract
Coating is referred as a covering of an object or substrate surface completely or partially which changes the properties of the substrate completely. Polymer coating is degraded when exposed to UV lights and thus failed to protect the layer. Therefore, to control the degradation of polymers, a new approach of adsorptive polymer coating has been employed. This chapter provides the brief view of adsorptive polymer coating and their coating techniques. This chapter also includes the comprehensive comparisons of coating techniques. Moreover, adsorptive polymer coating applications in various fields are also detailed. Furthermore, future perspective towards existing challenges provides a better direction and understanding to overcome these challenges in coming days.

Keywords: Coatings, adsorptive coating, coating techniques, coating applications

6.1 Introduction

The covering of an object usually a substrate is termed as coating which is spread over the substrate's surface. The applied coating may cover all over the surface completely or partially cover the substrate surface [1]. The morphological properties of an object, e.g., wear resistance, adhesion, wettability, and corrosion resistance, may be improved by applying coating. In

**Corresponding author*: aneela.pet.ceet@pu.edu.pk

Inamuddin, Rajender Boddula, Mohd Imran Ahamed and Abdullah M. Asiri (eds.) Polymer Coatings: Technology and Applications, (121–134) © 2020 Scrivener Publishing LLC

some cases, coating adds new properties to the materials such as electrical conductivity, magnetic response, and the formation of an essential part of the finished products [2].

In the past decades, its use has become ubiquitous for protective, functional, and decorative applications [3]. The protective and functional properties include anticorrosive, ice repelling or oleophobic, hydrophobic, flame retardancy, antifouling, self-healing, self-cleaning, and antibacterial, etc. [4]. Thus, to control the corrosion/degradation rate of materials, the modification is the basic approach to employ noble coatings as physical barrier.

6.2 Types of Coatings

In order to protect the materials from degradation/corrosion and get the desired properties, various types of coatings are applied. Some of common types of coatings are illustrated in Figure 6.1.

6.3 Polymer Coating

All coatings, when exposed to UV radiation, start degrading due to complex photochemical process. This degradation is enhanced by the weather conditions such as temperature, pollutants, and moisture, etc.; however, polymer coating do not degrade as they do not absorb UV radiation greater than 295 nm wavelength [5]. Thus, polymer has been extensively been employed as coating material in various fields. Polymer coating enhances mechanical

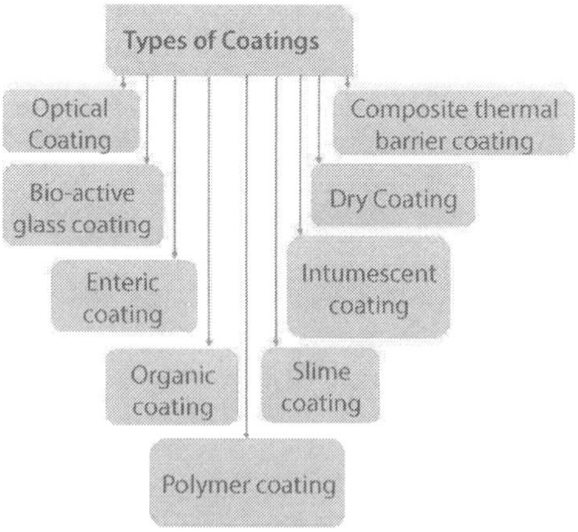

Figure 6.1 Various types of coatings.

strength, thermal stability, biomedical properties, separation characteristics, chemical resistance, and other various functional properties [6].

Polymer coating has various applications in the field of painting, storage media, semi-conductors, optical devices, fluorescent devices, and etc. [7]. Protection of material from closure stresses of the concealed materialization is a significant improvement of polymer coating. Polymer coatings also perform the protection of material from ambient conditions, abridged fragmentation, and protect from dusting [8].

6.4 Types of Polymer Coating

The concept of polymer coating suggests new routes in the vicinity of secure and durable materials and sustainability and improves the efficiency [9]. Polymer coatings are divided into different types as shown in Figure 6.2 depending on the superior properties in terms of stiffness, strength, thermal stability, cost, and light weight [10].

6.5 Adsorptive Polymer Coating

Adsorptive polymer has been proposed as an alternative coating materials in pharmaceutical, cosmetics, environmental, personal care products, food safety, and etc. [11–15]. Adsorptive polymer proposed large surface/volume ratio and present unique surface chemistry features, excellent porosity, good mechanical strength, and hydrophobic surface area. Increased consumer concern to food safety led to the development of adsorptive polymer coating, a renewable coating suitable for food, and pharmaceutical, etc. In food packaging, adsorptive polymer

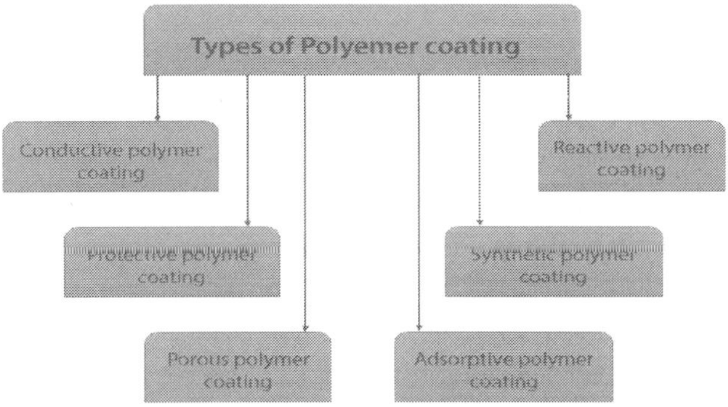

Figure 6.2 Types of polymer coating.

film perpetuate and insulate food from extrinsic chemical, physical, and microbiological elements [16, 17].

6.6 Materials

Numerous polymers have been used as coating material (shown in Figure 6.3). Polyethylene, polypropylene, polyethylene terephthalate, and nylon, etc., are common adsorptive polymers being widely used as coating material [18, 19]. Selection of polymers for coating depends on low price, low wettability, and excellent mechanical properties [20].

6.7 Adsorptive Polymer Coating Techniques

In adsorptive polymer coating, a polymer solution is filled in capillary that adsorb to the capillary surface that formed a dynamic coating or adsorbent is mixed with polymer. Various adsorptive coating methods have been established for coatings with thickness range of few nm to 10 nm, and numerous synthetic, natural, neutral, and charged polymers have been used in the pursuit of dynamic coating [21]. Some suitable coating techniques are shown in Figure 6.4.

6.7.1 Spray Coating

Spray coating is a facile and unique approach for applying a coat over a substrate. In this process, first suitable adsorbent is dissolved in polymer

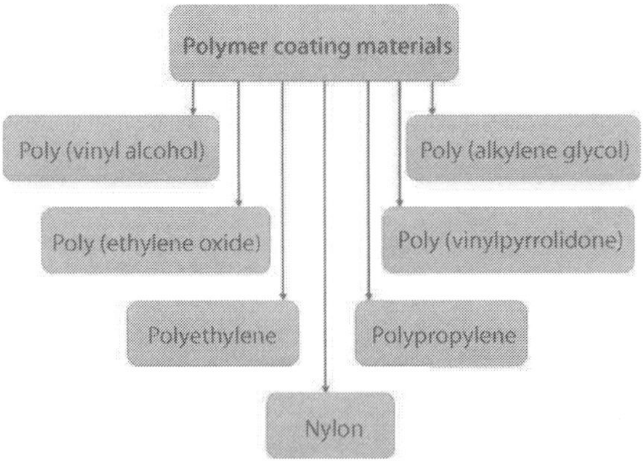

Figure 6.3 Common materials for adsorptive polymer coating.

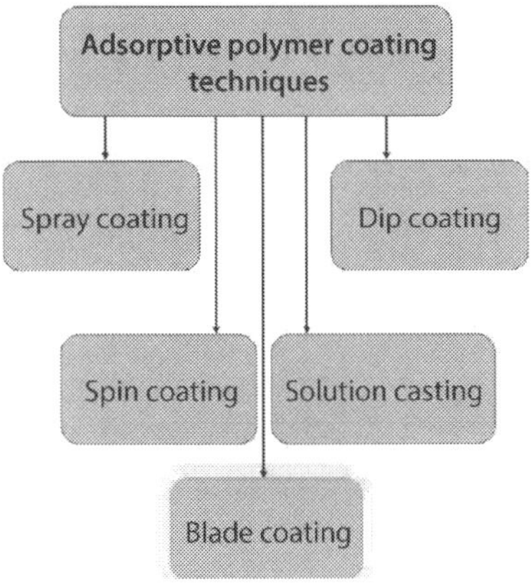

Figure 6.4 Adsorptive polymer coating techniques.

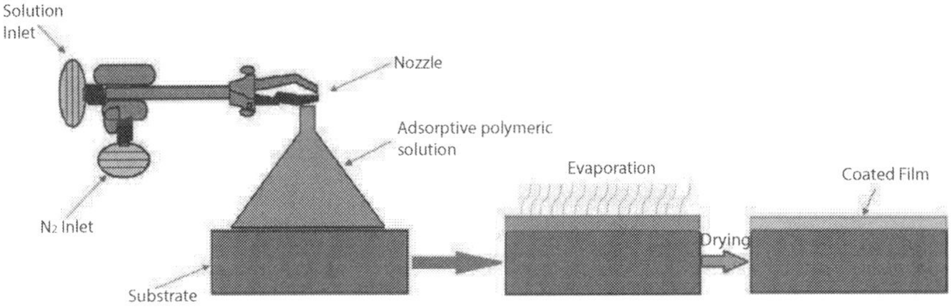

Figure 6.5 Spray coating technique on substrate.

and then dissolved or melted polymer is sprayed onto a substrate as shown in Figure 6.5 [22]. To generate continuous flow of droplets, the nozzle at the spray head atomized the desired material. For atomization process, the crucial process parameters such as surface tension, viscosity, nozzle design, and gas flow should be acknowledged [23].

6.7.2 Dip Coating

Dip coating technique is used to coat the three-dimensional object. In this technique, a solid object or substrate is submerged in polymer solution containing dissolved adsorbent as shown in Figure 6.6. The submerged object is withdrawn, solvent is evaporated, and object is dried at required temperature [24]. The important advantage of this technique is to obtain

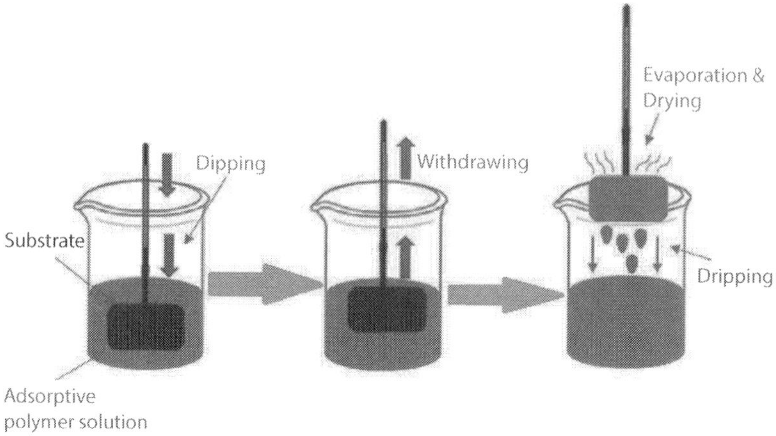

Figure 6.6 Dip coating technique on solid object.

uniform and high quality films. The controlling parameters such as submersion time, dipping cycle, withdrawal speed, solution composition, and concentration adjustment should be carefully acknowledged to obtain unique dip film thickness and required properties [25].

6.7.3 Spin Coating

To obtain high quality and fine films, spin coating technique has been employed. In this technique, centrifugal force spreads the coating material. A cramped quantity of adsorptive polymer film is pragmatic on the middle of the substrate as shown in Figure 6.7 [26]. Under centrifugal action, as the substrate gain required spinning speed, the coating material is spread

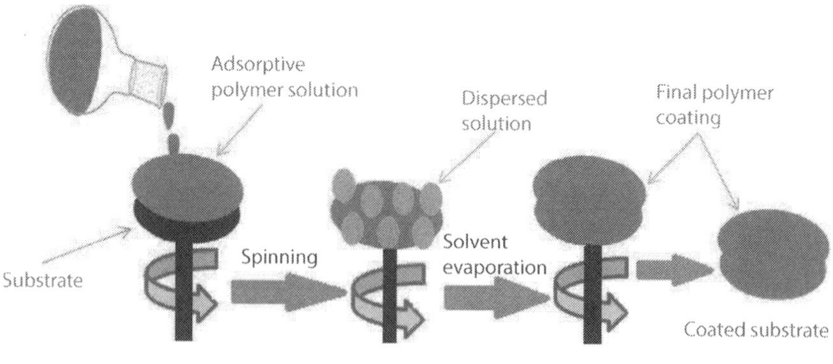

Figure 6.7 Spin coating on substrate.

on substrate. This is the most refined adsorptive polymer film forming technique, and covering instrument is named as spin coater [27].

6.7.4 Solution Casting

Solution casting is a simple coating approach. This method was developed to coat adsorptive polymer solution on a substrate. In this approach, the softened adsorbent in polymer solution is simply casted on the solid object and allowed to evaporate at desired temperature as shown in Figure 6.8. The obtained film on substrate or solid object acts as a coat and protect from external environment [28].

6.7.5 Blade Coating

To obtain thin film coating on substrate, blade coating technique is preferably used. In this technique, the dissolved selective adsorbent in polymer solution is layered on solid object with a high-pitched blade edge as shown in Figure 6.9. The volume of thaw or solution on surface of the material is controlled by the sharp knife edge [29]. Less volume of solution waste, consistency to a roll-to-roll fabrication, interlayer dissolution resistance, and large area uniformity are considered as the advantages of this technique and give well-defined coating. The fabrication parameters such as blade coating speed, blade gap, and solution concentration are properly adjusted to obtain optimized coating thickness [30–32].

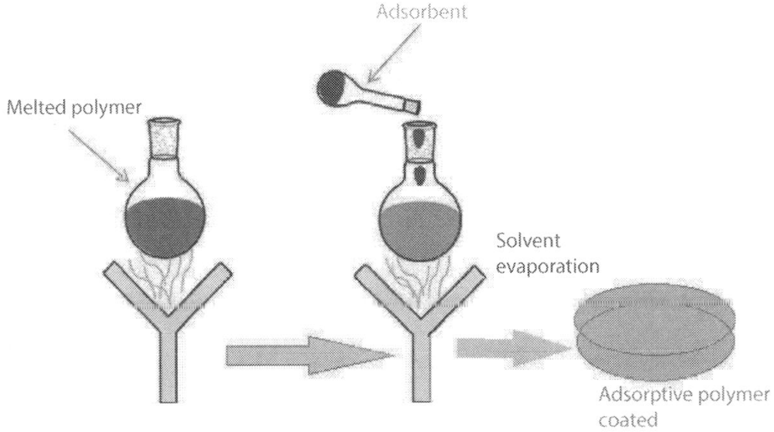

Figure 6.8 Solution casting technique.

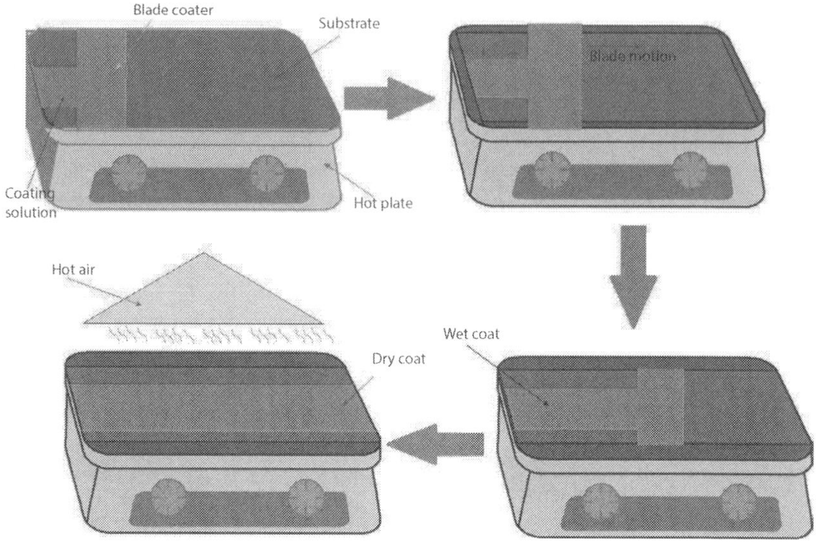

Figure 6.9 Blade coating technique on substrate.

In the absence of centrifugal action, polymer chains are free to move, which is an advantage of blade coating over spin coating. Therefore, there is no need of any kind of post treatment such as thermal and solvent annealing. Thermodynamically, more favorable morphology has been obtained through this technique due to slower evaporation of solvent [33].

6.8 Adsorptive Polymer Coating Applications

Adsorptive polymers are a viable candidate for coating of materials due to excellent thermal, chemical, and mechanical resistance and protect the material from deterioration. Due to these properties, a vast application of coating has been found in this regard. Some important and common applications are illustrated below.

6.8.1 UV Protection

When polymer coating is exposed to ultraviolet radiation, polymer degradation process accelerated due to breaking of polymer chains. The loss of surface gloss and significant deterioration of coating are resulted due to this degradation of chains as free radicals are produced which reduces the molecular weight of polymer [34]. However, adsorptive polymer material has ability to adsorb photon in ultraviolet region which makes adsorptive polymer as an innovative candidate to develop the stabilization and weathering of polymer coating [35, 36].

6.8.2 Biomedical

Due to biocompatibility, elasticity, corrosion protection, and mechanical strength, adsorptive polymer coating is considered as one of the promise candidate for variety of biomedical applications [37]. In drug delivery application, this type of coating is considered very effective [38].

6.8.3 Corrosion Protection

Due to contact of metals with environment, devastation/deterioration of metals occurs. Thus, to prevent the materials from deterioration, different approaches have been developed so far. However, adsorptive polymers have excellent corrosion inhibit properties [39, 40].

6.8.4 Mechanical and Wear Properties

Adsorptive polymers managed to enhance the impact, flexural, and tensile strength because of the excellent adsorbent interaction and act as an intelligent stress transfer through the interface [41]. Adsorptive polymer coating also protects materials from galling, physical damages and seizing. This type of coating improves the toughness wear resistance of polymer coating. Thus, adsorptive polymer coating stemmed in a outstanding development in wear and corrosion resistance [42].

6.8.5 Packaging

Packaging of materials is a very common and important application of adsorptive polymer coating. Anti-microbial packaging materials have been investigated to hinder microbial growth in food [43]. To design and control food packaging process, release kinetics, diffusion mechanism, polymer kinetics, etc., are the essential parameter to be considered. For liquid food packaging, adsorptive polymer coating was also being employed [44, 45].

6.9 Future Perspectives

In each of above mentioned technological areas, several challenges for adsorptive polymer coating still exist. Intelligent inhibiting adsorbent are also needed for ideal corrosion protection [46]. To obtain the better results and to control the coated film thickness, properties, morphology, and to obtain high performance for commercialization, a better understanding and relevant parameters optimization of processes needed [47].

Proper mixing of adsorbents in polymer and random aggregation of adsorbent is still a problem. Thus, this is indispensable to recognize the *in vitro* and *in vivo* approach for equalization of these coverings [48]. On large-scale application, quantitative analysis of uniformity is still lacking. Nevertheless, imminent exploration may undertake the complexities of these challenges and assist to gain excellent performance of materials and coatings [49].

References

1. Whitbourne, R.J. and Zhang, X., *Adherent, flexible hydrogel and medicated coatings*, Google Patents. STS Biopolymers, Inc., Henrietta, N.Y., 2000.
2. Féat, A. *et al.*, Coatings preventing insect adhesion: An overview. *Prog. Org. Coat.*, 134, 349–359, 2019.
3. Hansen, M.K., Larsen, M., Cohr, K.H., Waterborne paints. A review of their chemistry and toxicology and the results of determinations made during their use. *Scand. J. Work Environ. Health*, 13, 6, 473–485, 1987.
4. Ghosh, S.K., Functional coatings and microencapsulation: A general perspective. *Functional coatings*, 3, 1–28, 2006.
5. Nguyen, T. *et al.*, Degradation modes of crosslinked coatings exposed to photolytic environment. *J. Coat. Technol. Res.*, 10, 1, 1–14, 2013.
6. Guo, Q. *et al.*, Characterization of cross-linking depth for thin polymeric films using atomic force microscopy. *J. Appl. Polym. Sci.*, 132, 8, 1–6, 2015.
7. Bossi, A. *et al.*, Synthesis of controlled polymeric cross-linked coatings via iniferter polymerisation in the presence of tetraethyl thiuram disulphide chain terminator. *Biosens. Bioelectron.*, 25, 9, 2149–2155, 2010.
8. Zoveidavianpoor, M. and Gharibi, A., Application of polymers for coating of proppant in hydraulic fracturing of subterraneous formations: A comprehensive review. *J. Nat. Gas Sci. Eng.*, 24, 197–209, 2015.
9. Thakur, V.K. and Kessler, M.R., Self-healing polymer nanocomposite materials: A review. *Polymer*, 69, 369–383, 2015.
10. Balazs, A.C., Emrick, T., Russell, T.P., Nanoparticle Polymer Composites: Where Two Small Worlds Meet. *Science*, 314, 5802, 1107, 2006.
11. Gonçalves, F.P. *et al.*, Development of a Powdered Activated Carbon in Bar Adsorptive Micro-Extraction for the Analysis of Morphine and Codeine in Human Urine. *J. Chromatogr. Sci.*, 50, 574–81, 2012.
12. Asiabi, H. *et al.*, Electroplating of nanostructured polyaniline–polypyrrole composite coating in a stainless-steel tube for on-line in-tube solid phase microextraction. *J. Chromatogr. A*, 1397, 19–26, 2015.

13. Ye, N. *et al.*, Application of Graphene as Solid Phase Extraction Absorbent for the Determination of Parabens in Cosmetic Products by Capillary Electrophoresis. *Anal. Lett.*, 46, 13, 1991–2000, 2013.
14. Md Yusoff, M. *et al.*, An ionic liquid loaded magnetically confined polymeric mesoporous adsorbent for extraction of parabens from environmental and cosmetic samples. *RSC Adv.*, 7, 57, 35832–35844, 2017.
15. Alvarez-Rivera, G. *et al.*, Development of a multi-preservative method based on solid-phase microextraction–gas chromatography–tandem mass spectrometry for cosmetic analysis. *J. Chromatogr. A*, 1339, 13–25, 2014.
16. Trojanowicz, M., Analytical applications of carbon nanotubes: A review. *TrAC, Trends Anal. Chem.*, 25, 5, 480–489, 2006.
17. Alves, V.D., Costa, N., Coelhoso, I.M., Barrier properties of biodegradable composite films based on kappa-carrageenan/pectin blends and mica flakes. *Carbohydr. Polym.*, 79, 2, 269–276, 2010.
18. Neng, N., Silva, A., Nogueira, J., Adsorptive micro-extraction techniques—Novel analytical tools for trace levels of polar solutes in aqueous media. *J. Chromatogr. A*, 1217, 47, 7303–7310, 2010.
19. Ide, A.H. and Nogueira, J., New-generation Bar Adsorptive Microextraction (BAμE) Devices for a Better Eco-user-friendly Analytical Approach – Application for the determination of antidepressant pharmaceuticals in biological fluids. *J. Pharm. Biomed. Anal.*, 153, 126–134, 2018.
20. Hanani, Z.N. *et al.*, Effect of different fruit peels on the functional properties of gelatin/polyethylene bilayer films for active packaging. *Food Packag. Shelf Life*, 18, 201–211, 2018.
21. Horvath, J. and Dolník, V., Polymer wall coatings for capillary electrophoresis. *Electrophoresis*, 22, 4, 644–655, 2001.
22. Susanna, G. *et al.*, Airbrush spray-coating of polymer bulk-heterojunction solar cells. *Sol. Energy Mater. Sol. Cells*, 95, 7, 1775–1778, 2011.
23. Choi, D.Y. *et al.*, Annealing-free, flexible silver nanowire–polymer composite electrodes via a continuous two-step spray-coating method. *Nanoscale*, 5, 3, 977–983, 2013.
24. Brinker, C.J. *et al.*, Fundamentals of sol-gel dip coating. *Thin Solid Films*, 201, 1, 97–108, 1991.
25. Lu, Y. *et al.*, Continuous formation of supported cubic and hexagonal mesoporous films by sol–gel dip-coating. *Nature*, 389, 6649, 364–368, 1997.
26. Hall, D.B., Underhill, P., Torkelson, J.M., Spin coating of thin and ultrathin polymer films. *Polym. Eng. Sci.*, 38, 12, 2039–2045, 1998.
27. Chang, C.-C. *et al.*, Spin coating of conjugated polymers for electronic and optoelectronic applications. *Thin Solid Films*, 479, 1–2, 254–260, 2005.
28. Caruso, R.A. *et al.*, Modification of TiO2 network structures using a polymer gel coating technique. *Chem. Mater.*, 13, 3, 1114–1123, 2001.
29. Krebs, F.C., Polymer solar cell modules prepared using roll-to-roll methods: Knife-over-edge coating, slot-die coating and screen printing. *Sol. Energy Mater. Sol. Cells*, 93, 4, 465–475, 2009.

30. Schneider, A., Traut, N., Hamburger, M., Analysis and optimization of relevant parameters of blade coating and gravure printing processes for the fabrication of highly efficient organic solar cells. *Sol. Energy Mater. Sol. Cells*, 126, 149–154, 2014.
31. Tsai, P.-T. *et al.*, High-efficiency polymer solar cells by blade coating in chlorine-free solvents. *Org. Electron.*, 15, 4, 893–903, 2014.
32. Chang, Y.-H. *et al.*, Polymer solar cell by blade coating. *Org. Electron.*, 10, 5, 741–746, 2009.
33. Krebs, F.C., Fabrication and processing of polymer solar cells: A review of printing and coating techniques. *Sol. Energy Mater. Sol. Cells*, 93, 4, 394–412, 2009.
34. Katangur, P., Patra, P.K., Warner, S.B., Nanostructured ultraviolet resistant polymer coatings. *Polym. Degrad. Stab.*, 91, 10, 2437–2442, 2006.
35. Kotlík, P. *et al.*, Acrylic copolymer coatings for protection against UV rays. *J. Cult. Heritage*, 15, 1, 44–48, 2014.
36. Pathania, A., Arya, R.K., Ahuja, S., Crosslinked polymeric coatings: Preparation, characterization, and diffusion studies. *Prog. Org. Coat.*, 105, 149–162, 2017.
37. Abbasi, F., Mirzadeh, H., Katbab, A.-A., Modification of polysiloxane polymers for biomedical applications: A review. *Polym. Int.*, 50, 12, 1279–1287, 2001.
38. Wei, Q. and Haag, R., Universal polymer coatings and their representative biomedical applications. *Mater. Horiz.*, 2, 6, 567–577, 2015.
39. Sitaram, S.P., Stoffer, J.O., O'Keefe, T.J., Application of conducting polymers in corrosion protection. *J. Coat. Technol.*, 69, 866, 65–69, 1997.
40. Baumert, B., Stratmann, M., Rohwerder, M., The deformation response of ultra-thin polymer films on steel sheet in a tensile straining test: The role of slip bands emerging at the polymer/metal interface. *Z. Metallkd.*, 95, 447–455, 2004.
41. Shaffer, S.J. and Rogers, M.J., Tribological performance of various coatings in unlubricated sliding for use in small arms action components—A case study. *Wear*, 263, 7, 1281–1290, 2007.
42. Xia, X. *et al.*, Constitutive property evaluation of polymeric coatings using nanomechanical methods. *MRS Online Proc. Library Archive*, 522, 199–204, 1998.
43. Han, J.H., Antimicrobial food packaging, in: *Novel food packaging techniques*, 8, 50–70, 2003.
44. Rhim, J.-W. and Ng, P.K.W., Natural Biopolymer-Based Nanocomposite Films for Packaging Applications. *Crit. Rev. Food Sci. Nutr.*, 47, 4, 411–433, 2007.
45. López-Rubio, A. *et al.*, Overview of Active Polymer-Based Packaging Technologies for Food Applications. *Food Rev. Int.*, 20, 4, 357–387, 2004.

46. Rohwerder, M., Duc, L.M., Michalik, A., *In situ* investigation of corrosion localised at the buried interface between metal and conducting polymer based composite coatings. *Electrochim. Acta*, 54, 25, 6075–6081, 2009.
47. Williams, G., Coleman, A.J., McMurray, H.N., Inhibition of Aluminium Alloy AA2024-T3 pitting corrosion by copper complexing compounds. *Electrochim. Acta*, 55, 20, 5947–5958, 2010.
48. Arumugam, D.V.M., Theivanayagam, M.G., Manthiram, A., Rapid microwave-solvothermal synthesis of phospho-olivine nanorods and their coating with a mixed conducting polymer for lithium ion batteries. *Electrochem. Commun.*, 10, 903–906, 2008.
49. Krebs, F.C., Fyenbo, J., Jørgensen, M., Product integration of compact roll-to-roll processed polymer solar cell modules: Methods and manufacture using flexographic printing, slot-die coating and rotary screen printing. *J. Mater. Chem.*, 20, 41, 8994–9001, 2010.

7

Polyurethane Coatings

Nadia Akram*, Khalid Mahmood Zia, Nida Mumtaz, Muhammad Saeed, Muhammad Usman and Saima Rehman

Department of Chemistry, Government College University Faisalabad, Faisalabad, Pakistan

Abstract

Polyurethane belongs to a versatile category of polymers which includes a wide range of applications. It provides an excellent coverage to substrate for protecting it from wear and tear, abrasion, corrosion, and weathering. The complex chemistry of polyurethane has made its significant position among various polymers to dominate this important market. The significance lies in its leading role to replace conventional polymers by its unique formulations and development of both homogenous and heterogeneous emulsions. It provides a custom solution to décor and protect the surfaces and to probe the solution of the problems faced by polymer industry and allied industries. With the exception of few surfaces, PU is an eco-friendly dispersant to majority of surfaces which make it an excellent choice to be used with ease. Fast drying and easy curing without additional curing agents and variety of application methods enchant the consumer to use it on a range of surface.

Keywords: Polyurethane, formulation, coatings, protection

7.1 Introduction

The term "polyurethane" is employed for the polymers comprising a significant figure of urethane (ethyl carbamate) groups. In reaction between the hydroxyl group of the macrodiol/polyol and the isocyanate group, urethane linkage is formed, which is represented in Figure 7.1.

**Corresponding author*: nadiaakram@gcuf.edu.pk

Inamuddin, Rajender Boddula, Mohd Imran Ahamed and Abdullah M. Asiri (eds.) Polymer Coatings: Technology and Applications, (135–158) © 2020 Scrivener Publishing LLC

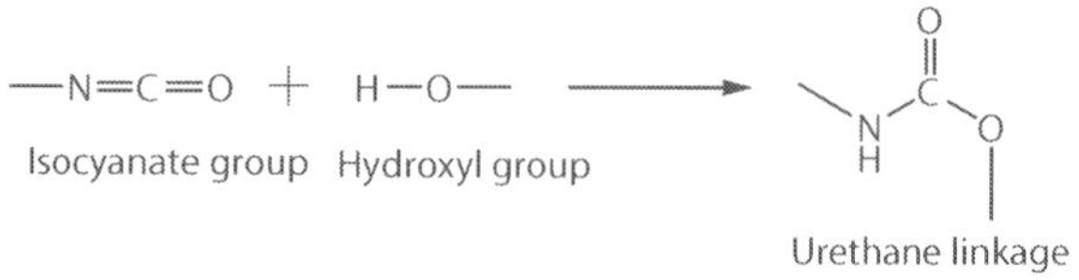

Figure 7.1 Representation of urethane linkage.

Like a conventional polymer, the name is derived due to repeated urethane linkages. Although, the usage of the common term polyurethane is illusive as polyurethane polymers neither belong to the class of polymers which consist of primarily urethane linkages nor they are synthesized by the monomeric polymerization of reactant urethane. The molecular chain in the major linkage groups is biurets, ethers, allophanate, and amides. The term is more frequently used for the expediency than accuracy [1]. The polymeric material of polyurethane has exclusive range of chemical and physical properties. Polyurethanes have gained attention in different industries, for example in adhesives, in the form of foams, as a thermoplastic elastomer and in coatings field [2]. Polyurethane has always been an emerging field of polymers and has enchanted the investigators to explore more and more about this material. It is an interesting system and its properties can be modified by the structural and chemical variation. Its broad spectrum ranges from soft flexible material to tough and rigid materials. It is hard like fiber glass, squishy like foam, protective like varnishes, bouncy like rubber, and sticky like glue.

These are widely distributed in the chemical, automobile, medicinal, packaging, computer, and athletics industries due to the fact that a large variety of raw material with the specific properties is available for polyurethane. In order to find out the best advantages of polyurethane polymeric material, the prime consideration is the enhancement of the steps of the process of synthesis [3].

Since its discovery, the polyurethane has been studied a lot by the researchers. They have developed the methods of its synthesis using linear and graft copolymers [4]. There are two convenient methods for the synthesis of polyurethanes, one shot method and the prepolymer method. Prepolymer method is considered efficient over one shot method due to its minimum chances of side reactions. Researchers have preferred prepolymer method over one shot method depending upon the requirement of the final product [5]. Polyurethane in various forms has been synthesized starting from rubber, elastomers, and foams, etc. They have been studied as thermoplastic elastomeric poly (propylene) system consisting of a lined structure contained numerous blocks of stereo regular or stereo irregular arrangements [6]. Kim and Lee have prepared the thermoplastic PUs by different hard part length as of a fixed molar ratio of polyols and diisocyanate. They have adopted different polymerization procedures and have studied the consequence of polymerization technique on mechanical and thermal properties of polyether-based thermoplastic polyurethanes [7]. Detailed reviews on the material have also been reported extensively. Wicks and coworkers have reviewed the polyurethane system involving water as a carrier medium. They have described that the chemistry and the raw components usage in this ground and has offered some factors that should give attention in order to make this field feasible for the industry [8]. Chattopadhyay and Webster have presented a detailed review on the thermal stability and flame retardancy of polyurethanes [9]. Most useful, common, and research materials are polyurethanes in the world. Polymeric materials have special groups like polyurethanes. In between the large number of plastic types, polyurethanes are different in numerous ways. Polyurethanes can be combined with variety of materials such as integral skin, paints, elastic fibers, liquid coatings, foams, insulator, and elastomers. During World War II in place of rubber, polyurethanes were first developed. Polyurethanes can be incorporated in different applications because of its versatility and appropriateness [10]. Polyurethanes have many superior properties like hardness, strength, modulus, and elongation. Because of these properties, polyurethanes are broadly used in various applications like building and construction applications, textiles, in biomedical, in automotive industry, etc. In polyurethanes, most repeating unit is urethane group. Along with urethane group, there are many other groups present in polyurethanes, for example, urea, ether, ester, or aromatic compounds. Synthesis of polyurethanes can be done by a wide range of sources and because this polyurethane gives a large no of specific applications. On the basis of required properties, polyurethanes can be classified into various groups like coating, elastomers, rigid, waterborne, flexible thermoplastic, adhesive, and sealants. Among the main applications of polyurethanes, very prominent polyurethane product is polyurethane foam. This product is used globally. There are several types of polyurethanes foams. All these types are of commercially significant. These are divided into these major types, for example, adhesives, rigid foams, surface coatings, elastomers, integral foams, and flexible foams [11]. Thermoplastic polyurethane as for its good strength, excellent abrasion resistive, brilliant chemical resistive, and have low smoke properties. This polymeric material is extensively used in the form of sheathing composite in particular determination, low voltage control cables, and also in the profile extrusions. It is proven that polyurethanes have fascinating shape-memory qualities. Shape-memory PU_S is abbreviated "S-PU_S". Shape memory polyurethanes are consist of polyester and polyether, molecular weight of 300–1,000. Shape memory polyurethanes are set in a definite shape above Glass transition temperature and cooled to a temperature below glass transition temperature. Shape memory cooled polyurethanes hold this particular shape as it is set. By heating shape memory polyurethanes to a temperature above glass transition temperature, the shape memory polyurethanes came back to their original shape.

Rigid polyurethane foams (RPFs) are three-dimensional polymeric substances, cross-linked, and possess cellular structures. Rigid polyurethane foams have low thermal conductivity or excellent mechanical resistance. Rigid polyurethane foams are used by means of insulating materials, refrigerator (domestic/commercial), in the field of building engineering, in automotive parts, and use in pipelines applications [12].

7.2 Chemistry of Polyurethane

Compounds which consist of hydroxyl groups mostly produce esters of substituted carbamic acid and urethanes. These consistently procedure portion of the end usage on the way to which the isocyanates are placed. The term polyurethane has turn out to be accepted by means of a generic term in order to cover the entire engineering products produced from isocyanates [13].

Polyurethanes are produced in presence of ultraviolet light activation, catalyst, chain extender, or any additives, by means of a chemical reaction among a di/poly isocyanate in addition of a diol or polyol, creating repeating urethane groups. In polyurethane backbone, urea, ester, aromatic ring and ether are present with urethane linkages [14]. The polyurethanes chemistry allows this class of polymers to assemble with further compounds that are mostly cooperatively termed to as reaction polymers. Such mixtures consist of epoxies, unsaturated polyesters, and phenolics [10].

Isocyanates stay reactive class, at room temperature isocyanate reacts by hydroxyl comparatively slowly. To some degree, the slow speed of isocyanate reaction is a reflection of phase unsuitability of the comparatively nonpolar or thicker isocyanate phase and comparatively polar or less-dense polyol phase. The reaction also remains slow at room temperature even if phase compatibilized with a surfactant [15]. Polyol molecules must certainly consist of two or more than two isocyanate groups (R–(N]C]O)n$2) and hydroxyl groups (R0–(OH)n$2), correspondingly. The shown qualities of the polyurethanes frequently rest on the categories of polyols and isocyanates by which they were prepared. Usually, soft flexible polymers can be obtained by the means of flexible long segments of polyols, although rigid or hard polymers can be prepared by a greater cross-linking amount. Stretchy polymeric substances might produce via elongated chains through low cross-linking, while hard polymers can be generated by the help of shorter chains within high cross-linking. A mixture of long chains by means of normal cross-linking, generate such polymeric material which are appropriate aimed at foam manufacturing. In polyurethanes, because of the presence of cross-linking, they frequently own an immeasurable molecular weight by means of a three-dimensional (3D) network build-up. Because of this motive, a small portion of polyurethanes can be mentioned to as a giant molecule.

This could also describe why typical polyurethanes frequently will not undergo melt and soft after they are heated. The addition of various additives into the isocyanates or polyols, and amendments to the processing settings, makes it probable to acquire an extensive series of characteristic features, which able them to be used for numerous applications. Various forms of polyols are available. By the use of various methods, all these polyols can be produced in laboratories [1].

Under the existence of catalyst by polyaddition reaction in alcohols and isocyanates, synthesis of polyurethane is carried out (Figure 7.2) [16]. This polyaddition reaction is nucleophilic addition in nature, in which different catalysts are used, for example, compound like organo tin, metals, and tertiary amines. Urethane formation is in fact an equilibrium reaction; the presence of a catalyst therefore also increases the rate of the reverse reaction at very high temperatures. Another important basic reaction involved in the polyurethane formation is the chain extension reaction which occurs between chain extender (diol or diamine) and isocyanate. The use of diol as chain extender generates urethane, while in final step of synthesis of polyurethane, diamine reaction produced urea linkages [17].

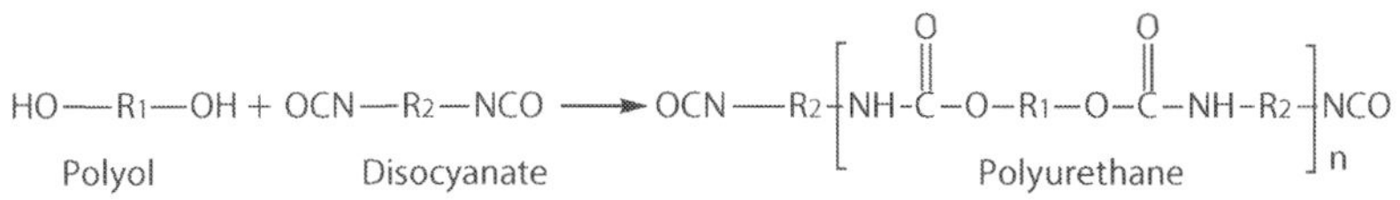

Figure 7.2 Reaction for the synthesis of polyurethane.

7.3 Formulation of PU Coating

7.3.1 Raw Material for Polyurethanes

7.3.1.1 Polyols

Polyols or macrodiols are reactive compounds which are usually found in the form of liquids, but are available in the solid form as well. These contain at least two functional groups in the form of hydroxyl or amino groups. A huge variety of polyols are accessible, but most of the polyols belong to two classes, hydroxyl terminated, and amine terminated polyols. So, depending upon the use of the starting polyol of polyether, polyester, or hydrocarbon materials, we can obtain ether-based, ester-based, or hydrocarbon-based polyurethanes as a finished product.

7.3.1.2 Polyether Polyols

This raw material has repeating linkage of R-O-R. They have two or more functional groups (hydroxyl groups) at the terminals. The reaction of epoxides like ethylene oxide or propylene oxide with the multifunctional initiator in the presence of a catalyst forms the polyether polyols. Common polyether diols are polyethylene glycol, polypropylene glycol, and poly (tetramethylene ether) glycol. Polyether polyols contribute about 90% to the industries. The hydroxyl groups of polyether can react with the other substances, e.g., isocyanates to develop urethane linkages. These are water soluble compounds and are available in a wide range of molecular weight 200 to 14,000 daltons. The physical properties of the material depend upon the chain length of the polymer. Their major contribution is in the polyurethane industry for the manufacturing of different products. These are mainly used in the form of polyethylene glycols or polypropylene glycols. Polypropylene glycol is similar to polyethylene glycol in most of its properties and is used in the same way as other polyether polyols (Figure 7.3).

Polyether polyols have always been used widely for the production of various polyurethanes. These polyethers have been used in the production

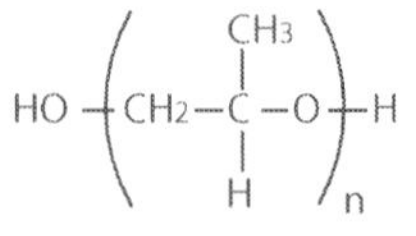

Figure 7.3 Chemical Structure of PPG.

of moisture curable and polyurethane resin, and these resins have been used as an adhesive on plastic and aluminum substrates [18]. While, polypropylene glycol with different molecular weights has been used for polyurethane plastic bandages [19].

Synthesis of polyurethane with cycloaliphatic diisocyanate has been carried out to study the structural investigation by the use of 1D and 2D NMR spectroscopy. In synthesis of polyurethanes, polyols are used. Polyols have mono disperse rigid segments [20]. Cross-linked polyurethane has also been formed in this connection. Blending of polyether and polyester polyols is also a common practice for the production of polyurethanes. Different synthesis protocols give different properties [21].

7.3.1.3 Hydrocarbon-Based Polyols

Hydroxyl-terminated polybutadiene HTPB is a commonly used hydroxyl compound in the polyurethane industry [22]. The complete hydrocarbon nature and unsaturation in the backbone make it a useful macrodiol for various polymeric products. Polybutadiene are very interesting compounds because of their very low T_g (−80°C) and lead to final materials with good viscoelastic properties [23]. HTPB produced strongest polyurethane. The low T_g and hydrocarbon-based backbone structure of polybutadiene impart better hydrolytic properties and higher flexibilities in the final product. However, the mechanical properties such as tensile strength, abrasion, and tear resistance are not considerably high for these materials [24]. Hydrocarbon-based polyurethanes develop phase segregation due to presence of a polar character of oligobutadiene. Consequently, some striking properties, for example, at low temperature retention of elastic behavior, highly resistive to acid or base hydrolysis, low water penetrability, and upgraded mechanical properties, have been established. Final product gain is much more beneficial properties in the presence of HTPB (Figure 7.4) [25].

The complete hydrocarbon nature of polybutadiene prevents formation of inter segment hydrogen bonding which are usually present in polyether or polyester-based polyurethanes favoring phase segregation [26].

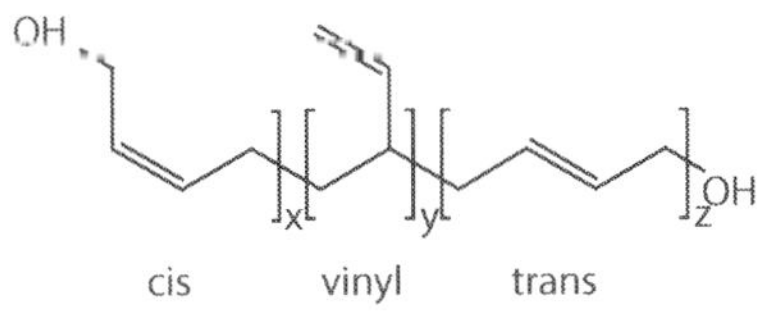

Figure 7.4 Chemical Structure of HTPB.

Polybutadiene in the form of a mixture with polystyrene blocks has been used to study the adhesive properties of the system in terms of peel and tack; the tackifiers have also been incorporated to check the effect as cross-linking agents [27]. The modification of the system has been used to improve the adhesive properties [28, 29]. The functionalization of HTPB is a common practice for the polyurethane formation and has been carried out to use in various properties as well [30]. Along with various properties of the material, the rheological properties have also been studied [31].

7.3.2 Isocyanates

The chemical structure of diisocyanate can be classified into two kinds. One is aromatic diisocyanate, and the other is aliphatic one. The PUs prepared with aliphatic diisocyanate does not show the change of color by UV light, heat, and aging [32]. With the help of resonance structure, high reactivity of the isocyanate group can be demonstrated (Figure 7.5). Lowest electron density is associated with carbon atoms, while higher electron density is at the oxygen atom. As an immediate consequence, the carbon atom has a positive charge, the oxygen atom has a negative one, and the nitrogen atom with an intermediate negative charge.

The free isocyanate contents influence the adhesive properties of the polyurethane [33]. The hydrophillic isocyanate has been reported to form stable and homogenous micelles in water [34]. The researchers have also found out that the self-curable polyurethane can be prepared by the use of aliphatic diisocyanate [35]. Polyurethane coatings were particularly used for impregnating paper and in the manufacturing of garments that remained resistive to mustard gas or rusting. For the purpose of coating of metals, high brightness textures used for airplanes and wood, polyurethanes coating behaves as chemically resistive. The initial industrial manufacturing of polyurethanes coatings started with various preparations for specific purposes. Initially, polyurethanes coatings were frequently used in adhesives, elastomers, rigid forms, and coatings. After that, relaxed and convenient pillows prepared by means of flexible foams which are commercially available. Furthermore, the development of elastic foams from inexpensive polyether polyols directed to various automotive or upholstery applications that are still applicable today. In coating, polyurethanes have a wide application,

Figure 7.5 Resonance structure of isocyanate groups.

because of their specific qualities: low temperature flexibility, corrosion, toughness, good mechanical strength, excellent abrasion. Polyurethane elastomer is one of the significant kinds of polyurethane. In different engineered products, polyurethane elastomers are incorporated. Their processing is easy because of the fact that they are malleable. Recycling possibility of polyurethane elastomers is high [1]. Some additives used in the synthesis of polyurethane and their advantages are listed in Table 7.1.

Table 7.1 Advantages of PU components.

Components	Advantages
Fillers	Fillers are used to enhance the properties of the material, for example, tensile strength and stiffness.
Catalysts	Catalysts are used to accelerate the rate of reaction between polyols and isocyanate and because of this even at lower temperature reaction can be carried out.
Smoke retardants	The main use of smoke retardants is the reduction of the generation of smoke during the burning of materials.
polyols	The main advantage of polyols is the contribution of flexible long segments, in result of which soft elastic polymers are produced.
Plasticizers	In order to decrease the hardness of the materials, plasticizers are used.
Isocyanate	Isocyanates are used to enhance the properties of polyurethanes, for example polyurethane curing and reactivity.
Pigments	Pigments are used in the production of polyurethanes colored materials, particularly for the purpose of aesthetic properties.
Blowing agents/ surfactants	Blowing agents/surfactants are used in the synthesis of polyurethanes foams, also used to control bubbles formation and cell structure of the foam during synthesis.
Flame retardants	Flammability of the material is reduced by the use of flame retardants.
Cross-linkers/ chain extenders	Cross-linkers/chain extenders are used to enhance the properties of the material for example enhance mechanical support and modification of structure.

7.3.3 Monomeric Diisocynate

In polyurethane products major value-added building blocks are monomeric diisocynate. In coating industry, they are widely used. There are two types of monomeric diisocynate aliphatic and aromatic diisocynate monomers. Both monomers have important differences. Aliphatic diisocynate is less reactive as com pare to the aromatic diisocynate. Urethane products prepared from aliphatic diisocyanates, if exposed to UV light, oxidize less while show more oxidation if prepared from aromatic diisocyanates. The furthermost significant monomers used in the polyurethane engineering are toluene diisocyanate. In the:

1. Toluene diisocyanate is used in the form of raw material in the synthesis of adducts and prepolymers in coatings industry.
2. To gain extensive industrial uses, diphenylmethane 4,4'-diisocyanate was first diisocynate monomers.
3. Hexamethylene diisocyanate is used for the synthesis of vast variety of polyisocyanate adducts. Hexamethylene diisocyanate is diisocynate monomer used in exterior coating [36].

7.3.4 Vegetable Oil-Based Polyurethane Coating

From various diisocyanates, chain extenders, and polyols, polyurethanes have been produced and properties of polyurethanes are examined. By means of petroleum source, most of the polyols are generated and used in the synthesis of polyurethanes. But because of environmental concern and high energy strains of the process of preparation have emphasis on the requirement for further appropriate or environmentally friendly replacements. This draws more academic and commercial attention renewable resources, for example, vegetable oils [10]. Many researches are dedicated to the production of polyols from various vegetable oils, for example, palm oil, rapeseed oil and soybean oil [13]. Vegetable oils primarily contain triglyceride molecules with various reactive places, such as ester, hydroxyl groups, and carbon-carbon double bonds [10]. It is necessary to functionalized vegetable oils for the production of polyurethanes. This functionalization is done by the help of double bond reaction, for example, hydroformylation, epoxidation, metathesis, and ester bond reaction [13]. Polyols attained by the source of vegetable oils can be able to partially replace polyols derived from the petroleum source, particularly crosslinked by various kinds of isocyanates for the manufacturing of polyurethanes [10].

In the composition of polyurethane foam, the usage of polyols from vegetable oil source makes it conceivable to expand the qualities of some prepared foams, specifically hydrophobicity. Polyurethanes from vegetable oil polyols can be characterized to have higher thermal stability as compare to those synthesis from propylene oxide polyols [13].

7.3.5 Water Borne Polyurethane Coating

For high-performance coating systems, solvent borne polyurethane sets high standard. The demand of environmentally friendly substitute and high performance of coating gave a new solution to the coating formulator and waterborne polyurethanes have been offered. To fulfill the performance demand of many coating applications, many waterborne technologies has been offered. Currently, as compare to solvent-based products, waterborne products because of their low toxicity dominate coating industry gradually. In mark by this tendency, in last decades, aqueous polyurethanes have grownup commercially because of environmental concern [36].

Bayhydrol Polyurethane Dispersions
Polyurethane dispersion can be described as a colloidal system in which particles of polyurethane are spread in a continuous phase (water). The structure of a polyurethane dispersion resin type is shown [36].

7.4 Applications of Polyurethane Coating

Polyurethane coatings have vast applications in various fields and some of which are listed below.

7.4.1 Multifunctional Polyurethane Coating

Presently elastic fibrous textile functional materials have gained much attraction because of wide range of applications they posses like antibacterial, UV blocking, electrical conductivity, fire retardant superhydrophobic properties. There are many approaches to impart these functional properties to the fabrics, for example, submission of molecular layer-by-layer self-assembly, pad-dry cure, solegel method, UV curing, magnetron sputter technique, and plasma-induced graft polymerization techniques. Far-infrared rays having a lengthier wavelength of 6e15 mm can penetrate 2e3 mm hooked on utmost living ingredients and at molecular level they have strongest rotational or vibrational properties, because this dilation of blood vessels takes place, or consequently enhancement of metabolism and blood microcirculation.

Far-infrared releasing ingredients absorbed energy from either heat of human body or from sun light. Far-infrared releasing ingredients can convert this energy into far-infrared rays having definite wavelength and, at that time, reemit these radiations in to humans. Far-infrared radioactivity gains much attention to claim of germanium or ceramics to textile materials, and these have nearby interaction with the human body, like cushions, sheet supplies, and clothing for health care and therapeutic determinations, because far-infrared radioactivity increases metabolism, enhance blood circulation, or encourages the regaining of exhausted muscles.

Researchers found that if we coat textiles with graphene graphene composites, we can gain a large number of important multifunctional properties, for example, thermal conductivity, anti-bacterial activity, and electrical conductivity. The graphene's potential is not investigated to enhance far-infrared emission. Consequently, as a functional unit to treat fabrics, graphene nanoplates applied. By doing this, far-infrared radiation and many other specific properties can be achieved by graphene [37].

7.4.2 Self-Cleaning of Polyurethane Coating

With passage of time, all surfaces are going to get fully contaminated in a natural environment. It consumes high efforts to clean these surfaces, surfactants used for cleaning also imparts negative impact on the environment. By the use of simple rain shower, dust from the leaves of many plants can be cleaned completely. The term superhydrophobic is used for these kinds of leaves. That is water dewdrops form spheres by very slight adhesion to the surface and then roll off very rapidly even at slight inclinations. On micro or nanometer scale, roughness is the main feature of superhydrophobic surfaces. Only on the tips of these structures, dirt and water droplets lie. Because of this, low adhesion forces are developed between the contaminating particles and rough surfaces. And consequently, by the help of rain shower, these contaminants can be washed out very effectively. In case of artificial superhydrophobic surfaces, self-cleaning qualities are not investigated yet. By means of chemical treatment, it is possible to hydrophobized silicon wafer specimen having different and even arrays of spikes. It can be used as a replicate of water repellant leaves. Likewise, hydrophobization of metal foils was carried out by the help of fluorinated agent. By the use of artificial fog, removal of contaminants is not possible because they have two levels of roughening. These surfaces can be cleaned perfectly when water drops have kinetic impact energy. Self-cleaning property of coating materials is of great importance to avoid and elimination of pollutants. Because of the presence of C-F bond in the fluorocarbon, they posses low surface energy and have excellent self-cleaning properties. Because of this, the use of fluorocarbons as having self-cleaning property gains much attention [1–3]. In case of coatings that are fluorine modified, possess surface free energy and their efficiency as a self-cleanable agent to eliminate the pollution are not much satisfactory because of having great content of fluorine. This is not best way to measure cleanable property of any coating. Because of this reason, a test named peel test was introduced to measure the ability of cleanliness of coating material. In pressure sensitive coatings, peel test is used to measure the content of adhesion, when we detached coated film from substrate [9]. By elimination of adhesion from the pressure sensitive adhesive and surface of fluorine coating, peel test is used to measure cleanability. Because of extensive range of applications in different kind of fields like cements, glass windows, and in textile industries, this technology has been widely used. On contrary of these beneficial aspects, these coating have many additional benefits, for example, by using this approach, there comes reduction in the cost, uses in snow preservation, and give shield from ecological legislation. There are two types of self-cleaning coating: hydrophobic and hydrophilic coatings. These both categories clean their selves by the use of water. Hydrophilic coatings clean themselves by spreading water over the surface by setting themselves, in this approach, breakage down of pollutant particles is carried out. Hydrophobic coatings clean themselves by the rolling of water droplets on the surface, in this way, dirt is removed. All types of coating are also associated with some limitations [7].

7.4.3 Self-Healing of Polyurethane Coating

Great attention is given to the self-healing materials because they posses excellent potential to reduce the effect of degradation and also diminish the cost of maintenance. These kinds of polymers or composites have the ability to heal themselves in case of any damage where and when necessary. This self-healing behavior is stimulated by biological systems in which self-healing is usually a common routine. Materials like autonomic healing ingredients posses very good potential in industrial system because of the quality that they react deprived of external intervention to ecological stimuli in a nonlinear or productive fashion. Such self-healing coatings are of interest which can prevent and cure rusting of the underlying surface. In order to achieve the self-healing properties, many approaches are used, some of which are reversible chemistry, polyionomers, encapsulation, hollow fibers, microvascular networks, monomer phase separation, and nanoparticle phase separation [38].

Majority of self-healing system have limitations like chemical and mechanical, which inhibit the use of these systems in coating. In case of self-healing systems, where environmental contamination can principally be avoided, it is necessary for a self-healing coating that it should be extremely stable to species existing in the environment. Because of air and water stability, self-healing chemistry is attractive.

And these self-healing systems remain stable even after exposure to high temperature. These properties enable the use of self-healing in thermal cure. Mechanical strength is of secondary importance in the field of coating system. Encapsulation of both of the phases (the healing agent and the catalyst) is beneficial in cases where the matrix can react with the healing agent [39].

Polyurethane polymer substance has a wide range of properties because of which this polymeric material used in the field of coating, such as they possess hardness or flexibility, they are good abrasion resistive, etc. Initially, this coating approach was formulated in presence of organic solvents but these synthesis protocols are substituted by the water-based approach throughout polymerization [1, 2]. One main drawback of water-based approach consists of polar groups, so there is a need of extra attention in order to stabilize water-based dispersions. Because of this factor, different properties reduce which inhibit the use of this dispersion in high application of high performance. However, by the use of cross-linking polymers, performance behavior of water-based dispersions can be increased [8, 9].

7.4.4 Nanodoped Polyurethane Coating

For the purpose of modification of physical characteristics, one of the best methods is material doping. For example, poor electrically conductive silicon (si) can be doped with arsenic or boron in order to make perfect hole and electron conductor correspondingly. The more attention is on self-heating property of electrically conductive composites, grounded on polymers filled with graphitic nanofillers is presently increasing for various applications, for example, anti-fog devices, anti- and de-icing systems (ADIS), or self-curing process. Although, it is required to expand the information about the electrothermal performance of such kind of materials [40].

7.5 Advantages of Polyurethane Coating

Polyurethane coatings have a large number of advantages. Some of these are discussed below.

7.5.1 Biodegradation of Polyurethane Coating

Polyurethanes are vulnerable to degrade themselves, even after the years of synthesis. There are differences in the degradation behavior of various samples of polyurethanes that were attributed to the large number of properties of polyurethanes like topology or chemical composition. In case of synthetic polymers, there exhibit the regularity which allows the chain of polymer to pack easily. In result, crystalline region is formed. This limits the polymer chain accessibility to degradative agents. In polyurethanes, microbial degradation depends on various qualities of the polymer, properties like crystallinity, molecular orientation, cross-linking, and chemical groups existing in the molecular chains which govern the approachability to degrading-enzyme structures. In microbes, esterase activity has been noted. This activity allows them to the use of polyurethanes. Because the esterase enzymes are used in ester bonds hydrolysis, the microbial degradation of polyester polyurethane is theorized. In the medicinal field, polyurethanes display strong resistive property to macromolecular oxidation, hydrolysis, and calcination. Elastomers of polyurethane are actually cast-off in place of other elastomers because of their higher toughness, good elasticity, oxidation, excellent resistance to tear, and humidity [41].

7.5.2 Antimicrobial Activity of Polyurethane Coating

Antimicrobial coating is used to reduce the growth of injurious bacteria. Because of the hygiene issues in everyday's practical life and in the medicinal field, antimicrobial coating gains much attention. While on the exposure with the atmosphere, or when used as coating material, or used as an adhesive, polyurethanes are expected to microbial attack. Blistering and disbanding of coating takes places because of the micro-organisms attack under different kinds of sever conditions of the environment. In marine technology, one of the main issues is bio-fouling, being natural phenomenon. It also causes an issue in the marine navigation. When organisms are accumulated on the surface of the submerged vessel of the reaction, this results in the higher fuel consumption and because of this speed is also reduce. Organisms like algae, tubeworms, and barnacle accumulates on the surface of submerged vessel.

To overcome these issues, anti-fouling paints are used to give the protection of the submerged surfaces against the marine bio-fouling. Anti-fouling paints can be described as the paints consist of biocidal species. Organotin compounds are the most effective ingredients of anti-fouling paints. TBT (tributyltin compounds) and its derivatives were found to be destructive molecules to aquatic eco-systems. Because of this reason, TBT-based paints/compounds are entirely forbidden. Compounds contain of mainly heterocyclic ring in their structure, and their derivatives are used in the field of the medicine. Coumarin complexes and their derivatives constitute a cluster of additional forty drugs, which are extensively used in the medicine field as a hypertensive, anticoagulant, immunomodulant agents, antiarrhythmic, and they own extraordinary properties in contradiction of bacteria and fungi. In addition to this, the above stated compounds (coumarin) have different substitutions thiazol rings. This shows good biological activities. Because of these biological activities, coumarin group chemistry gains much attention of synthetic chemists. Such compounds which consist of thiazol rings in their structure also belongs to the heterocyclic compounds. Because of their potential chemo- therapeutic properties, these kinds of groups possess extensive range of biological activities and have good impact in the field of the medicine. In all kinds of available penicillin's, the main integral part is thiazol nucleus which helped in the development of the treatment of the bacterial diseases. In the treatment of the microbial infections, thiazols also have a great impact. In different chemical compounds, the parent material is thiazol, which includes biocides, dyes, sulfur drugs, chemical reaction accelerator, and fungicides. 2-aminothiazol-derivatives are described to display important biological activities and are extensively used as medicines. In addition to this, coumarin thiazol is found to be the best anti-microbial and anti-cancer agent. By saturating bactericidal materials with the help of some anti-bacterial agents, bactericidal material is prepared. The efficiency of such structures is founded on the slow release of the anti-bacterial agents in their surroundings with passage of time. Anti-bacterial agents like quaternary ammonium compounds, antibiotics, halogens, and silver ions. Consequently, this approach is proved to be very effective approach. Active biocides exhausted with passage of time. Along this, leaching of heavy metal ions and toxic biocides into the environment is not anticipated from an environmental point of view [41].

7.5.3 Cloth Protection

To protect the surroundings where we used to live and work becomes even more complex day by day. Traveling from one part of the Earth to another part becomes easier even you can travel throughout the space. We belong to develop civilization where we can assess the incoming potential dangers around us, no matter where we are, all we need is anti-hazard environment.

Similarly, we also need protection in clothing industry. There is now an extensive range of cloth protection available, and continuous development of new raw materials helps to protect human beings from many hazards. Coatings are used in the form of laminations on the clothes in the pattern of continuous films to protect the well-being from the hazards of surrounding by the method of isolation. If we give treatment of water repellency to our fabric material, resistance to the wetting can be developed. Coating or protecting is broadly used in the protection of clothes, and to protect the films, in addition to this, the way or method by which these coatings are combined with the clothing assemblage, are of leading reputation to the manufacturing of a satisfactory clothing. Permethrin is an artificial made pyrethroid pesticide which associate to the compulsory properties of hot-feet, residual activity, repellency, and kill, and knockdown has extensively been used for periods of time, in cloth impregnation used as an arthropod interaction repellent. Consequently, permethrin is categorized by a higher level of potential in contradiction of a extensive variety of multiple repellent, vector insects or ticks, and toxicological effects on arthropods, rapid reactivity, excellent photostability resistance to weathering, and low mammalian toxicity. The combination of permethrin compound into human fabrics and exterior protecting devices essentially can be accomplished safely. Consequently, the use of permethrin is less expensive and easy to handle.

7.5.4 Anti-Scratch and Anti-Algal Coating

Presently, on the manmade exteriors such as roof tiles and buildings, the development of green algae is a main possible health and environmental problem. Because of this, it gains much attention. Because of the green algae, surface discoloration takes place which composed of corrosion and accelerated weathering. Additionally, anti-scratch and anti-algal coatings also have many other side effects, for example, diarrhea, asthma, respiratory failure, heart attack, liver cancer, itchy skin, fever, vomiting, seizures, paralysis, and DNA damage. If the toxins are consumed in large amount then results are even worse. It is necessary to control these green algae on the surface and it is also required to introduce such effective methods which are able to reduce the disadvantageous human health influences of these poisonous green algae.

Different core shell particles are synthesized; they cover the broader spectra of multiple applications. For example, silica-titania core-shell nanoparticle's utilization in nanopaint application by the use of anti-scratch quality of silica in the form of core material and anti-algal qualities of titania in the form of shell components in a single composite particle.

In coating industry, the use of Pus can increase the life span of the product and also increase product's appearance. Polyurethanes (Pus) are associated with the property of the strong bonding. Contrary, polyurethane sealants are associated with the property of providing tighter seals. Polyurethanes elastomers offer higher stress recovery and highly resistive to ecological conditions as these elastomers are much lighter than metals and can be into any desired shapes [42].

7.5.5 Flame Retardant Waterborne Polyurethane Coating

Polyurethanes have excellent chemical resistance, elastic properties, good hardness values, mechanical properties, and high abrasion. All these properties of polyurethanes increase its demand in the field of coating industry. Polyurethanes have strong impact to be used as fire hazards and ongoing public awareness about this impact forces the coating system to development of fire-retardant polyurethanes. By the addition of other reactive or non-reactive fire retardants, it is possible to enhance the flame resistance of a polyurethane coat. Addition of the reactive specie is preferable. As compare to the non-reactive class, reactive class of additives on the polymeric surface do not diffuse and, because of this reason, behave more effectively as resistive to combustion for a long period of time. On the lesser extent, they change the properties like (physical/mechanical) of the material in which they are incorporated. We can classify flame retardants as inorganic material or as an organic material. Inorganic materials include metal hydroxide, metN borates, and metal oxides. While organic materials include halogenated and phosphorous compounds. Inorganic flame retardants have poor effectiveness, lack of toxic byproducts, and they are also low in cost. That is why high level of materials are used for incorporation to reach the required flame-retardant level is required. This material can cause detrimental properties to other qualities of the polymer. Halogenated compounds are effective flame retardant but high temperature conditions are required to initiate the radical-trapping. Polyurethanes might not have as high temperature to start the activity of radical-trapping. Halogenated compounds also states hazardous because of the emission of the gas in the form of hydrogen halides. Continuous work is done to make such kind of flame-retardant product which does not include all these side effects. Compounds containing organic phosphorous are supposed to be very effective nontoxic flame-retardant additives [43].

7.6 New Innovations and Future of Polyurethane Coating

Most useful class of polymers is polyurethanes. Polyurethanes have numerous required properties that might be exploited by researchers for manipulative coatings. From vast range of applications, examples consist of:

- the usage of polyurethanes in clear coating for the glass,
- in aliphatic single-pack roof coatings,
- as the key binder in anticorrosive paint,
- as a scrap tyre binder for sporting surfaces,
- as an excellent quality, light-stable line marking paints,
- as encapsulants for weathered asbestos.

New isocyanates are commercially accessible now and they contribute a lot in the development of the technology of polyurethanes. Polyurethanes have good abrasion, cure well at low temperature, impact resistance, and excellent resistance to chalking. Aromatic polyurethanes are less expensive compare to the aliphatic polyurethanes. Because of this fact, this is more usual to have aromatic polyurethanes or epoxies, particularly used in base or tie-coats and use in primers [44].

7.6.1 Development in Biomaterials

The initial catastrophe failure of constituents, for example silk, metal, ivory, glass, and nylon brought significant principles into emphasis: thrombogenicity and durability. Then, focus was engaged at inert substances that slightly interrelate with blood and tissue. Polyethylene terephthalate and expanded polytetrafluoroethylene are presently the standard biomaterials of prosthetic vascular grafts. By the use of large period of time, properties that are examined state that both polyethylene terephthalate and expanded polytetrafluoroethylene grafts, exposed to achieve well at diameters of 6 mm, but neither substance has been appropriate for small-diameter of 4 mm applications. Biodegradable polymers might establish a momentary scaffold by which tissue in growth *in vivo* ultimately substitutes the prostheses and left behind a whole biological vascular conduit. The emergence of tissue-engineering technology has concurred the progress of a novel biologically viable vascular substitute possible, and it has been proved to be the possible solution for small-diameter vascular grafting.

7.6.2 Future of Paint Industry

While in the formation of high-quality coatings, because of the variable properties and diverse formulation, polyurethane binders are the preferred choice. A mature industry: coating and paint industry undergoes continuous change in the approaches/technologies. In the last 10 years, some market places, for example, automotive, have governed 95% change in the synthesis and composition of coating and paints, which they are used to apply on their products. In 2011, the worldwide marketplace share of polyurethane coatings, apart from architectural paints, can be predictable at nearly 15%. The configuration and medicine chemistry related with "compliant" coating technology are further discovered with respect to UV radiation curable coatings, high solids (solvent borne) coatings, powder coatings, and waterborne coatings. For future research and development, definite areas are being targeted. Polyurethane coating have the advantages of mechanical and chemical resistance and excellent weathering characteristics. These properties of the polyurethanes coating enable them to gain a leading market share in both the aerospace and auto refinish industries [45].

7.7 Conclusion

Polyurethane adhesives as a class can no longer be perceived as new raw materials. From a base of 217 million pounds, double-digit growth can no longer be expected. Even more significant growth of this industry will continue. Formulators are taking advantage of the tremendous flexibility of urethane chemistry in designing new adhesive products. Specialty niches such as waterbornes and reactive hot melts, for example, will continue to emerge and fuel growth. Exciting times lie ahead for innovative formulators of polyurethane adhesives.

References

1. Szycher, M., *Szycher Handbook of Polyurethanes*, p. 2, CRC Press, New York, 1999.
2. Ahmad, I., Zaidi, J.H., Hussain, R., Munir, A., Synthesis, characterization and thermal dissociation of 2-butoxyethanol-blocked diisocyanates and their use in the synthesis of isocyanate-terminated prepolymers. *Polym. Int.*, 56, 1521, 2007.

3. Hailu, K., Guthausen, G., Becker, W., König, A., Bendfeld, A., Geissler, E., *In-situ* characterization of the cure reaction of HTPB and IPDI by simultaneous NMR and IR measurements. *Polym. Test.*, 29, 513, 2010.
4. Baron, A., Hernandez, J.R., Ibarboure, E., Derail, C., Papon, E., A solventless castor oil-based PU adhesive for wood and foam substrates. *Int. J. Adhes. Adhes.*, 29, 1, 2009.
5. Heintz, A.M., Duffy, D.J., Hsu, S.L., Suen, W., Chu, W., Paul, C.W., Effects of reaction temperature on the formation of polyurethane prepolymer structures. *Macromolecules*, 36, 2695, 2003.
6. Lin, Y.G., Mallin, D.T., Chien, J.C.W., Winter, H.H., Dynamic mechanical measurement of crystallization-induced gelation in thermoplastic elastomeric poly (propylene). *Macromolecules*, 24, 850, 1991.
7. Kim, S.G. and Lee, D.S., Effect of polymerization procedure on thermal and mechanical properties of polyether based thermoplastic polyurethanes. *Macromol. Res.*, 10, 6, 365, 2002.
8. Wicks, Z.W.J., Wicks, D.A., Rosthauser, J.W., Two package waterborne urethane systems. *Prog. Org. Coat.*, 44, 161, 2002.
9. Chattopadhyay, D.K. and Webster, D.C., Thermal stability and flame retardancy of polyurethanes. *Prog. Polym. Sci.*, 34, 1068, 2009.
10. Akindoyo, J.O., Beg, M.D.H., Ghazali, S., Islam, M.R., Jeyaratnam, N., Yuvaraj, A.R., Polyurethane types, synthesis and applications-a review. *RSC Adv.*, 6, 115, 114453–114482, 2016.
11. Saunder, K.J., Detection of therapeutic radiation in three-dimensions. *Org. Poly. Chem.*, 358, 387, 1988.
12. Silva, M.C., Takahashi, J.A., Chaussy, D., Belgacem, M.N., Silva, G.G., Composites of Rigid Polyurethane Foam and Cellulose Fiber Residue. *J. App. Polymer. Sci.* vol. 117, pp. 3665–3672, 2010.
13. Buist, J.M. and Lowe, A., The chemistry of polyurethanes and their applications. *Ann. Occup. Hyg.*, 8, 2, 143–162, 1965.
14. Sharmin, E. and Zafar, F., *Polyurethane: An Introduction*, pp. 1–14, In Tech, Croatia, 2012.
15. Sonnenschein, M.F., *Introduction to Polyurethane Chemistry*, pp. 105–126, Sci. Tech. Mark. Tren, Hoboken, New Jersey, United States, 2015.
16. Bantu, B., Pawar, G.M., Decker, U., Wurst, K., Schmidt, M.A., Buchmeiser, R.M., CO_2 and Sn^{II} Adducts of N-Heterocyclic Carbenes as Delayed-Action Catalysts for Polyurethane Synthesis. *Chem. Eur. J.*, 15, 3103, 2009.
17. Vermette, P., Griesser, H.J., Laroche, G., Guidoin, R., *Biomedical applications of polyurethanes. (Tissue Engineering Intelligence Unit)*, CRC Press, Boca Raton, FL, 3, 2001.
18. Malucelli, G., Priola, A., Ferrero, F., Quaglia, A., Frigione, M., Carfagna, C., Polyurethane resin-based adhesives: Curing reaction and properties of cured systems. *Int. J. Adhes. Adhes.*, 25, 87, 2005.

19. Gogoi, R., Niyogi, U.K., Alam, M.S., Mehra, D.S., Study of effect of NCO/OH molar ratio and molecular weight of polyol on the physico-mechanical properties of polyurethane plaster cast. *World Appl. Sci. J.*, 21, 267, 2013.
20. Biemond, G.J.E., Braspenning, K., Gaymans, R.J.J., Polyurethanes with monodisperse rigid segments based on a diamine–diamide chain extender. *Appl. Polym. Sci.*, 107, 2180, 2008.
21. Kwak, Y.S., Kim, E.Y., Kim, H.D., Lee, J.B., Comparison of the properties of waterborne polyurethane-ureas containing different triblock glycols for water vapor permeable coatings. *Colloid Polym. Sci.*, 283, 880, 2005.
22. Sankara, G. and Nasar, A.S., The adsorption of gases on plane surfaces of glass, mica and platinum. *Pure Appl. Chem.*, 45, 721, 2008.
23. Boutevin, G., Ameduri, B., Robin, J.J., Boutevin, B., Joubert, J.P., Chemical modifications of functional polybutadienes and their derivatives Part 1. Use of meta isopropenyl-α, α'-methyl benzyl isocyanate (TMI). *Polym. Bull.*, 44, 239, 2000.
24. Vilar, W. and Akcelrud, L., Effect of HTPB structure on prepolymer characteristics and on mechanical properties of polybutadiene-based polyurethanes. *Polym. Bull.*, 35, 635, 1995.
25. Coutinho, F.M.B., Deplech, M.C., Alves, L.S., Anionic waterborne polyurethane dispersions based on hydroxyl-terminated polybutadiene and poly (propylene glycol): Synthesis and characterization. *J. Appl. Polym. Sci.*, 80, 566, 2001.
26. Desai, S., Thakore, I.M., Sarawade, B.D., Devi, S., Effect of polyols and diisocyanates on thermo-mechanical and morphological properties of polyurethanes. *Eur. Polym. J.*, 36, 711, 2000.
27. Kim, J.K., Kim, W.H., Lee, D.H., Adhesion properties of UV crosslinked polystyrene-block-polybutadiene-block-polystyrene copolymer and tackifier mixture. *Polym.*, 43, 5005, 2002.
28. Tyczkowski, J., Krawczyk, I., Woźniak, B., Martin-Martinez, J.M., Low-pressure plasma chlorination of styrene–butadiene block copolymer for improved adhesion to polyurethane adhesives. *Eur. Polym. J.*, 45, 1826, 2009.
29. O'Connor, A.E. and Willenbacher, N., The effect of molecular weight and temperature on tack properties of model polyisobutylenes. *Int. J. Adhes. Adhes.*, 24, 335, 2004.
30. Pant, C.S., Santosh, M.S.S., Banerjee, S.B., Khanna, P.K., Application of azide-containing molecules as modifiers of HTPB. *Propellants Explos. Pyrotech.*, 35, 1, 2013.
31. Courtois, B.J., Baroudi, I., Nouvel, N., Degrandi, E., Pensec, S., Ducouret, G., Chanéac, C., Bouteiller, L., Creton, C., Supramolecular soft adhesive materials. *Adv. Funct. Mater.*, 20, 1803, 2010.
32. Furukawa, M., Mitsuia, Y., Fukumaru, T., Kojio, K., Microphase-separated structure and mechanical properties of novel polyurethane elastomers prepared with ether based diisocyanat. *Polym.*, 46, 10817, 2005.

33. Silvaa, A.L.D., Martínez-Martínez, J.M., Bordadob, J.C.M.J., Influence of the free isocyanate content in the adhesive properties of reactive trifunctional polyether urethane quasi-prepolymers. *Int. J. Adhes. Adhes.*, 26, 355, 2006.
34. Tawa, T. and Ito, S., Preparation and reactions of hydrophilic isocyanate micelles dispersed in water. *Colloid Polym. Sci.*, 283, 731, 2005.
35. Sardon, H., Irusta, L., Santamaría, L.I.P., Berridi, M.J.F., Thermal and mechanical behaviour of self-curable waterborne hybrid polyurethanes functionalized with (3-aminopropyl) triethoxysilane (APTES). *J. Polym. Res.*, 19, 9956, 2012.
36. *The Chemistry of Polyurethane Coatings*, pp. 1–31, 412, 777, Baye. Mater. Sci., Pittsburgh, 2000.
37. Hu, X., Tian, M., Qu, L., Zhu, S., Han, G., Multifunctional cotton fabrics with graphene/polyurethane coatings with far-infrared emission, electrical conductivity, and ultraviolet-blocking properties. *Carbon*, 95, 625–633, 2015.
38. Huang, M. and Yang, J., Facile microencapsulation of HDI for self-healing anticorrosion coatings. *J. Mater. Chem.*, 21, 30, 11123–11130, 2011.
39. Cho, S., White, S., Braun, P., Self-healing polymer coatings. *Adv. Mater.*, 21, 139–159, 2009.
40. Doblas, A., Del Rosario, G., Prolongo, M.G., Prolongo, S.G., Electric heating performance of nanodoped polyurethane coatings. *Prog. Org. Coat.*, 135, May, 185–190, 2019.
41. Yagci, M.B., Bolca, S., Heuts, J.P.A., Ming, W., De With, G., Self-stratifying antimicrobial polyurethane coatings. *Prog. Org. Coat.*, 72, 3, 305–314, 2011.
42. Verma, J., Nigam, S., Sinha, S., Bhattacharya, A., Development of polyurethane based anti-scratch and anti-algal coating formulation with silica-titania core-shell nanoparticles. *Vacuum*, 153, 24–34, 2018.
43. Neisius, M., Liang, S., Mispreuve, H., Gaan, S., Phosphoramidate-containing flame-retardant flexible polyurethane foams. *Ind. Eng. Chem. Res.*, 52, 29, 9752–9762, 2013.
44. Xue, L. and Greisler, H.P., Biomaterials in the development and future of vascular grafts. *J. Vasc. Surg.*, 37, 2, 472–480, 2003.
45. Howarth, G.A., Polyurethanes, polyurethane dispersions and polyureas: Past, present and future. *Surf. Coat. Int.*, 86, 2, 111–118, 2003.

8

Electroactive Polymer Nanocomposite Coating

Ayesha Kausar

National Center For Physics, Quaid-i-Azam University Campus, Islamabad, Pakistan

Abstract

Shape memory polymers may relax to regain original shape through external heat, light, electricity, moisture, or pH. Electroactive polymers have capability to exhibit change in size or shape stimulated by electric field. Conducting polymers have been employed as successful electroactive materials. In this chapter, electroactive polymer coatings have been discussed regarding fabrication, features, and applications. Electroactive polymer coatings have shown light weight, high electrical conductivity, high durability, large bending deformation, rapid response, high stability, low voltage operation, high actuation, and anti-corrosion performance. Electroactive polymeric nanocomposites have been developed using conducting/nan-conducting polymers and nanoparticles such as graphene, carbon nanotube, and other nanoparticles. Electroactive nanocomposite actuator coatings have also been formed by incorporating nanoparticles in polymeric matrices. Electroactive materials have high surface area and exceptional electrical conductivity owing to beneficial interfacial interaction between electrons of conducting polymers and nanocarbon. Electroactive polymeric nanocomposite coatings exhibited homogeneous distribution of nanocarbon, high electrical conductivity, and electroactive shape recovery performance at low nanofiller loading. These nanocomposite coatings also possess mechanical and thermal stability. Electroactive polymers and nanocomposite coatings have been found beneficial for anti-corrosive coatings, antibacterial coatings, and films for sensors and actuators. More extensive research attempts are recommended to discover future applications of these coatings.

Keywords: Conducting polymer, nanocomposite, coating, actuator, nanofiller, conductivity, anti-corrosion, antibacterial

Email: asheesgreat@yahoo.com

Inamuddin, Rajender Boddula, Mohd Imran Ahamed and Abdullah M. Asiri (eds.) Polymer Coatings: Technology and Applications, (159–174) © 2020 Scrivener Publishing LLC

8.1 Introduction

Shape-memory polymer can be deformed and fixed into temporary shape, and permanent shape is usually recovered by external stimulus of electricity, heat, light, magnetic field, pH, moisture, etc. Shape memory polymers have gained significance in myriad of fields such as aerospace, aircraft, engineered structures, and biomedical. Conducting polymers are eminent high-performance materials owing to desired conductivity, strength, heat stability, and chemical inertness performance [1, 2]. Various conducting polymers have been exploited as electroactive materials such as polyaniline, polypyrrole, polythiophene, and derived forms [3, 4]. Electroactive polymers particularly have electroactivity better than their mechanical properties and thermal stability. Polymers have also been reinforced with conducting nanoparticles such as carbon nanotube, graphene nanosheets, and inorganic nanoparticles to attain electroactivity [5, 6]. In the fabrication of electroactivate shape memory nanocomposites, the shape-memory effect may depend on nanofiller content and surface-modification of nanoparticles. Conducting electroactive coatings have been used in superior corrosion protection of metallic substrates [7]. Electroactive polymeric nanocomposites have also been applied in microelectronics, sensors, e-textiles, and biomedical relevance [8–10]. Moreover, the electro-activate actuator coatings have been employed in controlling microaerial vehicles.

This chapter quotes recent developments in electro-induced shape recovery polymer coatings. In shape memory polymers filled with conducting nanofillers, electricity has been used as stimulus to enable resistive actuation. The electric triggering of shape memory polymer and nanocomposite coatings enlarges their technological potential from anti-corrosion to sensing and biomedical applications.

8.2 Electroactive Polymer

Electroactive polymers are materials proficient of transferring their size, shape, or volume when exposed to external stimulus of electric field [11]. The electrical energy is usually applied to mechanically deformed materials. The electrostatic forces in these polymers may lead to electromechanical change in shape of materials. Electroactive polymers are also named as shape memory polymers having low cost, lightweight, facile processing, fracture tolerance, easy programming, and high shape recovery ratio. Recently, the electro-activated shape memory polymers and composites have gained significance [12]. Electroactive materials with fine degradability, biocompatibility, robustness, and electroactivity are preferred. In electroactive polymeric nanocomposites, various types of nanofillers have been used. Incorporation of reinforcements may affect the conductive properties of these materials. The mechanisms of electrical conduction and shape recovery of electroactive polymers have been explored [10]. Electroactive polymers, composites, and nanocomposites have been employed in sensors, actuators, flexible circuits, and energy harvester [13]. Electroactive composites have usually fast response, bear large activation stresses, and may induce displacement under direct current (DC) activation, compared with the electroactive polymers. Moreover, electroactive nature of nanoparticle-reinforced materials has found widespread applications in tissue engineering and drug delivery. Electroactive nanocomposite materials with enhanced flexibility, scalability, advanced electrical field, and manufacturing possibilities have been researched [14, 15]. In scalability of these materials, potential errors dramatically increase to comply with the durability of architectural standards.

8.3 Electroactive Polymer and Nanocomposite Coating

In electroactive polymeric nanocomposite, the most important material used is conducting polymers. The electroactive polymeric nanocomposites have also been produced using non-conducting polymer and conducting nanofiller materials such as carbon nanotube, graphene, inorganic oxides, and metal nanoparticles [16, 17]. The electroactive nanocomposites have enhanced processability, electrical conductivity, specific capacitance, and mechanical strength [18]. Common conducting polymers used in electroactive materials are polyaniline, polypyrrole, polythiophene, and their derivatives [19]. Polyaniline (PANI) is an important conducting polymers with exceptional processing, electrical, optical, and environment features [20]. Shiri *et al.* [21] combined p-type conducting polymer with Gadolinium oxide nanoparticles through electropolymerization. The electroactive nanocomposite was used as supercapacitor electrode coating materials with specific capacitance and specific energy of 300 Fg^{-1} and 41.66 $Whkg^{-1}$, respectively. Electroactive polymer-based actuators have also gained attention in micro-electromechanical systems (MEMS) and biomedicine. Ru *et al.* [22] developed ionic electroactive polymeric actuators having large strain under low-voltage stimulation. Polymeric nanoporous carbon nanotube coating was fabricated with high conductivity of 1,000–2,000 S/cm. Actuator performance was evaluated in terms of generated strain of 1.26% and stress of 5.27 MPa. The increase in voltage by 1 V had enhanced the strain and stress by 0.336% and 1.407 MPa, respectively. Such actuators can be used as promising candidates for artificial muscles. The actuator has porous hierarchal structure to support electrical and mechanical properties for improved electrochemical and electromechanical properties. Prabhakar *et al.* [23] developed polyurethane coated with polyaniline and silver nanoparticle (AgNP) nanocomposite for biomedical application. The electroactive PANI/AgNP coating has shown reduced inflammatory response in human cells. The materials have shown potential application in electrochemical biosensors [24, 25]. The electroactive PANI/AgNP coatings have enhanced biocompatibility than the neat polyurethane. A major application of electroactive coatings is for the corrosion protection. Electroactive polymer nanocomposite coatings have enhanced the anti-corrosive abilities with the addition of nanoparticles, compared with the neat conducting polymers. Hence, electroactive polymeric nanocomposite coatings have superior conductivity, robustness, capacitance, and anti-inflammatory response compared with the unfilled polymers.

8.4 Applications of Electroactive Polymer Nanocomposite Coating

8.4.1 Electroactive Anti-Corrosive Coating

Electroactive polymer and nanocomposites have been employed as anti-corrosive coatings [26, 27]. Electroactive PANI, polyimide, and epoxy resin coatings have been developed on metallic electrodes and studied for corrosion protection [27, 28]. Mechanism of anti-corrosion of conducting polymers for underlying metal against the attack of aggressive species has been explored [29]. Huang *et al.* [30] prepared electroactive polyimide/nanoclay nanocomposite coatings for corrosion protection. The coating was formed on cold-rolled steel (CRS) electrode. The electrochemical measurements such as corrosion potential (E_{corr}), polarization resistance (R_p), and corrosion current (I_{corr}) were investigated in 5 wt.% NaCl electrolyte. Table 8.1 shows that nanocompsoite coating with higher nanoclay content had higher corrosion potential. The polarization resistance of highly loaded coating was also much greater than the uncoated CRS. Gas barrier properties and redox behavior of coatings were studied by gas permeability

Table 8.1 Electrochemical corrosion measurements of the materials [30]. Reproduced with permission from Elsevier.

Sample	E_{corr} **(mV)**	R_p **(kΩcm²)**	I_{corr} **(μA/cm²)**
CRS	−1020.9	2.47	9.14
Polyimide	−807.7	17.30	2.53
Polyimide/nanoclay 0.004 wt.%	−664.9	36.67	1.89
Polyimide/nanoclay 0.012 wt.%	−528.5	69.99	1.09

analysis and electrochemical cyclic voltammetry. The coatings possess excellent oxygen resistance and anti-corrosion properties. Yuan *et al.* [31] designed multifunctional coatings of polyaniline/functionalized carbon nanotube (PANI/fCNT) and ethylene tetrafluoroethylene (ETFE). The coatings revealed anti-corrosion, self-cleaning, and anti-wear properties. The ETFE coating with 6wt.% PANI/fCNT had superamphiphobic property with contact angles of 159°–167°. The superamphiphobic properties were retained upto 90 days in 3.5 wt.% NaCl. Chang *et al.* [32] prepared anti-corrosion electroactive polyimide/graphene nanocomposite coatings. The well-dispersed graphene nanosheets in coatings hindered the gas migration and corrosion resistance exponentially. Huang *et al.* [33] formed coatings of epoxy/amino-SiO_2 nanocomposite with electroactive amino-capped aniline trimer (ACAT). Corrosion potential, corrosion current, and electrochemical impedance spectroscopy studies on coatings were performed in 5 wt.% NaCl electrolyte. Electroactive epoxy/amino-SiO_2/ACAT coatings have shown enhanced oxygen barrier properties and corrosion resistance. Weng *et al.* [34] designed electroactive polyimide-TiO_2 (EPT) coatings. Increase in TiO_2 concentration revealed higher corrosion protection of these coatings. Electrochemical corrosion measurement of cold-rolled steel electrode coated with EPT was measured in 5 wt.% NaCl electrolyte. Figure 8.1 shows synthetic procedure of EPT. Firstly, poly(amic acid) solution was formed using 4,4′-(4,4′-isopropylidenediphenoxy) bis(phthalic anhydride). Figure 8.2a depicts redox catalytic properties of EPT. The mechanism was based on the formation of passivation protective metal oxide layers. The corrosion protection performance of EPT was found similar to the reported PANI coatings [35]. Figure 8.2b shows diffusion pathway of oxygen gas in the EPT coating formed on CRS electrode. The EPT nanocomposite coating with 0.5 wt.% TiO_2 revealed 21% reduction of O_2 permeability. Further, increase in TiO_2 loading resulted in more

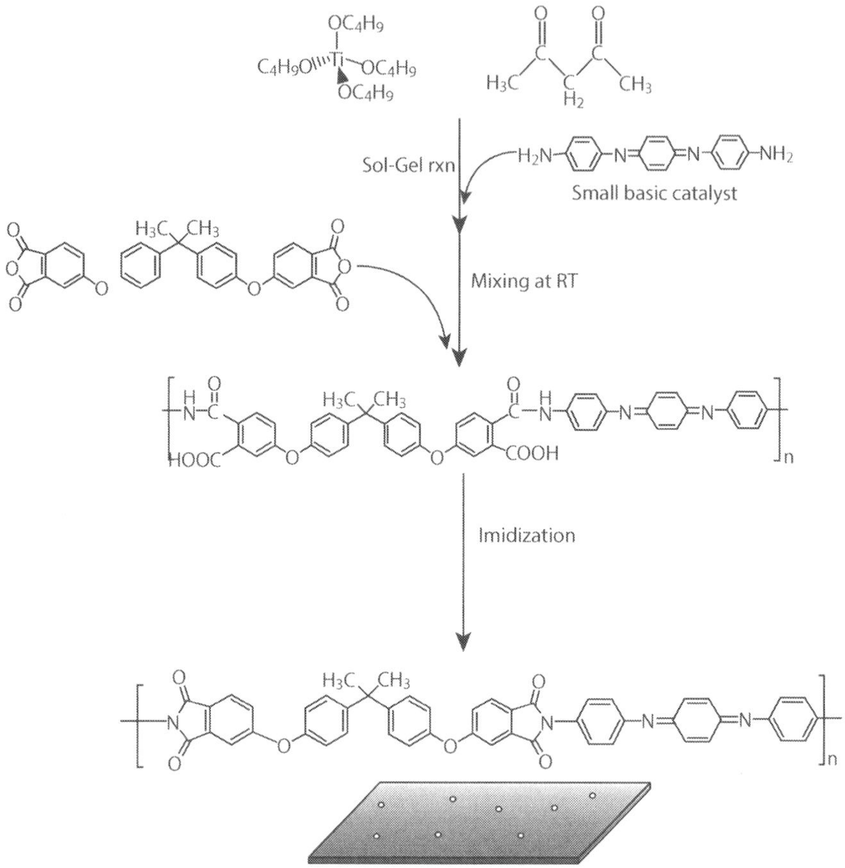

Figure 8.1 Schematic of synthesis of electroactive polyimide-TiO_2 [34]. Reproduced with permission from Elsevier.

enhanced molecular barrier properties of EPT coating. Hence, electroactive nanocomposite coatings have shown superior corrosion protection.

8.4.2 Electroactive Antibacterial Coating

Bio-hybrids have gained recent interest owing to synergistic properties of biomaterials with synthetic polymers [36–38]. For example, the natural biopolymer chitosan has been blended with conducting polythiophene, poly(3,4-ethylenedioxythiophene), and poly(4-styreneslfonic acid) to attain electroactive materials [39]. The bio-functionalization of conducting polymers may facilitate the electrical signal and actuation. Blending of conducting polymers with natural or biopolymers have enhanced

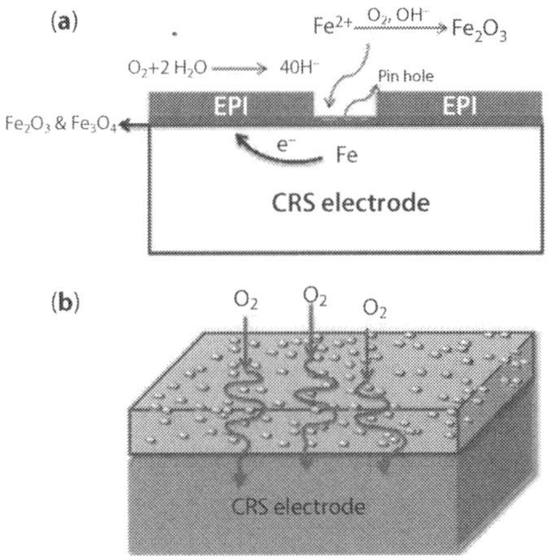

Figure 8.2 Schematic diagrams of (a) mechanism of CRS passivation by EPT coatings and (b) diffusion pathway of oxygen gas in the EPT [34]. Reproduced with permission from Elsevier.

electroactive antibacterial properties [40, 41]. Khan *et al.* [42] developed conductive bio-hybrid coatings of poly(3,4-ethylenedioxythiophene): poly(4-styreneslfonic acid) (PEDOT:PSS) with chitosan and polyvinyl alcohol (PVA). The biocidal potential of neat electroactive PEDOT:PSS, nanocomposite coatings, and chitosan were tested against S. *aureus* and gram negative E. *coli* bacterial strains using disk diffusion method (Figure 8.3). Pure PEDOT:PSS had no antibacterial activity against bacterial strains. The prepared bio-hybrid coatings were effective against

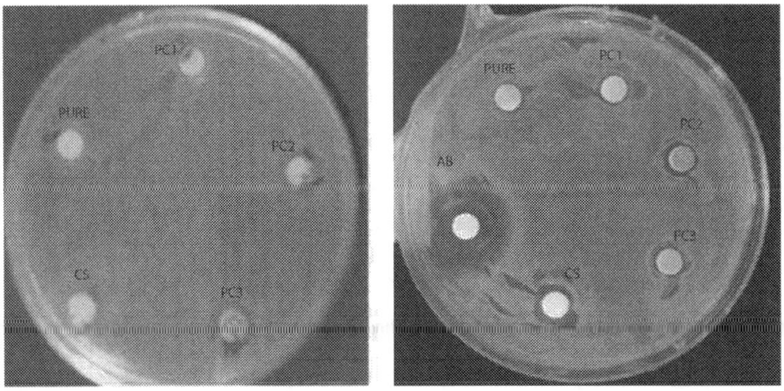

Figure 8.3 Inhibition zone of chemically synthesized pure PEDOT: PSS, bio-hybrid films PC1 (1:1), PC2 (1:2), and PC3 (1:3), and chitosan (CS) against gram positive bacteria and *S. aureus* [42]. Reproduced with permission from Elsevier.

gram positive bacteria. The bio-hybrid coatings with chitosan has much enhanced antibacterial property against *S. aureus* and gram negative *E. coli.* It was found that there were electrostatic interactions between amino groups of chitosan and anionic microorganisms leading to enhanced antibacterial effects. Gharibi *et al.* [43] designed polyurethane/siloxane-based electroactive wound dressing membranes. The physicochemical, mechanical, electrical, and biological properties of the coatings were found effective. Zhao *et al.* [44] formed poly(ethylene glycol)-*co*-poly(glycerol sebacate) as antibacterial, anti-oxidant, and electroactive dressing for cutaneous wound healing. The materials had shown good self-healing, biocompatibility, antibacterial activity, adhesiveness, conductivity, and electroactivity features. The electroactive dressing also possesses *in vivo* blood clotting capacity. Ahmed *et al.* [45] fabricated electroactive polyvinyl-*N*-carbazole (PVK) and single-walled carbon nanotube (SWCNT) coatings. The antibacterial effects were determined for *E. coli* and *Bacillus subtilis* strains. The hydrophobic and electrostatic interactions between conducting polymer and microorganisms synergistically inhibited the growth of microorganisms and bacterial strains and enhance the actuation performance.

8.4.3 Electroactive Coating for Sensors and Actuators

Polymer nanocomposites have been exploited in chemical and biological sensors [46, 47]. Polymers reinforced with nanoparticles such as graphene, carbon nanotube, and metal nanoparticles exhibited high surface area, electrical conductivity, and fast electron transfer rate. These nanocomposites have been employed in electronic sensors, biosensors, immunosensors, and electrochemical sensors. Recently, electroactive polymer nanocomposites have been employed in coatings of sensors and actuators. Kulkarni *et al.* [48] reported polyaniline wrapped-multi-walled carbon nanotube (PANI-MWCNT) coatings through *in situ* chemical polymerization. The synthesized nanocomposite coating was sensitive towards pH change of the solution. The materials were potentially important for optical pH sensing. Mahmoudian *et al.* [49] developed polypyrrole coating on nanospherical platinum. The coated materials was sensitive towards the mercury ion concentration in the range of 5 to 500 nM. Ionic or electrolytic polymers have also been studied as actuators. The driving force for such a actuators is the volume change due to oxidation/reduction reactions [50]. Such electroactive polymers have been utilized in artificial muscles. Conducting and electrolytic polymer coatings have found potential for sensors and actuators [51]. Their use in sensors and actuators depend on inherent conductivity

[52]. Park *et al.* [53] prepared poly(vinylidene fluoride) (PVDF) and carbon nanotube (CNT) nanocomposite coating on indium tin oxide (ITO). The coatings were investigated for self-sensor and micro-actuator applications. Figure 8.4 shows morphology of the surfaces of nanocomposite and ITO coating material. The ITO coating had non-entangled morphology compared with the nanocomposite. Figure 8.5 shows the effect of increasing nanofiller content on the activation displacement of nanocomposite and ITO coated material. With the increase in nanofiller content from 3 to 12 wt.%, the electrical conductivity and actuator displacement of ITO

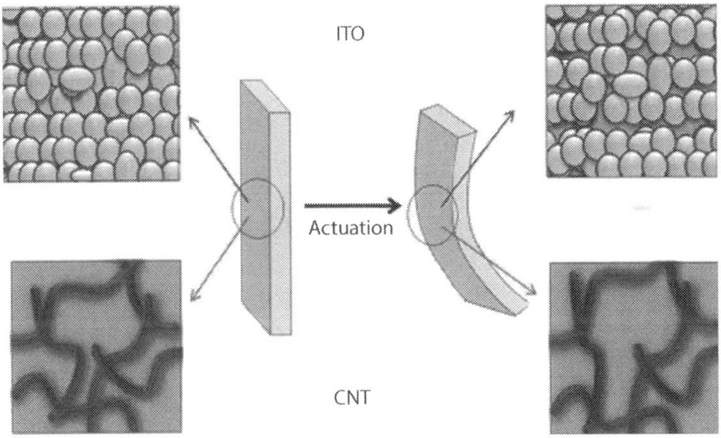

Figure 8.4 Sketches of proposed changes in layer morphology during actuation [53]. Reproduced with permission from Elsevier.

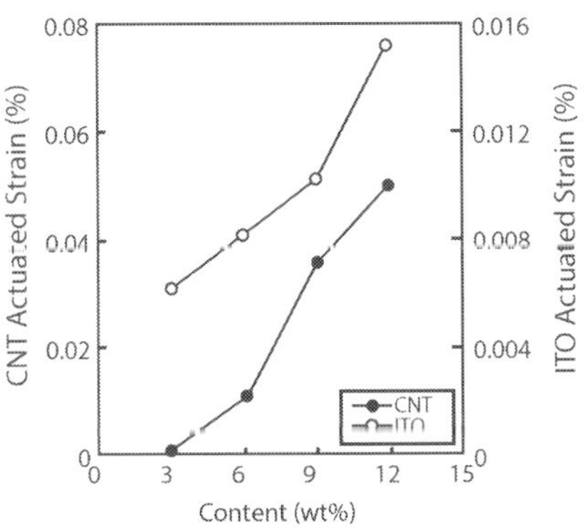

Figure 8.5 Actuated strain vs. nanofiller content for PVDF/CNT and ITO coated layers [53]. Reproduced with permission from Elsevier.

coated material were enhanced compared with the nanocomposite [54, 55]. Hence, PVDF/CNT nanocomposites coatings have shown high performance for self-sensors and micro-actuators.

8.5 Future and Summary

Electroactive polymer and nanocomposites have gained increasing academic and industrial interest. Development and characterization of new conducting polymers may bring about advancements in the field of electroactive materials [56, 57]. In this regard, conducting polymers as well as ionic/electrolytic polymers have been identified as electro-actuated materials. Both the conducting polymers and ionic polymers have been used as electroactive coatings. In the case of conducting polymers, transport mechanism is electronic in nature. Progress in this field depends upon the identification of new derivatives of polythiophene and polyaniline with enhanced electrical conductivity. Incorporation of conducting polymers in polymers may also enhance the electrical triggering and anti-corrosion performance of coatings. Advancements in ionically conducting polymers can be achieved through the dissolution of salts in macromolecular solids. For example, poly(ethylene oxide) has been mixed with lithium salts to get electroactive polymer. Similarly, other water soluble polymers can be mixed with salts. Amalgamation or modification of conducting polymers with biomaterials have led to unique actuation properties of coatings [37]. In this regard, new bio-hybrids had synergistic properties of conducting polymers and biomaterials. Field of electroactive polymers and nanocomposites needs to be researched to get high performance self-sensors and micro-actuator coatings.

This chapter describes various aspects of electrically triggered shape memory polymers and nanocomposites. The recent advances in this field depends upon the nature of polymer, fabrication strategies, conductivity properties, easy programming, fast recovery rate, and resistance to large activation stresses. Interaction between electro-activated polymer coating and substrate has also enhanced the anti-corrosion properties. Electroactive antibacterial properties of coatings have also been explored. Field of sensors and micro-actuators based on electro-activated polymer coatings were found to exhibit superior performance than non-electroactive polymers. Future advancements in the field of electroactive shape memory polymeric materials may enlarge their technological potential in several concealed high-performance areas.

References

1. Kausar, A., High-performance competence of polyaniline-based nanomaterials. *Mater. Res. Innovations*, 1–10, 2019.
2. Zamani, F.G., Moulahoum, H., Ak, M., Demirkol, D.O., Timur, S., Current Trends in the Development of Conducting Polymers-based Biosensors. *TrAC Trend. Anal. Chem.*, 118, 264–276, 2019.
3. Barbosa, J.R., Amorim, P.H., Mariana, C.D.O., Dornellas, R.M., Pereira, R.P., Semaan, F.S., Evaluation of 3D Printing Parameters on the Electrochemical Performance of Conductive Polymeric Components for Chemical Warfare Agent Sensing, in: *Developments and Advances in Defense and Security*, pp. 425–435, Singapore, Springer, 2020.
4. Wei, H., Cauchy, X., Navas, I.O., Abderrafai, Y., Chizari, K., Sundararaj, U., Liu, Y., Leng, J., Therriault, D., Direct 3D Printing of Hybrid Nanofibers-based Nanocomposites for Highly Conductive and Shape Memory Applications. *ACS Appl. Mater. Interfaces*, 11, 24523–24532, 2019.
5. Reddy, B. (Ed.), *Advances in Nanocomposites: Synthesis, Characterization and Industrial Applications*, BoD–Books on Demand, Intech, Rijeka, Croatia, 2011.
6. Dubey, N., Kushwaha, C.S., Shukla, S.K., A review on electrically conducting polymer bionanocomposites for biomedical and other applications. *Int. J. Polym. Mater. Polym. Biomater.*, 1–9, 2019.
7. Lei, Y., Qiu, Z., Liu, J., Li, D., Tan, N., Liu, T., Zhang, Y., Chang, X., Gu, Y., Yin, Y., Effect of Conducting Polyaniline/Graphene Nanosheet Content on the Corrosion Behavior of Zinc-Rich Epoxy Primers in 3.5% NaCl Solution. *Polymers*, 11, 850, 2019.
8. Ashtari, K., Nazari, H., Ko, H., Tebon, P., Akhshik, M., Akbari, M., Alhosseini, S.N., Mozafari, M., Mehravi, B., Soleimani, M., Ardehali, R., Electrically conductive nanomaterials for cardiac tissue engineering. *Adv. Drug. Deliv. Rev.*, 144, 162–179, 2019.
9. Zhang, R., Deng, H., Valenca, R., Jin, J., Fu, Q., Bilotti, E., Peijs, T., Carbon nanotube polymer coatings for textile yarns with good strain sensing capability. *Sens. Actuat. A: Phys.*, 179, 83, 2012.
10. Panda, S. and Acharya, B., Electronic Applications of Conducting Polymer Nanocomposites, in: *Proceedings of the Third International Conference on Microelectronics, Computing and Communication Systems*, pp. 211–220, Springer, Singapore, 2019.
11. Manghera, P., Rahman, F., Banerjee, S., Nazari, M., Tuff, W., Application of Electro-active Materials Toward Health Monitoring of Structures: Electrical Properties of Smart Aggregates, in: *Dynamics of Civil Structures*, vol. 2, pp. 7–15, Springer, Cham, 2020.
12. Cao, C., Gao, X., Conn, A., Elastomer membranes with magnetic repulsion, in: *Electroactive Polymer Actuators and Devices (EAPAD) XXI*, vol. 10966,

Y. Bar-Cohen (Ed.), Society of Photo-Optical Instrumentation Engineers (SPIE), USA, 2019.
13. M, Ortiz, IV, Shelton, J, Hueil, J, Swayze, Ethicon Inc, *Electroactive polymer-based articulation mechanism for circular stapler*, U.S. Patent 8, 317, 074, 2012.
14. Memon, I.A., Xudong, L., Xiaoqing, C., Zhiwei, J., Pervaiz, M., Weimin, Y., Haoyi, L., Sain, M., A review of electro-stimulated gels and their applications: Present state and future perspectives. *Mater. Sci. Eng. C*, 103, 109852, 2019.
15. Karri, S.N., Rao Nakka, N.M., Sesha Sainath, A.V., Palaniappan, S., Synthesis of homo/3-and 4-armed hydrophilic glycopolymers: To promote aniline to polyaniline-glycopolymers for fluorescence, electro active material, and electrostatic discharge applications. *J. Appl. Polym. Sci.*, 136, 48043, 2019.
16. Chen, Y., Han, M., Tang, Y., Bao, J., Li, S., Lan, Y., Dai, Z., Polypyrrole–polyoxometalate/reduced graphene oxide ternary nanohybrids for flexible, all-solid-state supercapacitors. *Chem. Commun.*, 51, 12377, 2015.
17. Ehsani, A., Vaziri-Rad, A., Babaei, F., Shiri, H.M., Electrosynthesis, optical modeling and electrocatalytic activity of Ni-MWCNT-PT nanocomposite film. *Electrochim. Acta*, 159, 140, 2015.
18. Shayeh, J.S., Norouzi, P., Ganjali, M.R., Studying the supercapacitive behavior of a polyaniline/nano-structural manganese dioxide composite using fast Fourier transform continuous cyclic voltammetry. *RSC Adv.*, 5, 20446, 2015.
19. Salehifar, N., Shayeh, J.S., Siadat, S.O.R., Niknam, K., Ehsani, A., Movahhed, S.K., Electrochemical study of supercapacitor performance of polypyrrole ternary nanocomposite electrode by fast Fourier transform continuous cyclic voltammetry. *RSC Adv.*, 5, 96130, 2015.
20. Shayeh, J.S., Ehsani, A., Ganjali, M.R., Norouzi, P., Jaleh, B., Conductive polymer/reduced graphene oxide/Au nano particles as efficient composite materials in electrochemical supercapacitors. *Appl. Surf. Sci.*, 353, 594, 2015.
21. Shiri, H.M. and Ehsani, A., Pulse electrosynthesis of novel wormlike gadolinium oxide nanostructure and its nanocomposite with conjugated electroactive polymer as a hybrid and high efficient electrode material for energy storage device. *J. Coll. Interface Sci.*, 484, 70, 2016.
22. Ru, J., Bian, C., Zhu, Z., Wang, Y., Zhang, J., Horiuchi, T., Sugino, T., Liu, X., Chen, H., Asaka, K., Controllable and durable Ionic electroactive polymer actuator based on nanoporous carbon nanotube film electrode. *Smart Mater. Struct.*, 28, 2019.
23. Prabhakar, P.K., Raj, S., Anuradha, P.R., Sawant, S.N., Doble, M., Biocompatibility studies on polyaniline and polyaniline–silver nanoparticle coated polyurethane composite. *Coll. Surf. B: Biointerfaces*, 86, 146, 2011.
24. Razzino, C.A., Sgobbi, L.F., Marciano, F.R., Lobo, A.O., Graphene-Based Sensors: Applications in Electrochemical (Bio) sensing, in: *Handbook of Graphene: Biosensors and Advanced Sensors*, 2019.
25. Sousa, C.P., Ribeiro, F.W., Oliveira, T.M., Correia, A.N., de Lima-Neto, P., Morais, S., Trends and Frontiers in Graphene-Based (Bio) sensors

for Pesticides Electroanalysis, in: *Handbook of Graphene: Biosensors and Advanced Sensors*, p. 59, 2019.
26. Huang, L., Hu, J., Lang, L., Wang, X., Zhang, P., Jing, X., Wang, X., Chen, X., Lelkes, P.I., MacDiarmid, A.G., Wei, Y., Synthesis and characterization of electroactive and biodegradable ABA block copolymer of polylactide and aniline pentamer. *Biomaterials*, 28, 1741, 2007.
27. Weng, C.J., Chang, C.H., Peng, C.W., Chen, S.W., Yeh, J.M., Hsu, C.L., Wei, Y., Advanced anticorrosive coatings prepared from the mimicked xanthosoma sagittifolium-leaf-like electroactive epoxy with synergistic effects of superhydrophobicity and redox catalytic capability. *Chem. Mater.*, 23, 2075, 2011.
28. Pillay, V., Tsai, T.S., Choonara, Y.E., du Toit, L.C., Kumar, P., Modi, G., Naidoo, D., Tomar, L.K., Tyagi, C., Ndesendo, V.M., A review of integrating electroactive polymers as responsive systems for specialized drug delivery applications. *J. Biomed. Mater. Res. A*, 102, 2039, 2014.
29. Tsai, T.S., Pillay, V., Choonara, Y.E., Du Toit, L.C., Modi, G., Naidoo, D., Kumar, P., A polyvinyl alcohol-polyaniline based electro-conductive hydrogel for controlled stimuli-actuable release of indomethacin. *Polymers*, 3, 150, 2011.
30. Huang, H.Y., Huang, T.C., Yeh, T.C., Tsai, C.Y., Lai, C.L., Tsai, M.H., Yeh, J.M., Chou, Y.C., Advanced anticorrosive materials prepared from amine-capped aniline trimer-based electroactive polyimide-clay nanocomposite materials with synergistic effects of redox catalytic capability and gas barrier properties. *Polymer*, 52, 2391, 2011.
31. Yuan, R., Wu, S., Yu, P., Wang, B., Mu, L., Zhang, X., Zhu, Y., Wang, B., Wang, H., Zhu, J., Superamphiphobic and electroactive nanocomposite toward self-cleaning, antiwear, and anticorrosion coatings. *ACS Appl. Mater. Interfaces*, 8, 12481, 2016.
32. Chang, K.C., Hsu, C.H., Lu, H.I., Ji, W.F., Chang, C.H., Li, W.Y., Chuang, T.L., Yeh, J.M., Liu, W.R., Tsai, M.H., Advanced anticorrosive coatings prepared from electroactive polyimide/graphene nanocomposites with synergistic effects of redox catalytic capability and gas barrier properties. *Exp. Polym. Lett.*, 8, 243, 2014.
33. Huang, T.C., Su, Y.A., Yeh, T.C., Huang, H.Y., Wu, C.P., Huang, K.Y., Chou, Y.C., Yeh, J.M., Wei, Y., Advanced anticorrosive coatings prepared from electroactive epoxy–SiO_2 hybrid nanocomposite materials. *Electrochim. Acta*, 56, 6142, 2011.
34. Weng, C.J., Huang, J.Y., Huang, K.Y., Jhuo, Y.S., Tsai, M.H., Yeh, J.M., Advanced anticorrosive coatings prepared from electroactive polyimide-TiO2 hybrid nanocomposite materials. *Electrochim. Acta*, 55, 8430, 2010.
35. Spinks, G.M., Dominis, A.J., Wallace, G.G., Tallman, D.E., Electroactive conducting polymers for corrosion control. *J. Sol. Stat. Electrochem.*, 6, 85, 2002.
36. Balint, R., Cassidy, N.J., Cartmell, S.H., Conductive polymers: Towards a smart biomaterial for tissue engineering. *Acta Biomater.*, 10, 2341, 2014.

37. Kiristi, M., Oksuz, A.U., Oksuz, L., Ulusoy, S., Electrospun chitosan/PEDOT nanofibers. *Mater. Sci. Eng. C*, 33, 3845, 2013.
38. Thakur, V.K. and Thakur, M.K., Recent Advances in Graft Copolymerization and Applications of Chitosan. *ACS Sustain. Chem. Eng.*, 2, 2637, 2014.
39. Guo, B., Glavas, L., Albertsson, A.C., Biodegradable and electrically conducting polymers for biomedical applications. *Prog. Polym. Sci.*, 38, 1263, 2013.
40. Wu, T., Pan, Y., Bao, H., Li, L., Preparation and properties of chitosan nanocomposite films reinforced by poly(3,4-ethylenedioxythiophene)-poly(styrenesulfonate) treated carbon nanotubes. *Mater. Chem. Phys.*, 129, 932, 2011.
41. Zare, E.N., Lakouraj, M.M., Mohseni, M., Biodegradable polypyrrole/dextrin conductive nanocomposite: Synthesis, characterization, antioxidant and antibacterial activity. *Syn. Met.*, 187, 9, 2014.
42. Khan, S. and Narula, A.K., Bio-hybrid blended transparent and conductive films PEDOT: PSS: Chitosan exhibiting electro-active and antibacterial properties. *Eur. Polym. J.*, 81, 161, 2016.
43. Gharibi, R., Yeganeh, H., Rezapour-Lactoee, A., Hassan, Z.M., Stimulation of wound healing by electroactive, antibacterial, and antioxidant polyurethane/siloxane dressing membranes: *In vitro* and *in vivo* evaluations. *ACS Appl. Mater. Interfaces*, 7, 24296, 2015.
44. Zhao, X., Wu, H., Guo, B., Dong, R., Qiu, Y., Ma, P.X., Antibacterial antioxidant electroactive injectable hydrogel as self-healing wound dressing with hemostasis and adhesiveness for cutaneous wound healing. *Biomaterials*, 122, 34, 2017.
45. Ahmed, F., Santos, C.M., Vergara, R.A.M.V., Tria, M.C.R., Advincula, R., Rodrigues, D.F., Antimicrobial applications of electroactive PVK-SWNT nanocomposites. *Environ. Sci. Technol.*, 46, 1804, 2012.
46. Khan, A., Abas, Z., Kim, H., Kim, J., Recent progress on cellulose-based electro-active paper, its hybrid nanocomposites and applications. *Sensors*, 16, 1172, 2016.
47. Shrivastava, S., Jadon, N., Jain, R., Next-generation polymer nanocomposite-based electrochemical sensors and biosensors: A review. *TrAC Trend. Anal. Chem.*, 82, 55, 2016.
48. Kulkarni, M.V. and Kale, B.B., Studies of conducting polyaniline (PANI) wrapped-multiwalled carbon nanotubes (MWCNTs) nanocomposite and its application for optical pH sensing. *Sens. Actuat. B: Chem.*, 187, 407, 2013.
49. Mahmoudian, M.R., Basirun, W.J., Alias, Y., A sensitive electrochemical Hg^{2+} ions sensor based on polypyrrole coated nanospherical platinum. *RSC Adv.*, 6, 36459, 2016.
50. Terasawa, N., Ono, N., Hayakawa, Y., Mukai, K., Koga, T., Higashi, N., Asaka, K., Effect of hexafluoropropylene on the performance of poly (vinylidene fluoride) polymer actuators based on single-walled carbon nanotube-ionic liquid gel. *Sens. Actuat. B: Chem.*, 160, 161, 2011.

51. Rathod, V., TRoy Mahapatra, D., Jain, A., Gayathri, A., Characterization of a large-area PVDF thin film for electro-mechanical and ultrasonic sensing applications. *Sens. Actuat. A: Phys.*, 163, 164, 2010.
52. Yang, L., Setyowati, K., Li, A., Gong, S., Chen, J., Reversible infrared actuation of carbon nanotube-liquid crystalline elastomer nanocomposites. *Adv. Mater.*, 20, 2271, 2008.
53. Park, J.M., Gu, G.Y., Wang, Z.J., Kwon, D.J., DeVries, K.L., Interfacial durability and electrical properties of CNT or ITO/PVDF nanocomposites for self-sensor and micro actuator applications. *Appl. Surf. Sci.*, 287, 75, 2013.
54. Tyson, B.M., Abu Al-Rub, R.K., Yazdanbakhsh, A., Grasley, Z., Carbon nanotubes and carbon nanofibers for enhancing the mechanical properties of nanocomposite cementitious materials. *J. Mater. Civil Eng.*, 23, 1028, 2011.
55. Huang, Y., Liang, J., Chen, Y., The application of graphene based materials for actuators. *J. Mater. Chem.*, 22, 3671, 2012.
56. Lyons, M.E. (Ed.), *Electroactive Polymer Electrochemistry: Part 1: Fundamentals*, Springer Science & Business Media, New York, USA 2013.
57. Scrosati, B. (Ed.), *Applications of electroactive polymers*, vol. 75, Chapman & Hall, London, 1993.

9

Conducting Polymer Coatings for Corrosion Resistance in Electronic Materials

U. Naresh[1], N. Suresh Kumar[2], D. Baba Basha[3], K. Chandra Babu Naidu[4*], M.S.S.R.K.N. Sarma[4], R. Jeevan Kumar[1], Ramyakrishna Pothu[5] and Rajender Boddula[6]

[1]*Department of Physics, Sri Krishnadevaraya University, Anantapur, India*
[2]*Department of Physics, JNTUA, Anantapuramu, India*
[3]*College of Computer and Information Sciences, Majmaah University Al'Majmaah, Saudi Arabia*
[4]*Department of Physics, GITAM Deemed to be University, Bangalore, India*
[5]*College of Chemistry and Chemical Engineering, Hunan University, Changsha, China*
[6]*CAS Center for Excellence in Nanoscience, National Center for Nanoscience and Technology, Beijing, China*

Abstract
Corrosion of the electronic material is the critical problem for many applications. Hence the researchers focused for the development of corrosion resistance coating materiel which has multifunctional properties such as electronic conductivity, better mechanical strength, more flexibility in the use and synthesis. In general, the polymers were used as coating material for the electronic and other materials. In the polymer types, the conducting polymers (CPs) got importance as it conduct electricity and have good multifunctional characteristics. Significantly the CPs such as PANI, PPy, and PEDOT:PSS and their derivatives are used as coating material to inhibit corrosion. In this paper we summarized the CPs and their com posites which have used for the corrosion resistance.

Keywords: Conducting polymers, corrosion, coating, PANI, PPy, PEDOT

**Corresponding author*: chandrababu954@gmail.com

Inamuddin, Rajender Boddula, Mohd Imran Ahamed and Abdullah M. Asiri (eds.) Polymer Coatings: Technology and Applications, (175–188) © 2020 Scrivener Publishing LLC

9.1 Introduction

The corrosion is the one of the source of the damage of the electronic materials and related peripheral electronic materials. Some of the few materials which are used in these days to fabricate electronic devices are more sensitive; hence, the materials affected themselves for the corrosion by oxidation. The demolition or worsening of electronic devices because of the nature and surroundings reaction is maybe referred as corrosion [1]. In general, the protection of the electronic materials from the corrosion did by the painting and coating of the organic materials (see Figure 9.1). The corrosion depends on the ability of coating which protects the metal surface when the metallic surface damaged in its life period. In recent days, conducting polymers for coatings are the one of the choice to protect electronic material defects from the corrosive nature due its conductivity and flexibility [2].

The corrosion environment for the electronic materials mainly happened by two basis that are first one is materials corrosion by itself and exterior world. Generally, hermit cavities are used for package or cover surface of the electronic materials to minimize first-type corrosion, in this case the air filtration process is used for the substances which are influenced

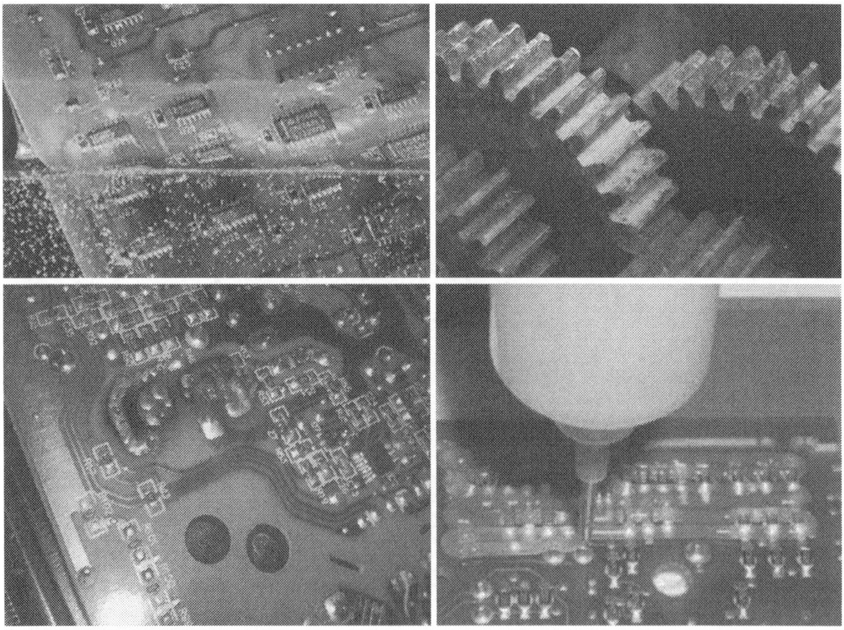

Figure 9.1 The photographs of corrosive electronic materials and removal process.

by corrosion due to external sources. The substances or materials which are exposed by external environment were well defined until now, even this substances corrosion practiced by photoactive packaging process [3]. The electronic devices mostly, in practice level of packaging restrictive packages, encapsulates and protective painting used to remove which the peripherals from internal and external world are available to degrade performance. These phenomena are also true for the components like printed boards, connectors, optical transmitter devices, electronic chips, which are kept in substrates, and many other same kind of devices [4–5]. The packaging materials itself are exposed to a nature of whose source is external and internal sources. The low molecular weight organic compounds interacting are the part of materials which are derived by the internal source due to its volatility and mobility the materials can diffuse on the metallic surface using other organic materials broadly used in electronic equipment fabrication [6]. Traegere *et al.* [7] made reviewed about the substances which are out gassed from organic materials those which were used in electronics. The researchers Weshler *et al.* [8] have discovered a number of components in insulation and corrosion resistance material. Those are present in both the vapor and condensed phase electronic machinery. On the other hand, the humans or atmosphere (external sources) made corrosion of the electronic substances. Gill and co-workers [9] have investigated and listed some compounds which are found in atmosphere with vapor pressure in specific range from internal.

In the exterior environment, there were many studies involved about corrosion studies. The researchers Vernon *et al.* [10] identified the products related to the sulfur dioxide and its oxidation states and humidity are the common responsible for corrosion in exterior world. He also gave results as the rate of corrosion increases with the absorption of humidity by material surface. The most materials have relative humidity value in the range of 70% to 80%. Rice *et al.* from United States studied rate of corrosion for the materials such as silver, nickel, cobalt, copper, and iron in different circumstances and concluded that the silver has low corrosion rate than the other. The coating to the materials depends on the corrosion which was happened by the interactions with surrounding environment. In the case of environmental interactions, generally the electronic materials get corrosion from the air interaction and some other devices more responsive to the dampness and ionic pollutant. All the elements such as metals and semiconductors which were used in the fabrication of electronic devices all are thermally active and oxidized easily at temperature 20°–40° (near in the room temperature). If we try to control the corrosion by decreasing oxidation through the self-limiting or restrained, then the materials go destructive. However, the electronic devices need the thin oxide films synthesized by the corrosion controlled conditions. In the periodic table, 3rd group and 5th group semiconductors which were commonly used for electronic materials significantly look as complex as the problems are arise at the synthesis of corrosion restrictive films of the 3rd and 5th group elements in the semiconductor technology and devices [11].

On the second major cause of the corrosion is interaction of water and other constituent. The attending of the moisture can give electrochemical effect; hence, the electronic devices performance surely be slow down or damaged permanently. The oxidation formation and the oxidation rate depend on the humidity in the atmosphere so it leads to deformation of structure and changes happened in the properties of electronic materials. The main components and connecters of electronic gadgets keep safe with the coating of noble metals like gold and platinum etc. The corrosion which comes from the gold coating gets creep in the surface of materials which leads to high contact resistance. Because the moisture is present, the atmosphere can increase the copper mixed oxide with sulfide in the case of gold-coated copper materials in the sulfide environment. Generally, in the case copper-gold alloys, the nickel barrier layer restrains the interdiffusion by eliminating pores and extents copper on contact surfaces between the copper and gold [12].

Generally, the corrosion constrained by the coating by the external sources such as polymer coating and sealing with respected sources. Commonly, the coating is referred as the painting on the surfaces of the electronic components, envelope with the shielding materials like the organic or inorganic elements. However, the electronic peripherals are coated with the polymers that are natural polymers or synthetic polymers. Nowadays, conducting polymers are used as the purpose of the coating due its electric conduction and flexibility [13]. Apart from the conducting polymers, there was natural polymers such as natural rubber or polyvinyl chloride, silicon, pollsters, etc. However, using the conducting polymers gives more advantages as it possesses conducting nature.

9.2 Conducting Polymers

9.2.1 Polyaniline (PANI)

Guptha *et al.* [14] prepared PANI-lignosulphonate (PANI-LGS) binder for the corrosion protection of aluminum 2024 alloys by chemical oxidative polymerization method and its performance is analyzed first time. They found the low content of PANI-LGS shows less dispersibility instead of high volume fraction of particles eliminates from the particles agglomeration. The maximum value of coating obtained at minimum PANI-LGS system and high volume microstructure were observed from the analysis with best adhesion after immersion then other coating formulation. The researchers also revealed that PANI distribution reduced the water and oxygen as a result the protection of coating increases effectively. The formation of complex of aluminum-sulfonate as giving the dependence where it required and also the thick layer of the oxide increases the corrosion protection. The authors also presented a conceptual chart in their study. When the aluminum alloy dipped into the corrosion environment the PANI-LGS system releases the doped LGS ions to prevent from the corrosion.

Ayad *et al.* [15] synthesized PANI ammonium hydroxide complex using chemical deprotonate method and PANI sulfamic, citric, succinic, and acetic acid complexes through the reprotonate method. The PANI oxidation ranges were estimated using the UV-Visible spectroscopy and the absorbance rate founded in between 360 nm and 438 nm. They analyzed using IUPAC procedure as taken PANI absorption as basis in the emeraldine state the doping concentration was estimated. The varied PANI concentration found as for sulfonic acid is 0.27, for succinic acid is 0.18, for citric acid 0.15 and for acetic acid is 0.13. The polarization calculations took for the estimation of corrosion nature. They terminated results as polarization resistance and productivity of corrosion seems to be as variation of PANI concentration. The increment of PANI concentration increases the polarization resistance then these results support the increment of coating concentration. The composite coatings gradual decreases of the polarization resistance values are observed after 300 hours. While for 150 hours, base coating demonstrates high decrement value of polarization; hence, it leads to the coating delimitation. From the above information, we can conclude that the PANI mixer coating decreases the delaminating.

Satyanarayana *et al.* [16] studied PANI tincture coated steel material in NaCl and HCl solution through EIS (electrochemical impedance spectroscopy). The studies reveal the experimental results as the wide corrosion resistance observed by coating PANI pigment coating in the acidic and neutral state. The studies indicate clearly the PANI coating highly protective in acidicmedium.

Olad *et al.* [17] synthesized the Zn nanoparticles incorporated PANI composite through the *in situ* polymerization process by energetic stirring process. The prepared PANI/Zn nanocomposite is used to study the electrical conductivity. The studies showed that the prepared PANI/Zn nanofilms conductivity is better than the pure PANI conductivity. The electro-active nature of PANI/Zn is confirmed through voltammetry. The more importantly, the corrosion study results are given as the anticorrosion of PANI/Zn nanocomposite on the iron materials clearly with kinetically and thermally as good as than the PANI.

Zhu *et al.* [18] synthesized graphene oxide and PANI composite through the *in situ* polymerization. The researchers observed that aniline polymerized graphene oxide shows water dispersibility. The GO/PANI composite coating is prepared and coated on mild steel for the corrosion studies. In the acidic environment (NaCl) using electro impedance spectroscopy and OCP, the corrosion properties are studied. The results are concluded as the PANI and GO have shown very good anticorrosive nature. The comparative anticorrosion study has made between the WEP composite and PANI/GO coated WEP composite, the researchers noticed that WEP/PANI/GO was shown better anti-corrosion than the WEP composites. The PANI works as electroactive protector and the GO sheets protected from the water obstacle performance.

Yanga *et al.* [19] synthesized nanofibers of PANI using conventional, interfacial polymerization and direct mixed reaction methods in sulfuric acidic media. The pure PANI nanofibers have uniform surface properties in the range of nanometers. The electrochemical corrosion studies of nanofibers which are synthesized through direct mixed reaction have shown very good corrosion protection for the kind of steel. As a whole, the studies revealed that PANI surface properties and nanostructure well anticorrosion achievement.

Paula *et al.* [20] synthesized the PANI microspheres by incorporating of the methyl methacrylate and butyacrylate, acrylic acid nanoparticles through the two-step process such as emulsion polymerization and core-shell formation on the aniline monomer. The prepared P (MMA/BA/AA)/PANI film-forming properties attained with acrylic waterborne formulations. The two-step synthesis process makes possible to get well semiconducting properties; this was confirmed by the UV-Visible spectroscopy and conductivity measurement. The corrosion properties were checked with the help of electrochemical impedance spectroscopy and also mechanical and thermal stability of the films looks better values. Finally the corrosion properties were analyzed as the passivation effect stimulated due to the incorporation of the PANI and instructed the enhancement protection to carbon steels.

The acrylic ester grafting epoxy (A-g-EP) with eCOOH functional groups was fabricated by the researchers Zhu *et al.* [21]. Then, the PANI and A-g-EP composite was got by via oxidation polymerization and *in situ* polymerization techniques of aniline on A-g-EP macromolecular chains.

The grafted epoxy and PANI-A-g-EP composite solution with 30 wt% concentration on the mild steel with the thickness of the coating was controlled to 25–30 μm. The PANI has served as the creating the conduction bridges in between A-g-EP chains it results the bupolaron existence so that the possibility of antistatic properties of PANI-A-g-EP revealed. The anticorrosion properties of PANI-A-g-EP composite coating are revealed by the electron transmission and passivation catalysis.

Zhua *et al.* [22] reported new method that simultaneous annealing conditions for the preparation of RGO-Fe_3O_4-PANI tracer nanocomposites should be formally reduced with aniline monomers on GO nanosheets but without the use of a secondary reducing reagent. And then, aniline RGO can be used in polymerization. π-π electronic interactions between RGO and PANI to promote metal passivation effect and also p-n junction between PANI and Fe_3O_4 to improve anti-corrosion performance. When 0.5 wt% of RGO-Fe_3O_4 in PANI nanocomposites were added to the water-borne acrylate amino solution, the resulting coating showed excellent anti-corrosion properties. The mechanism of counteraction of RGO-Fe_3O_4-PANI nanocomposites has been revealed to be an effective physical barrier due to the layered assembly of the microstructure. Increased passage catalysis and p-n junction have been established, which allows only one-way electron transport and limits oxygen depletion, thus effectively improving the corrosion resistance.

9.2.2 Polypyyrole (PPy)

Tunc *et al.* [23] prepared PPy film on the surface of the copper and brass metal through electrochemical method. The quantity of the copper ions incorporated at the time of synthesis to alter the PPy polymer film. The anticorrosion result analyzed as the PPy film could contribute protection in opposition to corrosion of brass and copper even at acidic environment at given period. Clear anticorrosion mechanism that was discussed by authors indicates the importance of the PPy for the corrosion resistance. However, the corrosion measurement revealed the fact that the coating of PPy is good on the copper then the brass.

Attarzadeh *et al.* [24] made an appealing study of the PPy polymer. The scientists used the Saccharinsodium to incorporate in the PPy through the electrodeposition for the coating of steel. At the time of synthesis the electro-deposition was varied by changing the inductance when the saccharin introduced then the maximum polymerization potential were observed. These PPy coatings expressed very good anti-corrosive properties. Interestingly, on the morphological properties the smooth surface was noticed at the saccharin concentration 0.5 g/l and it indicates the very good anti-corrosive nature of the PPy conducting polymer.

There were also some studies which related to the two or more conducting polymer contribution for the corrosion resentence. For example, Gouri *et al.* [25] created the PANI and PPy mixer coating on the surface of the aluminum substrate via electrochemical deposition with the help of the oxalic acid as a electrolyte. In the periodic voltammetry, an anodic peak is presented near 0.15 V in the minimum short time; the presented peak may due to the oxidation or adsorption of hydrogen atom. The composite coating periodic graphs also indicated an anodic peak power at 1.01 V versus SCE and the power peak of cathode near 0.7 V versus SCE. The FTIR analysis of coating reveals the composite coating of PPy structural vibration clearly. As well as the scanning electron microscopy images reflected the distinguished changes in the morphology of the polypyrrol and PANI coating.

The acidic nature of the surroundings is the main criteria for the efficiency of the coating polymers if the nature conditions are the acidic or base so that the corrosion also changes as per the instructions given by the

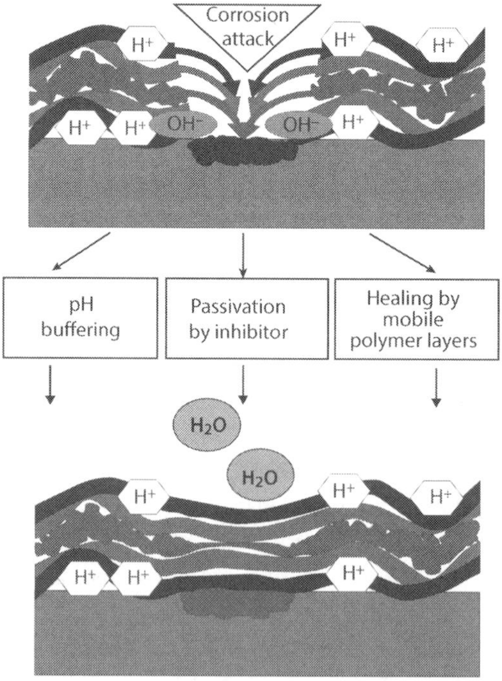

Figure 9.2 Influence of pH for the corrosion.

pH values (see Figure 9.2). The co-researchers of Jude *et al.* inspected this type of study. The researchers coated PPy on the steel rod at the conditions in both acidic and base through the polymerization of electrochemical. Fascinatingly iron oxalate layer formed in the acidic media with the potential of 1.2 versus SCE. As getting the positive value of the potential, the electrochemical polymerization of PPy raised on the substrate of passivation. The pH value of the reacted medium influenced y much on the passivation time and current density of the substrate. The passivation is spontaneous; the noncrystalline coatings were observed when the pH value is in maximum. Electropolymerization of PPy in the acid can only occurs afterword steel enclosed by the crystalline interphase.

Paliwoda *et al.* [26], group of scientists, study the MoO_{2-4} doped PPy conducting polymer and explained the corrosion properties. They observed clear inhibition effect on the delaminating process. Because of the wide pH value at the barrier of polymer and metal the PPy gets over oxidation. In the case of the Ppy/MoO_{2-4} coating, this buffering effect cannot seem to be observed. The corrosion resistive ions which were protected by the leaching and pH changes were not affected as PPy releasing on coating. However, the power of circumstances during the cathode delaminating process in the release behavior, especially the effect of small easy incorporable cations and the elevated pH, needs additional examination.

Bereket *et al.* [27] coated Poly (5-amino 1-naphthol) (PANAP) film on the top of the ppy polymer film which is synthesized through the electrochemical synthesis (cyclic voltammetry). Hence, the polymer films that are single PPy film and multilayer PPy/PANAP film were introduced to the comparative corrosion study, the results shown that PPy/PANAP multilayer film as good as the single PPy film for the corrosion protection of mild steel which is immersed in saline liquid. It is found that, if the immersive time is increases, then the protective behavior of synthesized single layer and multilayer PPY films were extremes. This was studied by the properties of auto undoping when the synthesized layers were immersed in the corrosive solution. Finally, Poly (5-amino 1-naphthol) (PANAP)covered PPy multilayer shows better anti-corrosion properties because Poly (5-amino 1-naphthol) (PANAP) contains structurally reactive –OH groups. The poly-pyrrole, alumina nanoparticles prepared through the polymerization process in the dihydroxybenzebene presents. The prepared nanocomposite was used to corrosion protection process as filler in the corrosion soundings. The dihydroxybenzene developed the better coating adhesion for the Al alloy metallic surface. The presence of dihydroxybenzene improved coatings adhesion to Al alloy surface [28].

Prissanaroon *et al.* [29] investigated the corrosion protective material for the stainless steel. The group of researchers used PPy conducting polymer for the coating material of the steel rod. The dodecylbenzenesulfonicacid (DBSA) and oxalic acid taken as a co doping systems to synthesize PPy films through the electro deposition.

9.2.3 Poly(3,4-ethylenedioxy thiophene): Polystyrene sulfonate (PEDOT:PSS)

Anmona *et al.* [30] made PEDOT:PSS coated microelectrode of gold materials for the application of long-term stable interface. These types of materials can also be used as iodine etching process because the surface morphology is important in the point of view etching time and iodine concentration. When the gold electrodes etched in iodine solution for period of time the PEDOT:PSS thick film showed more stability. The PEDOT:PSS voltammetry activation procedure reduces polarization and impedance of electrode at the time of stimulation process. The researchers were concluding that this PEDOT:PSS development is promising coating material for the electrode for stability for long life period.

Pranti *et al.* [31] taken a step for the development of the efficiency of coating material with PEDOT:PSS in the form of composite or film. They designed a material which has long life period and sophisticated PEDOT:PSS film for neural application. The PEDOT:PSS micrometer layer has careful etching process and 100% stability aluminum electrode and the material evaluated by the ultrasonic studies and vitro studies. The neural simulative applications of PEDOT:PSS has shown good consistency. The PEDOT:PSS film electrochemical modifications by the voltammetry studies improve the characteristics of coated electrodes. The polarization of the material decreases it leads to improvement of charge addition capacity of the PEDOT:PSS electrode. The increment of signal to noise ratio shows the decrement of electrode impedance. These two phenomena reflects the PEDOT:PSS is a good coating material for aluminum electrode.

Tim *et al.* [32] scrutinized that the PEDOT is a required material to stimulate the electrode or whether the conduction of the PEDOT changes with respect into its lifetime. For this purpose, the group of scientists has taken the realistic test environment for the PNS functional properties like charge injection, pulse width, and constant frequency. The experimental results shown as PEDOT are a kind of essayist material to stimulate electrode with extreme properties then the platinum for the subchronic studies. In their studies, they demonstrated how the coating fails or else conductivity is to be lost to suggest further studies of conducting polymer studies.

Zaid *et al.* [33] studied the neuronal sensing properties using the conducting polymer coating. They mainly focused on the charge density which is deposit greater than the 320 mC/cm^2 and which provide application in the neuronal stimulation and recording with PEDOT:PSS coating. The long period stimulation functions were observed in the case of PEDOT:PSS through maximum CIL and development in impedance at the time of biphasic stimulation. Conversely, for the implementations the material requires adhesion improvement. The biocompatible study of PEDOT:PSS coating expressed via unstencil and briefly of neurite outgrowth and neuronal possibility. However, there were no extra advantages observed from the macroporous than the moderate progress in electrochemical properties of PEDOT:PSS coating at the time of process.

Ghosh *et al.* [34] demonstrated the manufacture of highly durable electronic fabrics for flexible electromagnetic performance through an explicit method of dipping and drying. The conductivity of the PEDOT:PSS-treated PEG coated tissue was 51 cm every 1 dipping cycle (thickness of 0.38 mm). The increase in conductivity was dominated by the screening phenomenon (PEG 400) between PEDOT and PSS, which leads to the phase-separated structure and higher aggregation of crystalline PEDOT domains on the cotton fabric surface, confirmed by the smaller X angle and Lightning scattering study (SAXS). Efficiency was also increased with increasing coating cycles and a maximum of 46.80 dB reached just 20 dipping cycles. In addition, shielding and electrical performance was successfully increased with extended needle cycles of up to 82.7 S cm^{-1} and 65.6 dB. The special vodefiller mechanism that comes from 3D tomography effectively controls the electrical conductivity with shielding capability. High mechanical and chemical tolerance (thermal air aging, bending and turning, single-sided tape peeling test, prolonged exposure to sunlight, polar and non-polar solvent treatment, and boiling with water) Strength and durability proof (retention up to 95% EM Performance on each finished garment test).

PEDOT:PSS study showing clean steam electrodes coated with an electrically conductive polymer is stable for at least 7 weeks of continuous dipolar current stimulation, corresponding to a total of 4.2 billion dipolar current pulses. Analysis of impedance spectroscopy results showed no significant changes in impedance after this long-term evaluation. Additionally, volatile electrodes are not visible after soaking in PBS solution at body temperature for more than 10 months. However, impairment of the electrode sheath due to blackout in continuous motion shows the importance of stimulus protocol stability to protect PEDOT:PSS active Layer. Overall, our results suggest that PEDOT:PSS coated electrodes are a good candidate for long-term recording and stimulation of chronically organized neural interfaces [35]. Ding *et al.* [36] prepared xVS_2@PEDOT-PSS nanosheets through the methods such as solvothermal method VS_2 preparation and the immersion and freeze-drying process for the VS_2 incorporation into the PEDOT-PSS for the coating of anode materials in the LIB (lithium-ion battery).

9.3 Conclusion

The described studies are the considerable evidence for the use and importance of conducting polymers in the field of corrosion resistance of the electronic material. The Cps coated electrode shows better stability and maximum current density as well as anticorrosive nature. The CP, PANI, and its derivatives like PANI/Zn nanofilms conductivity and anticorrosive nature are better than the pure PANI. Whereas, the specified material like PANI and GO have shown very good anticorrosive nature. The PPy and their composite also show remarkable anticorrosive behavior for the electrodes which were synthesized different methods were revised. The PEDOT:PSS composite polymer widely used as coating material for the electronic devices.

References

1. Yao, B., Wang, G., Ye, J. *et al.*, Corrosion inhibition of carbon steel by PANI nanofibers. *Mater. Lett.*, 62, 1775–1778, 2011.
2. Hosseini, M.G., Sabouri, M., Shahrabi, T., Corrosion protection of mild steel by PPy phosphate composite coating. *Prog. Org. Coat.*, 60, 178–185, 2007.
3. Herrasti, P., Recio, F.J., Ocon, P. *et al.*, Effect of the polymer layers and bilayers on the corrosion behaviour of mild steel: Comparison with polymers containing Zn microparticles. *Prog. Org. Coat*, 54, 285–291, 2005.
4. Zhu, H., Zhong, L., Xiao, S. *et al.*, Accelerating effect and mechanism of passivation of PANI on ferrous metals. *Electrochim. Acta*, 49, 5161–5166, 2004.
5. Deng, Z., Smyrl, W.H., White, H.S., Stabilization of metal-metal oxide surfaces using electroactive polymer films. *J. Electrochem. Soc.*, 136, 2152–2158, 1989.
6. Bedioui, F., Voisin, M., Devynck, J. *et al.*, Electrochemistry of conducting PPy films containing cobalt porphyrin. *J. Electroanal. Chem. Interfacial Electrochem.*, 297, 257–269, 1991.
7. Traeger, R.K., Proc. Ekctron. *Compon. Conf.*, 27, 408, 1977.
8. Weschler, C.J. and Shields, H.C., Proceedings of the Annual Air Pollution Control Association Meeting, *Atmos. Environ.*, 33, 2301–2312, 1999.

9. Gill, P.S., Graedel, T.E., Weschler, C.J., Organic films on atmospheric aerosol particles, fog droplets, cloud droplets, raindrops, and snowflakes. *Rev. Geophys. Space Phys.*, 21, 903, 1983.
10. Vernon, W.H.J., A laboratory study of the atmospheric corrosion of metals. Part II.—Iron: The primary oxide film. Part III.—The secondary product or rust (influence of sulphur dioxide, carbon dioxide, and suspended particles on the rusting of iron). *Tram. Farada7 Soc.*, 31, 1668, 1935.
11. Machnikova, E., Pazderova, M., Bazzaoui, M. *et al.*, Corrosion study of PVD coatings and conductive polymer deposited on mild steel Part I: PPy. *Surf. Coat. Technol.*, 202, 1543–1550, 2008.
12. Camalet, J.L., Lacroix, J.C., Nguyen, T.D. *et al.*, Aniline electropolymerization on platinum and mild steel from neutral aqueous media. *J. Electroanal. Chem.*, 485, 13–20, 2000.
13. Comizzoli, R.B., Frankenthal, R.P., Milner, P.C., Sinclair, J.D., Corrosion of Electronic Materials and Devices. *Science*, 234, 340–345, 2000.
14. Gupta, G., Birbilis, N., Cook, A.B., Khanna, A.S., PANI-lignosulfonate/epoxy coating for corrosion protection of AA2024-T3. *Corros. Sci.*, 67, 256–267, 2013.
15. Salem, A.A. and Branimir, N., Grgur The influence of the PANI initial oxidation states on the corrosion of steel with composite coatings. *Prog. Org. Coat.*, 119, 138–144, 2018.
16. Sathiyanarayanan, S., Muthukrishnan, S., Venkatachari, G., Trivedi, D.C., Corrosion protection of steel by PANI (PANI) pigmented paint coating. *Prog. Org. Coat.*, 53, 297–301, 2005.
17. Olad, A., Haleh Rasoulil Enhanced Corrosion Protective Coating Based on Conducting olyaniline/Zinc Nanocomposite. *J. Appl. Polym. Sci.*, 115, 2221–2227, 2010.
18. Zhu, X., Ni, Z., Dong, L., Yang, Z., Cheng, L., Zhou, X., Xing, Y., Wen, J., Chen, M., *In-situ* modulation of interactions between PANI and graphene oxide films to develop waterborne epoxy anticorrosion coatings. *Prog. Org. Coat.*, 133, 106–116, 2019.
19. Yanga, X., c, Li, B., Wanga, H., Baorong Houc Anticorrosion performance of PANI nanostructures on mild steel. *Prog. Org. Coat.*, 69, 267–271, 2010.
20. Bertuoli, P.T., Baldissera, A.F., Zatter, A.J., Ferreira, C.A., Elaine, C.A., Armelin PANI coated core-shell polyacrylates: Control of film formation and coating application for corrosion protection. *Prog. Org. Coat.*, 128, 40–51, 2019.
21. Zhua, A., Wanga, H., Sunb, S., Zhanga, C., The synthesis and antistatic, anticorrosive properties of PANI composite coating. *Prog. Org. Coat.*, 122, 270–279, 2018.
22. Zhua, A., Shia, P., Sunb, S., Ruia, M., Construction of rGO/Fe3O4/PANI nano-composites and its corrosion resistance mechanism in waterborne acrylate-amino coating. *Prog. Org. Coat.*, 133, 117–124, 2019.
23. Tüken*, T. and Yazıcı, B., Erbil, M., The electrochemical synthesis and corrosion performance of PPy on brass and copper. *Prog. Org. Coat.*, 51, 152–160, 2004.

24. Attarzadeh, N., Raeissi, K., Golozar, M.A., Effect of saccharin addition on the corrosion resistance of PPy coatings. *Prog. Org. Coat.*, 63, 167–174, 2008.
25. Akundy, G.S., Rajagopalan, R., Jude, O., Iroh electrochemical Deposition of PANI–PPy Composite Coatings on Aluminum. *J. Appl. Polym. Sci.*, 83, 1970–1977, 2002.
26. Paliwoda- Porebska, G., Stratmann, M., Rohwerder, M., Potje-Kamloth, K., Lu, Y., Pich, A.Z., Adler, H.-J., On the development of poly-pyrrole coatings with self-healing properties for iron corrosion protection. *Corros. Sci.*, 47, 3216–3233, 2005.
27. Bereket, G. and Hur, E., The corrosion protection of mild steel by single layered poly pyrrole and multilayered poly pyrrole/poly(5-amino-1-naphthol) coatings. *Prog. Org. Coat.*, 65, 116–124, 2009.
28. Tallman, D.E., Levine, K.L., Siripirom, C., Gelling, V.G., Bierwagen, G.P., Stuart, G., Croll Nano composite of PPy and alumina nanoparticles as a coating filler for the corrosion protection of aluminum alloy 2024-T3. *Appl. Surf. Sci.*, 254, 5452–5459, 2008.
29. Prissanaroon N, W., Brack, P., Pigram, J., Liesegangco-J., doped, PPy coatings for stainless steel protection. *Surf. Rev. Lett.*, 13, 2 & 3, 319–327, 2006.
30. Pranti, A.S., Schander, A., Bödecker, A., Lang, W., *Highly Stable PEDOT:PSS Coating on Gold Microelectrodes with Improved Charge Injection Capacity for Chronic Neural Stimulation*, Presented at the Eurosensors 2017 Conference, Paris, France, 3–6 September 2017. *Proceedings*, 1, 4, 492, 2017, https://doi.org/10.3390/proceedings1040492.
31. Pranti, A.S., Schander, A.B., Odecker, A., Lang, W., PEDOT: PSS Coating on Gold Microelectrodes with Excellent Stability and High Charge Injection Capacity for Chronic Neural Interfaces. *Sensors and amp; Actuators: B. Chem.*, 275, 382–393, 2018.
32. Boretius, T., Schuettler, M., Stieglitz, T., On the Stability of Poly-Ethylenedioxythiopene as Coating Material for Active Neural Implants. *Artif. Organs*, Wiley Periodicals, Inc, 35, 3, 245–248, 2016.
33. Bartsch, H., Peipmann, R., Himmerlich1, M., Müller, J., Andreas Bund, Hartmut Witte PEDOT coating applied on thick film gold electrodes for increased miniaturization capability. *Prog. Org. Coat.*, 135, 545–554, 2019.
34. Ghosh, S., Ganguly, S., Remanan, S., Narayan Ch DasFabrication and investigation of 3D tuned PEG/PEDOT: PSS treated conductive and durable cotton fabric for superior electrical conductivity and flexible electromagnetic interference shielding. *Compos. Sci. Technol.*, 181, 107682, 2019.
35. Schander, A., Teßmann, T., Strokov, S., Stemmann, H., Kreiter, A.K., Lang, W., *In-vitro* Evaluation of the Long-term Stability of PEDOT:PSS Coated Microelectrodes for Chronic Recording and Electrical Stimulation of Neurons. *Conf. Proc. IEEE Eng. Med. Biol. Soc.*, 2016, 6174–6177, 2016.
36. Ding, Z., Zhang, Q., Chen, Y., Liu, G., Xin, X., He, H., Cai, B., Wu, J., Yao, X., PEDOT-PSS coated VS2 nano sheet anodes for high rate and ultrastable lithium-ion batteries. *New J. Chem.*, 43, 1681, 2019.

10

Polymer Coatings for Food Applications

Fatemeh Sadat Mostafavi[1]* and Davood Zaeim[2]

[1]Department of Food Science and Technology, University of Torbat-e Jam, Torbat-e Jam, Iran
[2]School of Food Science and Biotechnology, Zhejiang Gongshang University, Hangzhou, China

Abstract

In recent years, consumer demands, as well as obligatory regulations that have been passed by various national and international authorities, have pushed food manufacturers to produce tastier and healthier food products with longer shelf lives. Coating of food components is introduced as an effective approach to meet these demands and obligations. Biopolymers including polysaccharides, proteins, and waxes are the main ingredients used for creating edible coatings. They might be used individually or in combination. The coating process can be applied to a variety of food products ranging from very large objects to particles at the microscale or smaller. The term encapsulation is used for the latter. The objectives and techniques used for coating food products with large sizes are different from those used for encapsulating small-sized food materials. Dipping, enrobing, panning, and spraying are typical methods used for coating food products such as fruits and vegetables, meat and meat products, eggs, cheese, nuts, and fried food. While, spray-cooling, freeze-drying, emulsification, extrusion, and electrohydrodynamic atomization are the most common microencapsulation techniques. In this chapter, the main objectives, components and techniques used for coating food surfaces or encapsulating food components are discussed individually.

Keywords: Coating, encapsulation, hydrocolloids, waxes, shelf-life, polysaccharide, protein

**Corresponding author*: fatemehs.mostafavi@gmail.com

Inamuddin, Rajender Boddula, Mohd Imran Ahamed and Abdullah M. Asiri (eds.) Polymer Coatings: Technology and Applications, (189–232) © 2020 Scrivener Publishing LLC

10.1 Introduction

The idea of coating food products is not new. Edible coatings have been utilized for centuries to enhance food quality and shelf-life. The first recorded use was in the 12th century in China on oranges and lemons to give them a glossy appearance and retard their senescence [1]. Later, in England, the shelf-life of meat products was extended using lard or fats. This process was named "larding" [2]. In the last 20 years, coating food has become one of the most important applications of polymers in food science and great research efforts have been devoted to this field [3].

According to Saleh and Guigon [4], the food coating process is defined as the covering of edible materials with a surrounding layer being composed of a single or multiple GRAS (Generally Recommended as Safe) components with specific functions. The coating process can be applied to a variety of food products ranging from very large objects to particles at the micro or nanoscale. The term encapsulation is used for the latter [4, 5]. The objectives and techniques used for coating food products with large sizes are different from those used for encapsulating small-sized food materials. In this chapter, the major objectives, main components, and the techniques used for coating food surfaces or encapsulating food components are discussed individually.

10.2 The Main Objectives of Coating Food Surfaces

The main reasons for coating food surfaces could include the following.

10.2.1 Controlling Mass Transfer

Undesirable mass transfer among the food and its surrounding environment can contribute to loss of food quality [5]. Moisture absorption, surface dehydration, flavor and aroma gain or loss, fat migration, and gas transfer are some of the most important problems occurring due to mass transfer across the food. Edible coatings on the food surface can prevent these problems by acting as a barrier against moisture, gas, and ultraviolet. They can also act as a mass transfer barrier among particles within a food system [6].

10.2.2 Carrier of Functional Agents

Edible coatings can act as a carrier for functional ingredients to food systems to improve their stability, quality, or shelf-life [7]. These functional agents include antimicrobials, antioxidants, texturizers, nutraceuticals, probiotics, flavors, and colorants [8].

10.2.3 Physical Protection

Edible coatings applied on the food surface can protect them from physical damage caused by different mechanical forces applied during handling. This mechanical protection is especially important in fragile food products such as nuts, breakfast cereals, and snacks [3]. Han and Aristippos [9] reported improvement in the physical strength of cashew nuts against external forces with the aid of whey protein-based coating (Figure 10.1).

10.2.4 Sensorial Improvement

Usually, the food surface has a powerful impact on the consumers' first sensorial impressions. Therefore, the materials used for coating food can positively alter the organoleptic properties of the coated food [3]. Coatings can be used to give a glossy appearance or to provide a non-greasy or non-sticky surface when necessary [10]. They also can be used as an adhesive to add salt, sugar, colorants, seasonings, or flavors onto the surface of food products such as nut and confectionery [3, 11, 12].

10.3 Components of Edible Coatings

Forming a compact and uniform coating without pores and cracks on a food surface is important in preserving its quality. Moreover, a good adhesion should occur among the coating solution and the product surface [13].

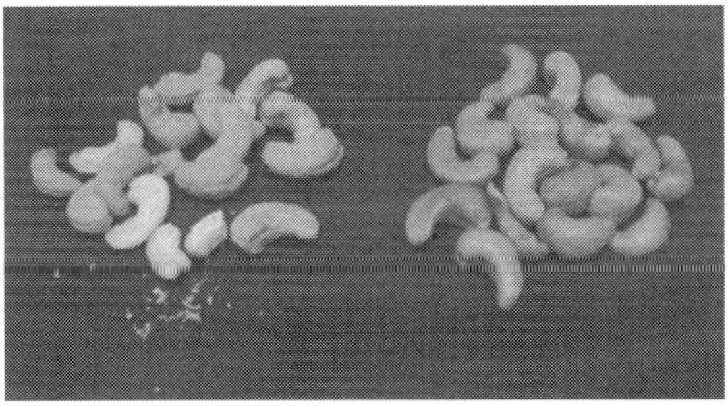

Figure 10.1 Whey protein-based edible coating protects the cashew nuts from physical damage. The figure has been reproduced with permission from Han and Aristippos [9].

The effectiveness and functionality of an edible coating mainly depend on its barrier against water vapor and gases, which depends on the surface properties of the food, its storage conditions, and the chemical composition of the coating materials [14]. Biopolymers including polysaccharides, proteins, and waxes are the main ingredients used for creating edible coatings. They could be used individually or in combination [15].

10.3.1 Polysaccharide

Polysaccharides, due to their favorable functional characteristics, are the most versatile polymers used for food coating applications. They have an oil-free and colorless appearance. Most of them are neutral; while some of them are negatively charged. Polysaccharide molecules can interact together through hydrogen bonding and create a continuous coating network that can act as a barrier against gas and oil transfer [16]. However, the coatings made from these types of biopolymers are hydrophilic, and therefore, have poor moisture barrier characteristics [17].

Polysaccharides-based coatings have been utilized for enhancing the shelf-life of several food products. High-fat food products such as nuts, sausages, and fillets can be protected from oxidation with the aid of appropriate polysaccharide-based coatings [18, 19]. Moreover, the barrier properties of these types of coatings can retard the respiration rate of fruits and vegetables and protect them from dehydration [20]. They can also be used for encapsulating bioactive agents [21]. The most important polysaccharides used for this purpose include cellulose derivatives, chitosan, starch derivatives, alginate, carrageenans, and pectin.

10.3.1.1 Cellulose Derivatives

Cellulose is the most abundant natural biopolymer on earth. It is a water-insoluble linear polysaccharide composed of D-glucose units linked through ß-1,4 glycosidic bonds. Methylcellulose (MC), hydroxypropyl cellulose (HPC), and hydroxypropylmethylcellulose (HPMC) are cellulose ethers obtained by partial substitution of hydroxyl groups in cellulose by ether functions. They are water-soluble and able to form a compact coating layer on the food surface [22]. These coatings have been reported to be tasteless, odorless, transparent, water-soluble, resistant to oil, with moderate gas, and moisture transmission properties [3]. They have been widely used for coating fruits [23, 24] or meat products [25, 26] and encapsulating bioactive agents [27].

10.3.1.2 Chitosan

Chitosan is the second most abundant natural biopolymer after cellulose. It is obtained by alkaline deacetylation of chitin. Chitosan coatings are characterized by having low oxygen permeability and excellent mechanical properties. Moreover, due to their cationic nature, they exhibit antimicrobial activity. Chitosan-based coatings have been widely used for enhancing the shelf-life of nuts [28], meat products [18, 29, 30], fruits and vegetables [31–33]. It has also been used as a delivery system for bioactive agents [34, 35].

10.3.1.3 Starch and Starch Derivatives

Starch is a mixture of amylopectin and amylose. Amylose is a linear biopolymer composed of α-1,4 anhydroglucose units; while, amylopectin is a highly branched biopolymer consisting of short α-1,4 chains linked by α-1,6 glucosidic branching points. Amylose is known to form continuous coating layers on the food surface, while coatings made from amylopectin have been reported to be brittle [36]. Starches and their derivatives have been widely used for coating cheese [37] fruits [38–40], bakery products [41], and encapsulating functional ingredients in food systems [42].

10.3.1.4 Seaweed Extracts

Alginates and carrageenans are two main seaweed extracts used for food coating applications. Alginates are linear biopolymers consisting of L-guluronic and D-mannuronic acid units; while carrageenan is a water-soluble polymer formed by alternate units of 3,6-anhydrogalactose and D-galactose joined by β-1,4- and α-1,3-glycosidic bonds [43]. These seaweed extracts can to form transparent, uniform and water-soluble coating layers with good gas and oil barrier properties, and high water vapor permeability. They have been widely used for coating meat products, fruits, and vegetables and also for encapsulating functional ingredients in food products [44].

10.3.1.5 Pectin

Pectin, one of the main components of the plant cell wall, is a complex anionic polysaccharide composed of poly α-1,4-galacturonic acids. Citrus peels and apple pomace are the main industrial sources for extracting this polysaccharide. It is widely used as a stabilizing, gelling, or thickening agent in food products [45]. Due to its edibility, biocompatibility, and selective gas permeability, it has been used for forming films and coating food products such as chicken fillets [46], fish [47], fruits [48], and vegetables [49]. It also has been widely used for encapsulating bioactive agents in food systems [50–52].

10.3.1.6 Other Polysaccharides

Other polysaccharides used for coating food commodities include different gums classified in three main groups named as plant exudates (acacia gum [53, 54], gum karaya [55], Persian gum, and tragacanth gum [13]), microbial gums (pullulan [56], xanthan gum [57], and gellan gum [58]), and seed gums (guar gum [59], locust bean gum [60], and basil gum [61]).

10.3.2 Proteins

Protein-based coatings are edible since food-grade proteins and additives are used. Hydrophobic interactions, disulfide bonds, and hydrogen bonds are the main interactions occurring while forming protein-based coatings. Any factor affecting these interactions might alter the mechanical and barrier characteristics of the final coating on the food surface [62].

Proteins used for coating food are derived from different plants or animals. Soybean and corn are two of the main sources of plant-based proteins; while casein, whey proteins, collagen, and gelatin are some of the most important animal-origin proteins used for coating the foods [3].

10.3.2.1 Collagen and Gelatin

Collagen is the most commercially successful edible protein used for coating food. It is a hydrophilic fibrous protein being the major constituent of connective tissue, skin, and tendon. Collagen-based coatings are mechanically strong, water-insoluble, and good oxygen barriers. However, they are not particularly good moisture barriers.

Collagen has been traditionally used for the production of edible sausage casings, replacing those previously made from natural gut. Collagen has also been recently used for coating fruits [63].

Gelatin protein results from controlled hydrolysis of collagen. Gelatin-based edible coatings are able to reduce the transfer of gas, oil, and moisture from the food surface [3]. Gelatin coatings containing natural antimicrobial agents have been previously used for improving the shelf-life of meat [64–66] and fruits [67, 68]. It has also been used for decreasing fat uptake in deep-fat fried foods [14].

10.3.2.2 Corn Zein

Zein is composed of prolamins which are a group of alcohol-soluble proteins found in corn endosperm. Zein has a hydrophobic nature, which is attributed to its high content of non-polar amino acids. Zein-based coatings are traditionally used for coating fruits, confectionery, nuts, and candies [3]. Recent researches show the use of zein-based coatings for enhancing the shelf-life of melon [69], guava [70], and egg [71], enhancing the survival of probiotic bacteria in gastric fluid [72], and controlling the release of bioactive components [73]. Zein coatings have also been reported to reduce oil uptake in deep-fat fried foods such as mashed potato balls [14]. Antimicrobial zein-based edible coatings incorporated with sorbic acid, calcium propionate, or nisin have been reported to be effective in controlling *Listeria monocytogenes* in cooked sweet corn [74] and ready-to-eat chicken [75].

10.3.2.3 Soy Protein

Soybeans contain much higher protein content in comparison with cereal grains. Most of the proteins in soybeans are water-insoluble [3]. Soy proteins-based coatings have been used for inhibiting lipid oxidation in nuts [76], retaining moisture, delaying lipid oxidation and deterioration in meat products [77–79], reducing fat-uptake in deep-fat fried foods [80], and enhancing the shelf-life of eggs [81] and fruits (such as fresh-cut pineapple [82], fresh-cut apples [83], Persian lime [84], artichoke [85]). It also has been used as a delivery system for bioactive compounds [86].

10.3.2.4 Whey Protein

Whey protein is a major type of milk protein, which remains soluble after precipitation of casein at pH = 4.6. They are heat-labile and globular naturally. The coatings made from this protein have a glossy and transparent appearance with good mechanical characteristics [3]. They are excellent barriers against lipid, oxygen, and aroma. Therefore, they have been widely used for improving the appearance and shelf-life of nuts [87, 88], eggs [89], confectionary [90], meat [91, 92], fruits and vegetables [11, 93, 94]. Whey protein-based coatings have also been widely used as a carrier for bioactive agents such as probiotic bacteria [95], lactoperoxidase, and α-tocopherol [96] for improving the functional properties of different food products.

10.3.2.5 Casein

Casein represents approximately 80% of the total milk proteins. Sodium and potassium caseinates are used in preference to calcium and magnesium caseinate as they are more soluble and have better functional characteristics. Casein molecules, due to their ability to interact together through hydrogen bonds, form compact coatings without needing further treatment. The casein coating layer on the food surface has been reported to be flavorless, transparent, flexible, and water-soluble [3, 97]. The casein edible coatings enriched with essential oils and bacteriocins have been reported to retard microbial contamination in fruits [98]. Casein has also been introduced as a promising agent for encapsulating sensitive food ingredients and bioactives [99, 100].

10.3.3 Lipids

Lipids are usually incorporated into food coatings to impart hydrophobicity and thus reduce moisture transfer. Different types of lipids are used for this purpose. The most commonly used hydrophobic materials in food coatings are waxes. Carnauba wax, beeswax, shellac wax, polyethylene wax, or candelilla wax are the most common waxes in the structure of food coatings [3, 101].

10.3.3.1 Shellac Wax

Shellac is a natural wax secreted by the lac insect. It has been approved by the FDA (Food and Drug Administration) for some foods since 1939. Due to its excellent coating ability and its shiny appearance, it has been used as an edible coating for fruits, vegetables, and confectionery products since the beginning of the 21st century [102, 103]. In recent researches, the combination of shellac wax and other biopolymers such as gelatin has been used for improving the shelf-life of various fruits [104, 105].

10.3.3.2 Carnauba Wax

Carnauba wax is obtained from the leaves of a Brazilian palm tree (*Copernica cerifera*). Refined carnauba wax is mostly composed of saturated long-chain alcohols and saturated wax acid esters containing 24–32 hydrocarbons. Therefore, it is an excellent barrier to moisture while poor barrier against gases [106]. Carnauba wax has been considered as GRAS and is allowed to be used in the formulation of coatings applied on fruits and vegetables, confectioneries, and chewing gums [107].

The main problems with using carnauba wax as a food coating material are its relatively poor gas barrier and its loss of gloss during storage. To overcome these drawbacks, it can be combined with shellac. Johnfresh™ is a commercial food coating wax composed of both carnauba wax and shellac [1].

10.3.3.3 Candelilla Wax

Candelilla wax is an exudate of the candelilla plant which mostly grows in the US and Mexico. It is considered as GRAS and has been used for coating different foods such as hard candy, chewing gum, fruits, and vegetables [108–111].

10.3.3.4 Beeswax

Beeswax is secreted by honeybees and is separated from the honey section by centrifuging. It is a complex mixture of hydrocarbons, monofunctional alcohols, and long-chain acids. It is considered a GRAS and that is permitted to be directly used in food with some limitations. The maximum permitted level is 0.005% in confections and frostings, 0.065% in chewing gum, 0.1% in soft candy, 0.04% in hard candy, and 0.002% in all other categories. It has been used in different food coatings for applying on the surface of fruits [2, 112, 113] and also for encapsulating bioactive agents [27].

10.3.4 Additives

The organoleptic, functional, nutritional, and technological characteristics of edible coatings can be modified with the aid of different additives.

Due to the high brittleness of the polymer-based coatings, the most frequently used additives in their structure are plasticizers. Plasticizers are low-molecular-weight components which locate between polymer chains. They decrease intermolecular forces, increase the mobility of the polymer chains, and therefore improve the film flexibility. Common plasticizers used for edible coatings are saccharides (glucose, fructose glucose syrups, and sucrose) and polyols (sorbitol, glycerol, polyethylene glycols, and glyceryl derivatives).

Coating functionality on the food surface depends on the adhesion between the coating and food surface. Surfactants such as Tweens, lecithin, or Spans are surface-active components being able to reduce the surface tension at the water-solid interface contributing to wettability improvement and better adhesion of the coating solution on the food surface.

They are also necessary for the stability of polysaccharide or protein coatings containing hydrophobic substances [3].

Texture enhancers, antibrowning and antimicrobial agents, nutraceuticals, aroma precursors, nutrients, probiotics, colorants, flavors, and nanoparticles are other additives used in the formulation of food coatings [1].

10.4 Application Methods of Edible Coating on Food Surface

For coating a food product, the coating solution should be applied on its surface and the coating form after solvent evaporation. Typical methods used for forming polymer-based edible coatings are dipping, enrobing, panning, and spraying [5]. The three latter methods are commonly used in the food industry; while dipping is usually employed for lab-scale coatings [114].

The dipping is best-suited for products with irregular shapes. In this method, the product is dipped into the coating solution for a specific duration of time and then drains off and dries. This method is simple and cheap and does not require sophisticated machinery equipment. However, the final formed coating might not be uniform enough. Therefore, in some cases, multiple dipping may be necessary to ensure full coverage [62, 115]. Moreover, difficulty in controlling the process, microorganism growth in the dipping tank, dilution of the coating solution, and residual of a high quantity of it at the end of the process are some of the major challenges of this method [114].

The enrobing method is more common for coating food products with flat surfaces in the meat and chocolate industry. During this process, the coating solution flows vertically to the food product. Gravitational and viscous forces are important in the coating process. For controlling the coating thickness on the final product, the viscosity of the coating solution is a key parameter that should be adjusted properly [114, 116].

The panning technique is usually used for coating confectionery products. In this method, the product is placed in a large rotating pan. The coating material is sprayed or poured into the pan. The pan rotates and the coating material evenly distributes over the food surface. Finally, forced air is used to dry the coating [62, 117].

The spraying technique is preferred for coating products with a large surface area. In this method, food is placed on a plate which rotates under a hand-held spraying gun. This process continues until achieving the desired coating thickness. Finally, a uniform coating layer with controlled thickness is formed over the food surface. However, when using the spray-coating technique, the bottom surface of the food should be coated separately after drying the initial coating layer [62, 115].

10.5 Food Applications of Edible Coatings

10.5.1 Fruits and Vegetables

Fruits and vegetables are living tissues that undergo respiration, transpiration, and ethylene production after harvest, and thus, major physical, chemical, and microbiological changes occur during storage contributing to changes in their appearance, flavor, texture, and nutritional aspects [10].

Coating fruits and vegetables with biopolymer-based coatings can retard the transfer of gas, water vapor, and volatile compounds through their surface and thus provide them with a modified atmosphere which reduces the rate of respiration and senescence, retains moisture, reduces aroma loss, and delays color changes and microbial contamination during storage [13, 33]. Vu, Hollingsworth [118] enhanced the shelf-life of strawberries by coating them with chitosan-based solutions (Figure 10.2).

Edible materials used for coating fruits and vegetables are composed of hydrocolloids, (consisting of either proteins or polysaccharides) or hydrophobic compounds (e.g., waxes). Successful fruit coatings are usually composed of a mixture of hydrocolloids and hydrophobic compounds [1]. Since now, different polysaccharides (e.g., chitosan [31], starch [38], carboxymethyl cellulose [119], gellan [120]), proteins (e.g., soy protein isolate [82], zein [69], gelatin [67], gluten [121], whey protein isolate and calcium caseinate [122]), and waxes (such as carnauba

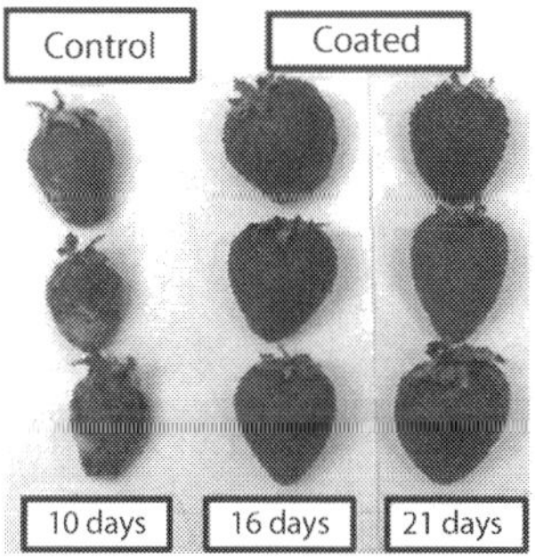

Figure 10.2 Uncoated and coated strawberries with chitosan-based coatings after 10, 16, and 21 days storage at 4°C. The figure has been reproduced with permission from Vu, Hollingsworth [118].

Table 10.1 Applications of polymer-based edible coatings for improving the quality and extending the shelf-life of different fruits and vegetables.

Fruit type	Main polymer	Additives	Results	Reference
Apple	Starch	Glycerol and calcium chloride	Decreased weight loss and improved appearance	[227]
	Chitosan	Stevia extract	Reduced the total yeast and mold contamination	[32]
	Soy-protein	Glycerol and ferulic acid	Controlled weight loss and firmness	[83]
Strawberry	Chitosan and CMC		Maintained volatile aroma and firmness	[228]
Tomato	Alginate and chitosan	*Flourensia cernua* extract	Inhibited microbial growth and reduced weight loss and gas exchange rate	[229]
	Starch	Sorbitol	Retarded ripening and deterioration rate of tomatoes up to 20 days	[230]
	Pullulan	*Laminaria japonica*-derived oligosaccharides	Reduce respiratory intensity, vitamin C loss, weight loss and softening, increased titratable acidity, and the overall likeness of fruit	[231]

(*Continued*)

Table 10.1 Applications of polymer-based edible coatings for improving the quality and extending the shelf-life of different fruits and vegetables. (*Continued*)

Fruit type	Main polymer	Additives	Results	Reference
Banana	Soybean protein isolate	Cinnamaldehyde and zinc oxide	Reduced ripening rate, weight loss, enhanced firmness, titratable acidity, and sensory quality	[232]
	Starch	Sucrose esters	Retarded the senescence, reduced weight loss, and chlorophyll degradation, resulted in a 40% extension in banana shelf-life	[20]

wax [123, 124], beeswax [112], candelilla wax [109], or shellac [125]) have been used for this purpose. Fruit coatings can also be potential carriers of additives that help to preserve or even improve the quality of the product. Antimicrobial compounds (e.g., potassium sorbate) [126], anti-browning agents (e.g., ascorbic acid) [127], nutrients (e.g., vitamin E) [128], texture enhancers (e.g., calcium chloride) [129], probiotics (e.g., *Bifidobacterium lactis*) [130], aroma precursors (e.g., linoleic acid) [131], and flavors and colorants (e.g., apple puree) [132] are different additives incorporated into edible coatings applied on the surface of fruits and vegetables. Table 10.1 shows the components of some of the coatings recently applied on apple, strawberry, tomato, and banana that successfully enhanced their shelf lives. As it is observed, improving appearance, decreasing weight loss, retarding ripening, inhibiting microbial growth, and maintaining the firmness, aroma, and nutritional compounds are the main achievements resulted from coating these fruits and vegetables.

10.5.2 Meat and Meat Products

Meat, due to its high moisture content and the availability of nutritional and functional components on its surface, is susceptible to shrinkage, development of microorganisms (especially *Listeria monocytogenes*), chemical deterioration, enzymatic autolysis, and lipid oxidation which contribute to discoloration and changes in flavor, aroma, and other organoleptic characteristics during processing, handling, and storage [133, 134]. Moreover, in the case of freezing meat for achieving a longer shelf-life, they become susceptible to freeze burn and dehydration during storage. These changes result in substantial economic losses for producers as well as a considerable risk for consumers' health [135, 136]. Edible coatings are able to retard these changes. Volpe, Siano [137] improved the shelf-life of fish fillet with the aid of carrageenan-based coatings (Figure 10.3).

Polymer-based edible coatings act as barriers against the migration of oxygen, moisture and carbon dioxide. They also enable incorporation of functional natural extracts to prolong shelf-life, improve sensory and nutritional quality and retain the market value of meat and its products [138]. Various types of coating formulations have been used for this purpose [135]. They are usually composed of a polymeric basis with different functional additives incorporated into it.

Among the different biopolymers, chitosan is the most commonly used polymer for coating meat, poultry, and seafood. It is used as a polymeric basis as well as antimicrobial and antioxidant agents [139–152]. Other previously used polysaccharides and proteins for coating meat and meat products include carboxymethyl cellulose [133, 153–155], pectin [46, 156], starch [156–158], alginates [155, 156, 159–162], carrageenan [156], tragacanth [163], salep [163], whey protein [91, 134, 164–166], soy protein [77, 167], gelatin [156, 168, 169], sodium caseinate [170], egg albumin [167] and zein [171].

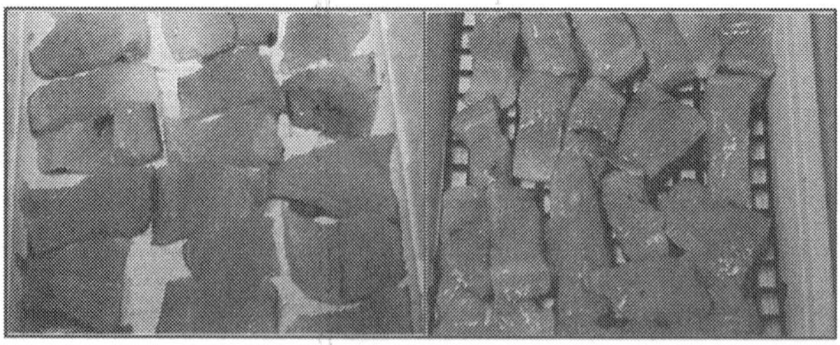

Figure 10.3 Uncoated (left) and coated with carrageenan (right) fish fillets after 7 days of storage at 4°C. The figure has been reproduced with permission from Volpe, Siano [137].

Polysaccharide- and Protein-based coatings, due to their hydrophilic nature, have selective permeability for oxygen and carbon dioxide and resist lipid migration and therefore, are able to extend the shelf-life of meat products by preventing oxidative rancidity, dehydration, shrinkage, and surface browning and improving juiciness and texture of the meat product. However, they usually have insufficient moisture barrier effects and low mechanical resistance, and therefore, might become fragile under mechanical stresses [172].

Lipids are efficient barriers against moisture migration. The main lipids used for coating meat products are natural waxes (such as paraffin wax and beeswax), acetoglycerides, resins, fatty acids, and both vegetable (e.g., sunflower oil [46]) and animal oils (e.g., krill oil [173]). They are either used as separate coatings or incorporated into protein- and polysaccharide-based coatings to make composite coatings [174]. Lipid incorporation can improve hydrophobicity, cohesiveness, and flexibility of the coating. This can help to extend the freshness and the stability of color, aroma and tenderness as well as the microbiological quality of fresh and processed meat products. However, it should be noted that these types of coatings may confer an undesirable taste to meat products [172].

There are also other additives that have been previously incorporated into edible coatings used for meat and meat products. They include organic acids (e.g., gallic [143], malic [164], benzoic, lactic, propionic, acetic and sorbic acids), antioxidants (e.g., tea polyphenols and vitamin C [160], bamboo vinegar [152]) polypeptides, fatty acid esters, and bacteriocins (e.g., lactoperoxidase [175], lauric arginate ester [144], peroxidase, lysozyme [165, 166], nisin [140, 144, 159, 161, 164], sodium lactate [156] and sodium diacetate [139]); plant essential oils (e.g., essential oils extracted from oregano [134, 151, 162], rosemary [154, 159], clove [134, 147], ginger [170], *Satureja khuzestanica* [150], *Thymus vulgaris* [176], *Satureja thymbra* [153], *Origanum virens* [91], *Ziziphora clinopodioides* [133, 155], *Mentha spicata* [133], *Zataria multiflora* [142, 149], cinnamon [46, 159, 161], nettle [177], garlic [46] and lemongrass); plant extracts (e.g., sour lemon peel extract [157], grape seed extract [145, 164], apple peel extract [155]), flavoring and coloring agents (e.g., curcumin [46], pomegranate juice [142]), nanoparticles (e.g., zinc oxide nanoparticle [155]) and probiotic bacteria are the most common.

Table 10.2 shows some of the recent researches carried out on coating meat and meat products. As it is observed, retarding moisture loss, inhibiting microbial growth and lipid oxidation, maintaining the color during storage and thus improving the shelf-life are the main achievements resulted from coating meat and meat products.

Table 10.2 Applications of polymer-based edible coatings for improving the quality and extending the shelf-life of meat and different meat products.

Product	Coating polymer	Additives	Results	Reference
Chicken fillets	Whey Protein Isolate	Oregano or clove essential oils	Inhibited microbial contamination and enhanced the storage time from 6 to 13 days	[134]
	CMC	*Ziziphora clinopodioides* and *Mentha spicata* essential oils	Reduced the number of test pathogens and increased the shelf-life	[133]
	Chitosan	Pomegranate juice and *Zataria multiflora* essential oil	Decreased total viable counts of microorganisms, decreased peroxide value, Thiobarbituric acid reactive substance values and protein oxidation, improved sensory scores	[142]
Pork	Chitosan	Gallic acid	Inhibited lipid oxidation and myoglobin oxidation	[143]
	Chitosan	Nisin + gallic acid	Inhibited microbial growth, decreased protein and lipid oxidation, helped to maintain color	[140]
Beef	Chitosan	*Zataria multiflora* essential oil and sumac extract	Inhibited lipid oxidation and microbial growth	[149]

(Continued)

Table 10.2 Applications of polymer-based edible coatings for improving the quality and extending the shelf-life of meat and different meat products. (*Continued*)

Product	Coating polymer	Additives	Results	Reference
Fish	CMC	*Satureja thymbra EO*	Controlled oxidative deterioration, extended the shelf-life up to 35%	[153]
	Sodium alginate	Vitamin C and tea polyphenols	Retarded the decay and water loss, reduced chemical spoilage and increased the overall sensory quality	[160]
Turkey Frankfurter	Whey protein isolate	Grape seed extract, nisin, malic acid, and ethylenediamine tetraacetic acid	Inhibited microbial growth	[164]
Cooked pork sausages	Chitosan	Clove essential oil	Retarded microbial growth and sensory changes and significantly enhanced shelf-life	[147]
Cold-smoked fish	Whey protein isolate	Lysozyme	Retarded microbial growth	[165, 166]
Cooked pork chops	Chitosan	Bamboo vinegar	Retarded lipid oxidation maintained the color and improved shelf-life	[152]
Portuguese sausages	Whey protein concentrate	*Origanum virens* Essential oils	Inhibited the total microbial load, prevented lipid oxidation and color fading, extended the shelf-life	[91]

10.5.3 Bakery Products

The main aims of coating bakery products are to prevent microbial growth and to slow down moisture migration across their surface and thus to maintain their quality during storage [3]. Edible coatings can prevent moisture absorption in bakery products with low moisture content such as biscuits and crackers and also reduce the rate of moisture loss in bakery products with higher moisture content such as bread and cake.

Bartkiene, Bartkevics [178] improved the shelf-life of wheat bread by coating it with a starch-based coating containing lactobacilli and cranberry extract as natural antifungal agents. Saraiva, Naponucena [179] coated cake with starch-based coatings incorporated with citric acid and potassium sorbate as antimicrobial agents. They reported that coating maintained the quality parameters including their moisture, hardness, and elasticity and inhibited the growth of yeasts and molds during storage.

10.5.4 Cheese

During the storage of cheese, contamination by microorganisms and moisture loss may increase the hardness of the final product and undesired organoleptic properties [180]. Coatings can act as a barrier against carbon dioxide, oxygen, and moisture loss preventing dehydration of cheese during the storage and controlling the maturation rate. They can also act as carriers of flavoring, nutritional agents, antimicrobial components (e.g., nisin, natamycin, lysozyme, essential oils, and organic acids), colorants, or sweetener agents [181, 182].

Proteins (e.g., zein [183] and whey protein [184]) and polysaccharides (e.g., chitosan [181, 185], alginate [186] and agar [187]) have been widely used for coating cheese. Water solubility is the main advantage of these types of polymers. Only a few studies showed the potential use of waxes for coating cheese. Previous researches show that beeswax is the one already applied to the cheese with success. It has been reported that hot-melting methods or organic solvents should be used during applying wax-based coatings to cheese and this has limited their application for cheese coating [182].

10.5.5 Nuts

Nuts are a generic term for almonds, peanuts, walnuts, pistachio, ginkgo nuts, pine nuts, etc. [76]. They usually add a desirable crunch and a specific flavor to many foods or might be consumed individually as snacks.

Nuts are rich in polyunsaturated fatty acids which can decrease the risk of stroke and cardiovascular disease in humans. However, due to the high content of polyunsaturated lipids, they are susceptible to lipid oxidation and the formation of off-flavors (oxidative rancidity) [188]. The other most common forms of spoilage of nuts are lipid migration in food systems, fungal growth, and absorption of foreign odors and moisture contributing to texture changes and loss quality [3].

Edible coatings with hydrophobic character can prevent moisture absorption of nuts and help to maintain their crispiness during the storage. On the other hand, coatings with hydrophilic character can act as oxygen barriers which limit oil oxidation and fungal growth and also resist against lipid migration from nut surface. Chitosan [28, 189–191], cassava starch [192], sodium alginate [193], whey protein isolate [88, 188, 194], and soy protein isolate [76, 192] have been recently used for this purpose. They might also contain antioxidant (e.g., *Laurus nobilis*) and antimicrobial agents to enhance the shelf-life of nuts [191]. However, coating nuts, which have hydrophobic surfaces, with hydrophilic coatings is quite difficult due to the differences in the chemical nature of the two compounds. During this process, the dewetting of the coating solution by the nut surface occurs which results in cracking and shrinking of the coating material during drying. Previous researches have shown that increasing the roughness of the nuts surface and addition of surfactants to the coating solution are effective in facilitating the wettability and coverage of nuts with the coating materials [188].

10.5.6 Eggs

Fresh eggs are excellent protein sources. Nevertheless, they are highly perishable and can lose their carbon dioxide and moisture through microscopic pores in their shell contributing to weight loss, interior quality deterioration, and major economic losses. In addition, different microorganisms penetrate through the pores and contaminate the internal content [195].

Edible coatings can maintain egg quality by preventing microorganism, moisture, and gas migration through the eggshell and enhancing shell strength [115]. Coatings based on either polysaccharides (e.g., chitosan [196–198], CMC [199], MC, and HPMC [200]), proteins (e.g., whey protein [71, 89, 201], zein [71], soy protein and wheat gluten [199]), or waxes (e.g., Shellac [202]) have been used for this purpose. However, it has been reported that among the different types of coatings, polysaccharide-based coatings are preferred for coating eggshells due to their greater transparency, lower thickness, and higher flexibility [115]. Hydrophobic additives could be added to decrease the moisture loss and enhance the wettability of the coating solution on the egg surface. Suppakul, Jutakorn [200] used MC and HPMC-based coatings containing polyethylene glycol, stearic acid, and palmitic acid for improving the shelf-life of eggs. Wardy, Torrico [203] preserved the internal quality of eggs using chitosan-soybean oil emulsions.

10.5.7 Fried Food

In recent years, there has been an increasing demand for consumption of fried food. However, after frying, these products may absorb a large amount of fat which contributes to the high risk of cardiovascular diseases, obesity, and certain cancers [204].

During the frying process, in the first 60 s, moisture molecules rapidly evaporate from the surface of the product. Then, the water migrates from the center of the food to its surface. This results in forming empty voids on the food surface that can be easily filled with fat molecules contributing to an oil concentrated crust (Figure 10.4) [205].

Edible coatings make compact and protective layers on the food surface with fewer pores, which prevent the migration of moisture molecules from the surface of the food. Therefore, the moisture molecules on the food surface cannot be replaced by the oil [206]. If the coating material has gelling abilities, during heating, a thermo-revisable gel forms around the food, which captures water and stops the migration of water out and oil inside the product contributing to a moist, tender, and soft core with a uniform

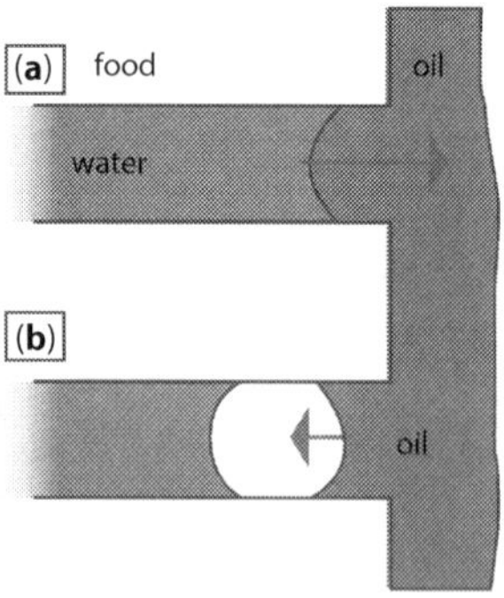

Figure 10.4 Water migration from the center of the food to its surface (a) and fat migration to the food crust (b). The figure has been reproduced with permission from Mellema [226].

crispy crust [207]. Besides, coatings can decrease the oil absorption of the food surface by decreasing the surface hydrophobicity. Therefore, it is preferred to choose a hydrophilic coating material such as proteins (e.g., gelatin, egg white, sodium caseinate, wheat gluten, soy protein, and whey protein) or polysaccharides (corn starch, cellulose derivate, carrageenan, gums, and pectin) for this purpose. It has been mentioned that coatings with high viscosity (such as gums) are much easily applicable to the food surface in comparison with those with low viscosities (such as proteins). These coatings can either act as a lipid barrier after forming thin films around the food or by their gelation properties during frying. Moreover, functional components can be added to the coating materials which can improve the nutritional value of the fried food and also control the transfer of fat and moisture across the surface of fried food [207].

Fry Shield™-calcium pectinate is one of the commercially available coatings used for reducing oil uptake of food products such as frying potatoes, fish, and vegetables. The spraying and dipping techniques are usually used for this purpose [207].

10.6 Microencapsulation of Bioactive Components in Food Systems

Bioactive components in food including proteins (e.g., enzymes or peptides), lipids (e.g., conjugated linoleic acid, fat-soluble vitamins, and omega-3 fatty acids), minerals (e.g., Fe, Se, and Zn), or live cells (e.g., probiotics) play significant roles in our well-being. There are also various molecules that determine organoleptic properties of food such as flavor and color. Most of these components cannot be incorporated directly into the food systems due to their sensitivity to processes or incompatibility with other ingredients. For example, probiotic cells, as functional food components, usually lose their viability and benefits during common food processes. Fat-soluble vitamins such as vitamin A or D are other examples of bioactive components that require encapsulation. They cannot be directly added to a water-based formulated juice due to the phase separation of immiscible liquids.

Microencapsulation is defined as a process in which a semipermeable polymeric membrane or matrix encloses sensitive materials leading to the formation of special microparticles or microcapsules that can release their content in a particular location under a controlled manner [208, 209]. This process is introduced as an approach to meet the nutritional or technological challenges in the production of formulated foods.

The main objectives of encapsulating bioactive components in food systems include:

- Protecting sensitive materials against evaporation, unwanted chemical interactions, and immigration of some molecules in food;
- Improving stability and activity of bioactive components during food processing, storage, and consumption;
- Masking off-flavors;
- Facilitating the handling and utilization of food ingredients;
- Delivering active component to their target site in the body, e.g., mouth, stomach, and colon, and releasing them controllably.

On the other hand, microencapsulation leads to an increase in complexity and cost of food chain production. In addition, microcapsules may have undesirable impacts on the food appearance and texture leading to a negative effect on its marketability. Although this process has not been prevalently used by food producers, current demands for tastier and healthier foods made microencapsulation an interesting topic to academics and industry owners.

10.6.1 Terminology

Key terms of microencapsulation and delivery systems for food systems include:

Delivery system: All mechanisms and processes engaged with microencapsulation, delivery, and release of active components [210].

Active component: A sensitive food component that is encapsulated in a polymeric microparticle or microcapsule. This component could be a bioactive molecule, live cell, or volatile compound. Molecular characteristics (e.g., molecular weight, spatial structure, density, charge, polarity, and physical state) and sensitivity of active component to ambient condition determine the type of wall materials and the technique being used for microencapsulation. The active component is also called "active agent", "fill", "core material", "internal phase", or "payload phase" [210, 211].

Wall material: A single or a combination of food-grade components including polysaccharides, proteins, fats, or surfactants that enclose active component. Wall material is also called "carrier material", "shell", "matrix", "membrane", "encapsulant", or "external phase" in different sources. Wall material may have a homogenous or heterogeneous structure depending on its constituents and fabrication method [210, 211].

Delivery process: The process of carrying an encapsulated active component into a specific site in the body like mouth, nose, stomach, small, or large intestine.

Controlled release: The process of releasing active components in the proximity of their target site with a specific concentration-time profile. Common terms used in the context of controlled release are:

- Targeted release: Release of the active component in a specific location within the body such as mouth, stomach, or colon;
- Triggered release: Release of active component in response to specific mechanical stress like chewing or environmental stimulators like alternation in pH, ionic strength, temperature, or enzymatic activity;
- Burst release: Releasing a major part of active components over a short period;
- Sustained-release: Gradual release of active components during a long period under a relatively constant rate [210].

10.6.2 Structure of Microcapsules

The structure of microcapsules can be classified into two main groups including mono-core and poly-core structures (Figure 10.5). In the mono-core structure, the active component is usually coated by a single- or multi-layer wall. However, in the poly-core structure, several active components are confined in a matrix or polymeric network within a single microparticle. The poly-core structure is more resistant to mechanical stresses than the mono-core type. However, the active component may be located on the surface of the particle and get exposed to the surrounding environment. In the poly-core structure, the active component distributes in the

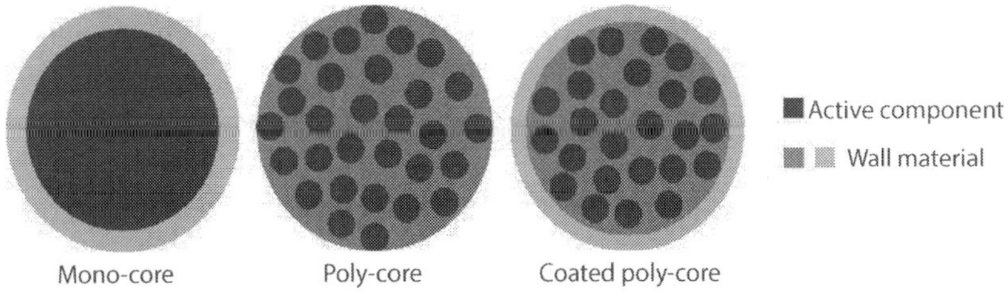

Figure 10.5 Different structures of microcapsules.

wall material network in the form of small emulsion droplets, live cells, and discrete or aggregated active molecules. A combination of these two structures leads to a coated poly-core structure, which has benefits of both.

10.6.3 Materials for Microencapsulation

The wall material or matrix should be able to protect the active component against chemical, physical, or biological destructions. Besides they should prevent the release of the active component before arriving at their target site. Numerous synthetic and natural polymers are prevalently used for medical and pharmaceutical purposes in delivery systems. However, wall materials for food applications should be GRAS and prevent releasing active component during processing, transportation, and storage of food. Thus, many studies are currently being carried out by researchers to use various food-grade biopolymers in targeted delivery systems. Among all, proteins and polysaccharides are the most common biopolymers because they are biocompatible, biodegradable, and able to release their content controllably in the presence of environmental stimulators [212, 213]. In the food industry, biopolymers have been used as fat-replacers in low-calorie products for many years, because they resemble organoleptic and textural characteristics of fat droplets in food matrices. Recently, their applications in microencapsulation and targeted delivery systems have attracted much attention from food researchers.

The most diverse biopolymers being used as wall material for microencapsulation of bioactive compounds and live cells in food systems are polysaccharides. Starch and its derivatives (e.g., amylose, amylopectin, dextrins, maltodextrins, and polydextrose), cellulose and its derivatives (e.g., CMC and HPC), plant exudates (e.g., galactomannans, pectins, tragacanth gum, and soluble soybean polysaccharides), seaweed gums (e.g., alginates and carrageenans), and animal or microbial polysaccharides (e.g., chitosan, xanthan, and dextran) have been used for microencapsulation in food systems. Proteins are also employed for microencapsulation in food systems. Whey protein concentrate, whey protein isolate, egg albumin, collagen, gelatin, zein, and casein have been widely studied as proteinous wall materials. Acacia gum, which is a complex of glycoproteins and polysaccharides, has been also used for microencapsulation of food flavors and live cells in food systems [214]. Lipids such as fatty acids, fatty alcohols, waxes (e.g., beeswax or carnauba wax), glycerides, and phospholipids are also suitable for microencapsulation of lipophilic active components in food.

10.6.4 Microencapsulation Techniques

One of the most crucial steps in preparing an efficient delivery system is the microencapsulation process, which should be chosen and performed regarding functional properties and molecular characteristics of the active component and wall material. Up to now, various strategies have been used to microencapsulate bioactive materials for food applications, including aerosol-assisted techniques (e.g., spray-drying, spray-cooling, and electrospraying), freeze-drying, emulsification techniques, and extrusion. In this section, a brief overview of the most common microencapsulation techniques is presented.

10.6.4.1 Spray Drying

Spray drying is the oldest microencapsulation technique in the food industry so that 80%–90% of food products containing encapsulated components are currently prepared by this method [208]. In this technique, the active component is thoroughly blended with the wall material, sprayed into a chamber, and exposed to hot airflow leading to sudden evaporation of the solvent. Then, the dried microparticles containing the active component are collected in a cyclone [215]. Spray-drying has many advantages over other techniques. It is a low-cost, flexible, simple, and continuous process and is able to fabricate microparticles with lower than 40 μm diameter. However, the application of high temperatures in the spray-drying method may damage sensitive components [216]. Moreover, the internal condition of the drying chamber is not uniform and also controlling the size of microparticles is relatively difficult. It should be noted that some researchers believe spray-drying is actually a technique of material immobilization, not an encapsulation method. Because, not only some active component may locate on the surface of microparticles and get exposed to the surrounding, but also they may release instantly in an environment which contains their solvent [208, 215].

10.6.4.2 Spray Cooling

Spray-cooling or spray-chilling is basically similar to the spray-drying in which a mixture of the active component and wall material is sprayed into a closed chamber. However, in this technique, a very cold airflow inside the chamber solidifies droplets into microparticles. Spray-cooling, as one of the cheapest encapsulation methods, is usually employed for lipid matrices.

Some disadvantages of this method are the low encapsulation efficiency of core material in microparticles and gradual expulsion of active component from the matrix (due to crystalline structure and polymorphic rearrangement of lipids) [216]. Lipids with high melting points are often used as wall material in this method. Spray-cooling has been previously used for microencapsulation of enzymes, flavors, minerals, proteins, and probiotics [215, 217].

10.6.4.3 Freeze-Drying

Freeze-drying has been used to produce probiotic powders for many years. However, its application for microencapsulation of active materials is relatively new in food products [216]. In this process, active component and wall material are dispersed in water, frozen at –40°C to –90°C and dried under vacuum through the phase transition of ice to vapor. The final product is obtained in the form of a highly porous brittle cake, which may be ground into microparticles by mechanical milling [211]. Since the drying process is carried out at a very low temperature, the functional properties of the active component are preserved better than the spray-drying method. However, formation and growth of ice crystals in the freezing step may injure some active components such as live cells and osmotic stress during drying may lead to loss of viable cells. To diminish the deleterious effects of the freezing step, cryoprotectant compounds such as sugars, polyols, amino acids, carbohydrate polymers, and proteins may be added to the initial mixture. Cryoprotectants are usually accumulated inside cells and modulate osmotic pressure on both sides of the cell membrane [216]. The main disadvantages of the freezing-drying technique are high energy consumption and long processing time. The final product usually has a very porous structure. Therefore, the protective effect of the matrix sharply decreases if the long-term release of the active component is required [208].

10.6.4.4 Emulsification

Emulsification is the process of dispersing a fluid into an incompatible or immiscible second fluid. It can be used to produce microcapsules by adding the active component into the first fluid [215]. In this method, the dispersed phase (a mixture of wall material and active component) is added to a high volume of the continuous phase (usually an edible oil) and is homogenized to form a water-in-oil emulsion. Upon the formation of emulsion, the water-soluble wall material becomes insoluble through the formation of crosslinks. Consequently, solid microcapsules are formed in the continuous phase that is separated by filtration. The size of microcapsules is usually adjusted by controlling homogenization speed, which can produce particles with 25 μm to 2 mm diameter [216, 218]. Usually, by increasing the homogenization speed, the encapsulation efficiency of the active component decreases to some extent. Smaller particles can be formed by adding emulsifiers to reduce the interfacial tension of the two phases. Emulsification can be easily used for scale-up production of microcapsules. However, the complexity of the process and impossibility of continuous production have put some limitations on its application. Since expensive edible oils such as soybean, sunflower, rapeseed, or corn oils are used as a continuous phase for food applications, the cost of the process is high. Moreover, the residue of oil in the microcapsules limits the application of this method for low-calorie foods [219, 220].

10.6.4.5 Extrusion

Extrusion is the most widely used method for particle formation by hydrocolloids. In this method, the active component is mixed with a hydrocolloid solution, which is actually the wall material. The mixture is slowly pumped into a narrow nozzle that forms pendant droplets at the lower end of the nozzle. These droplets fall into a hardening bath one by one and instantly form spherical particles. Hardening bath may only contain ice water for lipid coatings. However, it often contains multivalent ions (e.g., Ca^{2+} and Mg^{2+}) or enzymes (e.g., transglutaminase) that crosslink polymer chains and make hydrogel networks. The shape and size of the particles depend on the inner diameter of the nozzle, surface tension of the hydrocolloid solution, and distance between the nozzle and the hardening bath [220]. In contrast to spray-drying or spray-cooling, there is more control over droplet formation in the extrusion technique. In addition, uniform particles can be formed by vibrating the nozzle. Moreover, by using a coaxial needle (Figure 10.6) as the extrusion nozzle, it is possible to produce multi-layer particles with heterogeneous structures like coated poly-core structure in a single-stage process. These nozzles provide an opportunity to use incompatible materials without direct contact with each other in the structure of a particle; for example, encapsulation of live cells in wall material containing organic solvents [216].

Extrusion is a simple and low-cost method. Additionally, the gentle formation of particles guarantees high efficiency of active component encapsulation. The main disadvantages of this method are the large-size (2–5 mm) and the slow-rate of particle formation, which makes it difficult to increase the scale of production [220]. Some modifications including the application of multi-nozzle systems, rotating disc atomizers, and cutting jet technique have been performed to increase particle production. Among all, jet-cutter

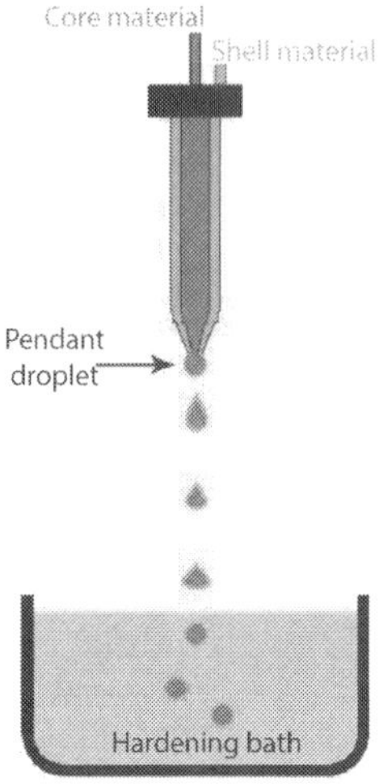

Figure 10.6 Extrusion process using a coaxial nozzle.

technology has the most potential to produce microparticles on the industrial scale [208]. The extrusion method can be carried out under aerobic or anaerobic conditions. The latter is a major advantage for the encapsulation of obligatory anaerobic microorganisms or materials sensitive to oxidation. In this case, the extrusion apparatus is placed in a sterile isolated chamber and oxygen in the chamber is replaced with nitrogen. Up to now, extrusion has been used for microencapsulation of aromas, enzymes, proteins, and living cells.

10.6.4.6 Electro-Hydrodynamic Atomization

Elcctro-hydrodynamic atomization or electrospray is a simple technique conceptually in which a conductive polymeric solution is sprayed into micro- or nano-droplets by strong electrical forces. First, the polymeric solution containing the active component is pumped into a narrow tube, which is connected to a charged electrode. As a pendant droplet is formed, electrical repulsions start competing with viscoelastic force and surface tension of the solution. This leads to the formation of the Taylor cone structure (Figure 10.7). Once electrical forces dominate, the solution is atomized into ultra-small charged droplets. These droplets are forcibly pulled toward a collector, which is connected to the opposite electrode. The collector can be a solid surface or a liquid bath. In the first case, dry microparticles are formed instantly, which is called dry-electrospray; while in the latter case, hydrogel microparticles are usually obtained through the wet-electrospray process (Figure 10.7) [214, 221]. Focusing and controlling

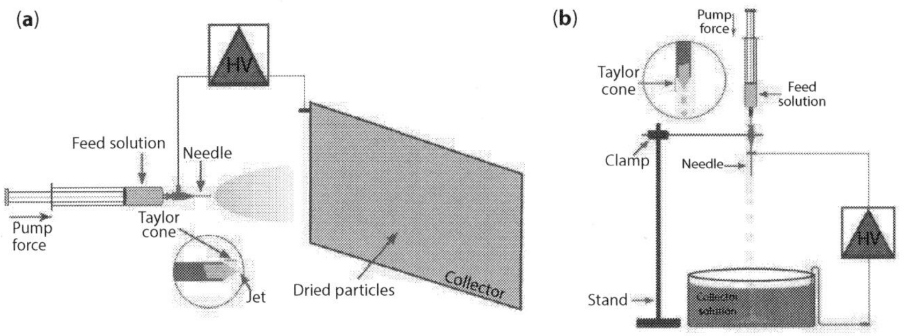

Figure 10.7 Experimental setup for electro-hydrodynamic atomization. Dry-electrospray in horizontal setup (a) and wet-electrospray in vertical setup (b).

the motion of flying charged droplets by external electric fields are easier than uncharged droplets formed by conventional spray systems such as spray drier. Moreover, microparticles produced by electrospray are almost uniform and their deposition efficiency is higher than uncharged ones. Since the process is usually carried out in the ambient condition, it does not involve high temperature, which makes it a suitable choice for microencapsulation of heat-sensitive active components such as live cells and antioxidants [222]. Charged microparticles are adequately scattered while flying toward the collector due to electrical repulsions, which prevents agglomeration of the resultant microparticles [223]. Also, it is not necessary to make emulsion before electrospraying and micro-droplets can be formed from fairly viscose solutions [224].

Low productivity of electro-hydrodynamic atomization and blocking the feeding needle are considered as two important drawbacks of the method. The scale of the production can be increased considerably by making some modifications in the injection device, for example, application of porous cylindrical injection devices, multi-nozzle sprayers, or multiple spikes needleless systems. Moreover, blocking the feeding nozzle can be prevented by using needleless systems [225]. Totally, electro-hydrodynamic atomization has a flexible experimental setup, which can be simply modified based on the users' demand.

10.7 Conclusions

In this chapter, coating and microencapsulation of food components were introduced as effective approaches for the production of healthier foods with a longer shelf-life and better organoleptic properties. A successful coating or microencapsulation process strongly depends on the materials being used for covering the food object. However, the technique used for this purpose plays a significant role. Thus, more studies are required to be carried out for finding new functional edible biopolymers or combination of biopolymers for coating or microencapsulation of food components. Moreover, some innovations are demanded to improve the efficiency of coating/microencapsulation techniques to reduce the cost of process and productivity.

References

1. Olivas, G.I. and Barbosa-Cánovas, G., Edible films and coatings for fruits and vegetables, in: *Edible films and coatings for food applications*, pp. 211–244, Springer, New York, US, 2009.
2. Velickova, E. *et al.*, Impact of chitosan-beeswax edible coatings on the quality of fresh strawberries (Fragaria ananassa cv Camarosa) under commercial storage conditions. *LWT-Food Sci. Technol.*, 52, 2, 80–92, 2013.
3. Fernández-Pan, I., Ignacio, J., Caballero, M., Biopolymers for edible films and coatings in food applications, in: *Biopolymers–New materials for sustainable films and coatings*, pp. 233–254, 2011.
4. Saleh, K. and Guigon, P., Chapter 7 Coating and encapsulation processes in powder technology, in: *Handbook of Powder Technology*, A.D. Salman, M.J. Hounslow, J.P.K. Seville (Eds.), pp. 323–375, Elsevier Science B.V, New York, US, 2007.
5. Pavlath, A.E. and Orts, W., Edible films and coatings: Why, what, and how?, in: *Edible films and coatings for food applications*, pp. 1–23, Springer, New York, US, 2009.
6. McHugh, T.H. and Krochta, J.M., Water vapor permeability properties of edible whey protein-lipid emulsion films. *J. Am. Oil Chem. Soc.*, 71, 3, 307–312, 1994.
7. Han, J.H., *Innovations in food packaging*, Elsevier, London, UK, 2005.
8. Falguera, V. *et al.*, Edible films and coatings: Structures, active functions and trends in their use. *Trends Food Sci. Technol.*, 22, 6, 292–303, 2011.
9. Han, J.H. and Aristippos, G., 15 - Edible films and coatings: A review, in: *Innovations in Food Packaging*, J.H. Han (Ed.), pp. 239–262, Academic Press, London, 2005.
10. Tahir, H.E. *et al.*, Recent developments in gum edible coating applications for fruits and vegetables preservation: A review. *Carbohydr. Polym.*, 224, 115141, 2019.
11. Javanmard, M., Shelf life of apples coated with whey protein concentrate-gellan gum edible coatings. *J. Food Biosci. Technol.*, 1, 1, 56–62, 2012.
12. Klinjapo, R. and Krasaekoopt, W., Chapter 14 - Microencapsulation of Color and Flavor in Confectionery Products, in: *Natural and Artificial Flavoring*

Agents and Food Dyes, A.M. Grumezescu and A.M. Holban (Eds.), pp. 457–494, Academic Press, London, UK, 2018.

13. Mostafavi, F.S., The surface characteristics of biopolymer-coated tomato and cucumber epicarps: Effect of guar, Persian and tragacanth gums. *J. Food Meas. Charact.*, 13, 1, 840–847, 2019.
14. Ananey-Obiri, D. *et al.*, Application of protein-based edible coatings for fat uptake reduction in deep-fat fried foods with an emphasis on muscle food proteins. *Trends Food Sci. Technol.*, 80, 167–174, 2018.
15. Baldwin, E.A., Hagenmaier, R., Bai, J., *Edible coatings and films to improve food quality*, CRC Press, Florida, US, 2011.
16. Dang, K.T., Singh, Z., Swinny, E.E., Edible coatings influence fruit ripening, quality, and aroma biosynthesis in mango fruit. *J. Agric. Food Chem.*, 56, 4, 1361–1370, 2008.
17. Mostafavi, F.S. *et al.*, Preparation and characterization of tragacanth–locust bean gum edible blend films. *Carbohydr. Polym.*, 139, 20–27, 2016.
18. Mehdizadeh, T. and Langroodi, A.M., Chitosan coatings incorporated with propolis extract and Zataria multiflora Boiss oil for active packaging of chicken breast meat. *Int. J. Biol. Macromol.*, 141, 401–409, 2019.
19. Martin Xavier, K.A. *et al.*, Functionality of chitosan in batter formulations for coating of fish sticks: Effect on physicochemical quality. *Carbohydr. Polym.*, 169, 433–440, 2017.
20. Thakur, R. *et al.*, A starch edible surface coating delays banana fruit ripening. *LWT*, 100, 341–347, 2019.
21. Fathi, M., Martín, Á., McClements, D.J., Nanoencapsulation of food ingredients using carbohydrate based delivery systems. *Trends Food Sci. Technol.*, 39, 1, 18–39, 2014.
22. Zhao, G. *et al.*, Biodegradable and transparent cellulose film prepared eco-friendly from durian rind for packaging application. *Food Packag. Shelf life*, 21, 100345, 2019.
23. Nadery Dehsheikh, F. and Taghian Dinani, S., Coating pretreatment of banana slices using carboxymethyl cellulose in an ultrasonic system before convective drying. *Ultrason. Sonochem.*, 52, 401–413, 2019.
24. Tesfay, S.Z. and Magwaza, L.S., Evaluating the efficacy of moringa leaf extract, chitosan and carboxymethyl cellulose as edible coatings for enhancing quality and extending postharvest life of avocado (*Persea americana Mill.*) fruit. *Food Packag. Shelf life*, 11, 40–48, 2017.
25. Ruan, C. *et al.*, Effect of sodium alginate and carboxymethyl cellulose edible coating with epigallocatechin gallate on quality and shelf life of fresh pork. *Int. J. Biol. Macromol.*, 141, 178–184, 2019.
26. Nouri Ala, M.A. and Shahbazi, Y., The effects of novel bioactive carboxymethyl cellulose coatings on food-borne pathogenic bacteria and shelf life extension of fresh and sauced chicken breast fillets. *LWT*, 111, 602–611, 2019.
27. Ngamekaue, N. and Chitprasert, P., Effects of beeswax-carboxymethyl cellulose composite coating on shelf life stability and intestinal delivery of holy

basil essential oil-loaded gelatin microcapsules. *Int. J. Biol. Macromol.*, 135, 1088–1097, 2019.
28. Sabaghi, M. *et al.*, Active edible coating from chitosan incorporating green tea extract as an antioxidant and antifungal on fresh walnut kernel. *Postharvest Biol. Technol.*, 110, 224–228, 2015.
29. Dong, C. *et al.*, Effects of edible chitosan coating on Harbin red sausage storage stability at room temperature. *Meat Sci.*, 159, 107919, 2019.
30. Ebadi, Z. *et al.*, The shelf life extension of refrigerated Nemipterus japonicus fillets by chitosan coating incorporated with propolis extract. *Int. J. Biol. Macromol.*, 139, 94–102, 2019.
31. Jiang, Y. *et al.*, Electrostatic spraying of chitosan coating with different deacetylation degree for strawberry preservation. *Int. J. Biol. Macromol.*, 139, 1232–1238, 2019.
32. Karagöz, Ş. and Demirdöven, A., Effect of chitosan coatings with and without Stevia rebaudiana and modified atmosphere packaging on quality of cold stored fresh-cut apples. *LWT*, 108, 332–337, 2019.
33. Zhang, W. *et al.*, Different molecular weights chitosan coatings delay the senescence of postharvest nectarine fruit in relation to changes of redox state and respiratory pathway metabolism. *Food Chem.*, 289, 160–168, 2019.
34. Ahmed, F. *et al.*, Recent progress in biomedical applications of chitosan and its nanocomposites in aquaculture: A review. *Res. Vet. Sci.*, 126, 68–82, 2019.
35. Safdar, R. *et al.*, Potential of Chitosan and its derivatives for controlled drug release applications – A review. *J. Drug Delivery Sci. Technol.*, 49, 642–659, 2019.
36. Thakur, R. *et al.*, Starch-based films: Major factors affecting their properties. *Int. J. Biol. Macromol.*, 132, 1079–1089, 2019.
37. González-Forte, L.d.S., Amalvy, J.I., Bertola, N., Corn starch-based coating enriched with natamycin as an active compound to control mold contamination on semi-hard cheese during ripening. *Heliyon*, 5, 6, e01957, 2019.
38. Patel, C. and Panigrahi, J., Starch glucose coating-induced postharvest shelf life extension of cucumber. *Food Chem.*, 288, 208–214, 2019.
39. Basiak, E. *et al.*, Dynamic behaviour of starch-based coatings on fruit surfaces. *Postharvest Biol. Technol.*, 147, 166–173, 2019.
40. Sapper, M., Bonet, M., Chiralt, A., Wettability of starch-gellan coatings on fruits, as affected by the incorporation of essential oil and/or surfactants. *LWT*, 116, 108574, 2019.
41. Eom, H. *et al.*, Development of a starch/gum-based edible coating for rice cakes to retard retrogradation during storage. *LWT*, 97, 516–522, 2018.
42. Martínez-Ortiz, M.A. *et al.*, Effect of using microencapsulated ascorbic acid in coatings based on resistant starch chayotextle on the quality of guava fruit. *Sci. Hortic.*, 256, 108604, 2019.
43. Tavassoli-Kafrani, E., Shekarchizadeh, H., Masoudpour-Behabadi, M., Development of edible films and coatings from alginates and carrageenans. *Carbohydr. Polym.*, 137, 360–374, 2016.

44. Dehghani, S., Hosseini, S.V., Regenstein, J.M., Edible films and coatings in seafood preservation: A review. *Food Chem.*, 240, 505–513, 2018.
45. Espitia, P.J.P. *et al.*, Edible films from pectin: Physical-mechanical and antimicrobial properties-A review. *Food Hydrocolloids*, 35, 287–296, 2014.
46. Abdou, E.S., Galhoum, G.F., Mohamed, E.N., Curcumin loaded nanoemulsions/pectin coatings for refrigerated chicken fillets. *Food Hydrocolloids*, 83, 445–453, 2018.
47. Nisar, T. *et al.*, Physicochemical responses and microbiological changes of bream (*Megalobrama ambycephala*) to pectin based coatings enriched with clove essential oil during refrigeration. *Int. J. Biol. Macromol.*, 124, 1156–1166, 2019.
48. Santagata, G. *et al.*, Pectin-honey coating as novel dehydrating bioactive agent for cut fruit: Enhancement of the functional properties of coated dried fruits. *Food Chem.*, 258, 104–110, 2018.
49. Sucheta, *et al.*, Composite edible coatings from commercial pectin, corn flour and beetroot powder minimize post-harvest decay, reduces ripening and improves sensory liking of tomatoes. *Int. J. Biol. Macromol.*, 133, 284–293, 2019.
50. Nguyen, S. *et al.*, Studies on pectin coating of liposomes for drug delivery. *Colloids Surf. B: Biointerfaces*, 88, 2, 664–673, 2011.
51. Niu, Z. *et al.*, Protection of native lactoferrin under gastric conditions through complexation with pectin and chitosan. *Food Hydrocolloids*, 93, 120–130, 2019.
52. Rehman, A. *et al.*, Pectin polymers as wall materials for the nano-encapsulation of bioactive compounds. *Trends Food Sci. Technol.*, 90, 35–46, 2019.
53. Rakshit, M. and Ramalingam, C., Gum acacia coating with garlic and cinnamon as an alternate, natural preservative for meat and fish. *Afr. J. Biotechnol.*, 12, 4, 406–413, 2013.
54. Ali, A. *et al.*, Gum arabic as a novel edible coating for enhancing shelf life and improving postharvest quality of tomato (*Solanum lycopersicum* L.) fruit. *Postharvest Biol. Technol.*, 58, 1, 42–47, 2010.
55. Sarpong, F. *et al.*, Comparative study of enzymes inactivation and browning pigmentation of apple (*Malus domestica*) slices by selected gums during low temperature storage. *J. Food Biochem.*, 42, 6, e12681, 2018.
56. Wu, S. and Chen, J., Using pullulan-based edible coatings to extend shelf life of fresh-cut 'Fuji' apples. *Int. J. Biol. Macromol.*, 55, 254–257, 2013.
57. Sharma, S. and Rao, T.R., Xanthan gum based edible coating enriched with cinnamic acid prevents browning and extends the shelf life of fresh-cut pears. *LWT-Food Sci. Technol.*, 62, 1, 791–800, 2015.
58. Tapia, M. *et al.*, Use of alginate- and gellan-based coatings for improving barrier, texture and nutritional properties of fresh-cut papaya. *Food Hydrocolloids*, 22, 8, 1493–1503, 2008.
59. Yu, L. *et al.*, Effect of guar gum with glycerol coating on the properties and oil absorption of fried potato chips. *Food Hydrocolloids*, 54, 211–219, 2016.
60. Rojas-Argudo, C., Del Río, M., Pérez-Gago, M., Development and optimization of locust bean gum (LBG)-based edible coatings for postharvest storage of 'Fortune' mandarins. *Postharvest Biol. Technol.*, 52, 2, 227–234, 2009.

61. Hashemi, S.M.B. *et al.*, Basil-seed gum containing Origanum vulgare subsp. viride essential oil as edible coating for fresh cut apricots. *Postharvest Biol. Technol.*, 125, 26–34, 2017.
62. Dangaran, K., Tomasula, P.M., Qi, P., Structure and function of protein-based edible films and coatings, in: *Edible films and coatings for food applications*, pp. 25–56, Springer, New York, US, 2009.
63. Lima, Á.M. *et al.*, New edible coatings composed of galactomannans and collagen blends to improve the postharvest quality of fruits - Influence on fruits gas transfer rate. *J. Food Eng.*, 97, 1, 101–109, 2010.
64. Jridi, M. *et al.*, Effects of active gelatin coated with henna (L. inermis) extract on beef meat quality during chilled storage. *Food Control*, 84, 238–245, 2018.
65. Kaewprachu, P. *et al.*, Gelatin films with nisin and catechin for minced pork preservation. *Food Packag. Shelf life*, 18, 173–183, 2018.
66. Sun, X. *et al.*, Preservative effects of fish gelatin coating enriched with CUR/βCD emulsion on grass carp (*Ctenopharyngodon idellus*) fillets during storage at 4 °C. *Food Chem.*, 272, 643–652, 2019.
67. Khorram, F., Ramezanian, A., Hosseini, S.M.H., Shellac, gelatin and Persian gum as alternative coating for orange fruit. *Sci. Hortic.*, 225, 22–28, 2017.
68. Kan, J. *et al.*, Development of active packaging based on chitosan-gelatin blend films functionalized with Chinese hawthorn (*Crataegus pinnatifida*) fruit extract. *Int. J. Biol. Macromol.*, 140, 384–392, 2019.
69. Boyacı, D. *et al.*, Development of flexible antimicrobial zein coatings with essential oils for the inhibition of critical pathogens on the surface of whole fruits: Test of coatings on inoculated melons. *Food Packag. Shelf life*, 20, 100316, 2019.
70. Santos, T.M. *et al.*, Enhancing storage stability of guava with tannic acid-crosslinked zein coatings. *Food Chem.*, 257, 252–258, 2018.
71. Caner, C. and Yüceer, M., Efficacy of various protein-based coating on enhancing the shelf life of fresh eggs during storage. *Poultr. Sci.*, 94, 7, 1665–1677, 2015.
72. Laelorspoen, N. *et al.*, Microencapsulation of Lactobacillus acidophilus in zein–alginate core–shell microcapsules via electrospraying. *J. Funct. Foods*, 7, 342–349, 2014.
73. Wang, X. *et al.*, Encapsulation and binding properties of curcumin in zein particles stabilized by Tween 20. *Colloids Surf. A: Physicochem. Eng. Asp.*, 577, 274–280, 2019.
74. Carlin, F. *et al.*, Utilization of Zein Coating and Sorbic Acid to Reduce Listeria monocytogenes Growth on Cooked Sweet Corn. *J. Food Sci.*, 66, 9, 1385–1389, 2001.
75. Janes, M.E., Kooshesh, S., Johnson, M.G., Control of Listeria monocytogenes on the Surface of Refrigerated, Ready-to-eat Chicken Coated with Edible Zein Film Coatings Containing Nisin and/or Calcium Propionate. *J. Food Sci.*, 67, 7, 2754–2757, 2002.
76. Kang, H.-J. *et al.*, Inhibitory effect of soy protein coating formulations on walnut (*Juglans regia* L.) kernels against lipid oxidation. *LWT-Food Sci. Technol.*, 51, 1, 393–396, 2013.

77. Guerrero, P. *et al.*, Application of soy protein coatings and their effect on the quality and shelf life stability of beef patties. *RSC Adv.*, 5, 11, 8182–8189, 2015.
78. Shon, J., Eo, J.-H., Eun, J.-B., Effect of soy protein isolate coating on quality attributes of cut raw han-woo(Korean cow) beef, aerobically packaged and held refrigerated. *J. Food Qual.*, 33, s1, 42–60, 2010.
79. Kurt, Ş. and Kılınççeker, O., Performance optimization of soy and whey protein isolates as coating materials on chicken meat. *Poultr. Sci.*, 90, 1, 195–200, 2011.
80. Liberty, J.T., Dehghannya, J., Ngadi, M.O., Effective strategies for reduction of oil content in deep-fat fried foods: A review. *Trends Food Sci. Technol.*, 92, 172–183, 2019.
81. Xu, L. *et al.*, Internal quality of coated eggs with soy protein isolate and montmorillonite: Effects of storage conditions. *Int. J. Food Prop.*, 20, 8, 1921–1934, 2017.
82. Yousuf, B. and Srivastava, A.K., Impact of honey treatments and soy protein isolate-based coating on fresh-cut pineapple during storage at 4 °C. *Food Packag. Shelf life*, 21, 100361, 2019.
83. Alves, M.M., Gonçalves, M.P., Rocha, C.M.R., Effect of ferulic acid on the performance of soy protein isolate-based edible coatings applied to fresh-cut apples. *LWT*, 80, 409–415, 2017.
84. González-Estrada, R.R. *et al.*, Antimicrobial soy protein based coatings: Application to Persian lime (*Citrus latifolia Tanaka*) for protection and preservation. *Postharvest Biol. Technol.*, 132, 138–144, 2017.
85. Ghidelli, C. *et al.*, Novel approaches to control browning of fresh-cut artichoke: Effect of a soy protein-based coating and modified atmosphere packaging. *Postharvest Biol. Technol.*, 99, 105–113, 2015.
86. Tang, C.-H., Nanostructured soy proteins: Fabrication and applications as delivery systems for bioactives (a review). *Food Hydrocolloids*, 91, 92–116, 2019.
87. Gounga, M. *et al.*, Effect of whey protein isolate–pullulan edible coatings on the quality and shelf life of freshly roasted and freeze-dried chinese chestnut. *J. Food Sci.*, 73, 4, E155–E161, 2008.
88. Han, J. *et al.*, Coating of peanuts with edible whey protein film containing α-tocopherol and ascorbyl palmitate. *J. Food Sci.*, 73, 8, E349–E355, 2008.
89. Caner, C., Whey protein isolate coating and concentration effects on egg shelf life. *J. Sci. Food Agric.*, 85, 13, 2143 2148, 2005.
90. Henriques, M., Gomes, D., Pereira, C., Whey protein edible coatings: Recent developments and applications, in: *Emerging and traditional technologies for safe, healthy and quality food*, pp. 177–196, Springer, New York, US, 2016.
91. Catarino, M.D. *et al.*, Development and performance of whey protein active coatings with Origanum virens essential oils in the quality and shelf life improvement of processed meat products. *Food Control*, 80, 273–280, 2017.
92. Brink, I., Šipailienė, A., Leskauskaitė, D., Antimicrobial properties of chitosan and whey protein films applied on fresh cut turkey pieces. *Int. J. Biol. Macromol.*, 130, 810–817, 2019.

93. Cisneros-Zevallos, L. and Krochta, J.M., Whey Protein Coatings for Fresh Fruits and Relative Humidity Effects. *J. Food Sci.*, 68, 1, 176–181, 2003.
94. Feng, Z. *et al.*, Edible coating based on whey protein isolate nanofibrils for antioxidation and inhibition of product browning. *Food Hydrocolloids*, 79, 179–188, 2018.
95. Odila Pereira, J. *et al.*, Impact of whey protein coating incorporated with Bifidobacterium and Lactobacillus on sliced ham properties. *Meat Sci.*, 139, 125–133, 2018.
96. Shokri, S. and Ehsani, A., Efficacy of whey protein coating incorporated with lactoperoxidase and α-tocopherol in shelf life extension of Pike-Perch fillets during refrigeration. *LWT-Food Sci. Technol.*, 85, 225–231, 2017.
97. Mostafavi, F., Tehrani, M., Mohebbi, M., Rheological and sensory properties of fat reduced vanilla ice creams containing milk protein concentrate (MPC). *J. Food Meas. Charact.*, 11, 2, 567–575, 2017.
98. López Aguayo, M.d.C. *et al.*, Effect of different activated coatings containing enterocin AS-48 against Listeria monocytogenes on apple cubes. *Innovative Food Sci. Emerg. Technol.*, 35, 177–183, 2016.
99. Elzoghby, A.O., Abo El-Fotoh, W.S., Elgindy, N.A., Casein-based formulations as promising controlled release drug delivery systems. *J. Controlled Release*, 153, 3, 206–216, 2011.
100. Ranadheera, C.S. *et al.*, Utilizing unique properties of caseins and the casein micelle for delivery of sensitive food ingredients and bioactives. *Trends Food Sci. Technol.*, 57, 178–187, 2016.
101. Debeaufort, F. and Voilley, A., Lipid-based edible films and coatings, in: *Edible films and coatings for food applications*, pp. 135–168, Springer, New York, US, 2009
102. Hagenmaier, R.D. and Shaw, P.E., Permeability of shellac coatings to gases and water vapor. *J. Agric. Food Chem.*, 39, 5, 825–829, 1991.
103. Limmatvapirat, S. *et al.*, Modification of physicochemical and mechanical properties of shellac by partial hydrolysis. *Int. J. Pharm.*, 278, 1, 41–49, 2004.
104. Soradech, S. *et al.*, Utilization of shellac and gelatin composite film for coating to extend the shelf life of banana. *Food Control*, 73, 1310–1317, 2017.
105. Jo, W.-S. *et al.*, Quality and microbial safety of 'Fuji' apples coated with carnauba-shellac wax containing lemongrass oil. *LWT-Food Sci. Technol.*, 55, 2, 490–497, 2014.
106. de Freitas, C.A.S. *et al.*, Carnauba wax uses in food – A review. *Food Chem.*, 291, 38–48, 2019.
107. Hall, D.J., Edible coatings from lipids, waxes, and resins, in: *Edible coatings and films to improve food quality*, pp. 79–101, CRC Press, Boca Raton, FL, 2012.
108. De León-Zapata, M.A. *et al.*, Edible candelilla wax coating with fermented extract of tarbush improves the shelf life and quality of apples. *Food Packag. Shelf life*, 3, 70–75, 2015.

109. De León-Zapata, M.A. *et al.*, Changes of the shelf life of candelilla wax/tarbush bioactive based-nanocoated apples at industrial level conditions. *Sci. Hortic.*, 231, 43–48, 2018.
110. Kowalczyk, D. *et al.*, Effect of carboxymethylcellulose/candelilla wax coating containing potassium sorbate on microbiological and physicochemical attributes of pears. *Sci. Hortic.*, 218, 326–333, 2017.
111. Oregel-Zamudio, E. *et al.*, Effect of candelilla wax edible coatings combined with biocontrol bacteria on strawberry quality during the shelf life. *Sci. Hortic.*, 214, 273–279, 2017.
112. Formiga, A.S. *et al.*, Use of edible coatings based on hydroxypropyl methylcellulose and beeswax in the conservation of red guava 'Pedro Sato'. *Food Chem.*, 290, 144–151, 2019.
113. Oliveira, V.R.L. *et al.*, Use of biopolymeric coating hydrophobized with beeswax in post-harvest conservation of guavas. *Food Chem.*, 259, 55–64, 2018.
114. Zhong, Y., Cavender, G., Zhao, Y., Investigation of different coating application methods on the performance of edible coatings on Mozzarella cheese. *LWT-Food Sci. Technol.*, 56, 1, 1–8, 2014.
115. Eddin, A.S., Ibrahim, S.A., Tahergorabi, R., Egg Quality and Safety with an Overview of Edible Coating Application for Egg Preservation. *Food Chem.*, 296, 29–39, 2019.
116. Bhat, Z.F., Kumar, P., Kumar, S., Effect of skin, enrobing and refrigerated storage on the quality characteristics of chicken meat balls. *J. Food Sci. Technol.*, 50, 5, 890–899, 2013.
117. Kramer, M.E., Structure and function of starch-based edible films and coatings, in: *Edible films and coatings for food applications*, pp. 113–134, Springer, New York, US, 2009
118. Vu, K.D. *et al.*, Development of edible bioactive coating based on modified chitosan for increasing the shelf life of strawberries. *Food Research Int.*, 44, 1, 198–203, 2011.
119. Shahbazi, Y., Application of carboxymethyl cellulose and chitosan coatings containing Mentha spicata essential oil in fresh strawberries. *Int. J. Biol. Macromol.*, 112, 264–272, 2018.
120. Tomadoni, B. *et al.*, Gellan-based coatings incorporated with natural antimicrobials in fresh-cut strawberries: Microbiological and sensory evaluation through refrigerated storage. *LWT*, 97, 384–389, 2018.
121. Tanada-Palmu, P.S. and Grosso, C.R.F., Effect of edible wheat gluten-based films and coatings on refrigerated strawberry (*Fragaria ananassa*) quality. *Postharvest Biol. Technol.*, 36, 2, 199–208, 2005.
122. Vu, K.D. *et al.*, Development of bioactive coatings based on γ-irradiated proteins to preserve strawberries. *Radiat. Phys. Chem.*, 81, 8, 1211–1214, 2012.
123. Motamedi, E. *et al.*, Performance of carnauba wax-nanoclay emulsion coatings on postharvest quality of 'Valencia' orange fruit. *Sci. Hortic.*, 240, 170–178, 2018.

124. Chen, H., Sun, Z., Yang, H., Effect of carnauba wax-based coating containing glycerol monolaurate on the quality maintenance and shelf life of Indian jujube (*Zizyphus mauritiana Lamk.*) fruit during storage. *Sci. Hortic.*, 244, 157–164, 2019.
125. Bai, J., Hagenmaier, R.D., Baldwin, E.A., Coating selection for 'Delicious' and other apples. *Postharvest Biol. Technol.*, 28, 3, 381–390, 2003.
126. Valencia-Chamorro, S.A. *et al.*, Antimicrobial edible films and coatings for fresh and minimally processed fruits and vegetables: A review. *Crit. Rev. Food Sci. Nutr.*, 51, 9, 872–900, 2011.
127. Lee, J. *et al.*, Extending shelf life of minimally processed apples with edible coatings and antibrowning agents. *LWT-Food Sci. Technol.*, 36, 3, 323–329, 2003.
128. Martín-Belloso, O., Rojas-Graü, M.A., Soliva-Fortuny, R., Delivery of flavor and active ingredients using edible films and coatings, in: *Edible films and coatings for food applications*, pp. 295–313, Springer, New York, US, 2009.
129. Amal, S. *et al.*, Improving strawberry fruit storability by edible coating as a carrier of thymol or calcium chloride. *J. Hort. Sci. Ornam. Plants*, 2, 2, 88–97, 2010.
130. Tapia, M. *et al.*, Alginate-and gellan-based edible films for probiotic coatings on fresh-cut fruits. *J. Food Sci.*, 72, 4, E190–E196, 2007.
131. Maya-Meraz, I.O. *et al.*, Production of Volatiles in Fresh-Cut Apple: Effect of Applying Alginate Coatings Containing Linoleic Acid or Isoleucine. *J. Food Sci.*, 79, 11, C2185–C2191, 2014.
132. Rojas-Graü, M.A., Soliva-Fortuny, R., Martín-Belloso, O., Edible coatings to incorporate active ingredients to fresh-cut fruits: A review. *Trends Food Sci. Technol.*, 20, 10, 438–447, 2009.
133. Ala, M.A.N. and Shahbazi, Y., The effects of novel bioactive carboxymethyl cellulose coatings on food-borne pathogenic bacteria and shelf life extension of fresh and sauced chicken breast fillets. *LWT*, 111, 602–611, 2019.
134. Fernández-Pan, I., Carrión-Granda, X., Maté, J.I., Antimicrobial efficiency of edible coatings on the preservation of chicken breast fillets. *Food Control*, 36, 1, 69–75, 2014.
135. Ustunol, Z., Edible films and coatings for meat and poultry, in: *Edible films and coatings for food applications*, pp. 245–268, Springer, New York, US, 2009.
136. Sánchez-Ortega, I. *et al.*, Antimicrobial edible films and coatings for meat and meat products preservation. *Sci. World J.*, 1–18, 2014.
137. Volpe, M.G. *et al.*, Active edible coating effectiveness in shelf life enhancement of trout (Oncorhynchusmykiss) fillets. *LWT-Food Sci. Technol.*, 60, 1, 615–622, 2015.
138. Berizi, E. *et al.*, Microbial, chemical, textural and sensory properties of coated rainbow trout by chitosan combined with pomegranate peel extract during frozen storage. *Int. J. Biol. Macromol.*, 106, 1004–1013, 2018.
139. Jiang, Z., Neetoo, H., Chen, H., Efficacy of freezing, frozen storage and edible antimicrobial coatings used in combination for control of Listeria

monocytogenes on roasted turkey stored at chiller temperatures. *Food Microbiol.*, 28, 7, 1394–1401, 2011.

140. Cao, Y., Warner, R.D., Fang, Z., Effect of chitosan/nisin/gallic acid coating on preservation of pork loin in high oxygen modified atmosphere packaging. *Food Control*, 101, 9–16, 2019.
141. Abdallah, M.R. *et al.*, Improving the sensory, physicochemical and microbiological quality of pastirma (A traditional dry cured meat product) using chitosan coating. *LWT*, 86, 247–253, 2017.
142. Bazargani-Gilani, B., Aliakbarlu, J., Tajik, H., Effect of pomegranate juice dipping and chitosan coating enriched with Zataria multiflora Boiss essential oil on the shelf life of chicken meat during refrigerated storage. *Innovative Food Sci. Emerg. Technol.*, 29, 280–287, 2015.
143. Fang, Z. *et al.*, Effect of gallic acid/chitosan coating on fresh pork quality in modified atmosphere packaging. *Food Chem.*, 260, 90–96, 2018.
144. Guo, M. *et al.*, Antimicrobial films and coatings for inactivation of Listeria innocua on ready-to-eat deli turkey meat. *Food Control*, 40, 64–70, 2014.
145. Hassanzadeh, P. *et al.*, Effect of functional chitosan coating and gamma irradiation on the shelf life of chicken meat during refrigerated storage. *Radiat. Phys. Chem.*, 141, 103–109, 2017.
146. Kanatt, S.R. *et al.*, Effects of chitosan coating on shelf life of ready-to-cook meat products during chilled storage. *LWT-Food Sci. Technol.*, 53, 1, 321–326, 2013.
147. Lekjing, S., A chitosan-based coating with or without clove oil extends the shelf life of cooked pork sausages in refrigerated storage. *Meat Sci.*, 111, 192–197, 2016.
148. Mohan, C. *et al.*, Effect of chitosan edible coating on the quality of double filleted Indian oil sardine (Sardinella longiceps) during chilled storage. *Food Hydrocolloids*, 26, 1, 167–174, 2012.
149. Langroodi, A.M. *et al.*, Effects of sumac extract dipping and chitosan coating enriched with Zataria multiflora Boiss oil on the shelf life of meat in modified atmosphere packaging. *LWT*, 98, 372–380, 2018.
150. Pabast, M. *et al.*, Effects of chitosan coatings incorporating with free or nano-encapsulated Satureja plant essential oil on quality characteristics of lamb meat. *Food Control*, 91, 185–192, 2018.
151. Petrou, S. *et al.*, Chitosan dipping or oregano oil treatments, singly or com bined on modified atmosphere packaged chicken breast meat. *Int. J. Food Microbiol.*, 156, 3, 264–271, 2012.
152. Zhang, H. *et al.*, Antioxidant and antimicrobial effects of edible coating based on chitosan and bamboo vinegar in ready to cook pork chops. *LWT*, 93, 470–476, 2018.
153. Choulitoudi, E. *et al.*, Antimicrobial and antioxidant activity of Satureja thymbra in gilthead seabream fillets edible coating. *Food Bioprod. Process.*, 100, 570–577, 2016.

154. Choulitoudi, E. *et al.*, Edible coating enriched with rosemary extracts to enhance oxidative and microbial stability of smoked eel fillets. *Food Packag. Shelf life*, 12, 107–113, 2017.
155. Rezaei, F. and Shahbazi, Y., Shelf life extension and quality attributes of sauced silver carp fillet: A comparison among direct addition, edible coating and biodegradable film. *LWT*, 87, 122–133, 2018.
156. Neetoo, H., Ye, M., Chen, H., Bioactive alginate coatings to control Listeria monocytogenes on cold-smoked salmon slices and fillets. *Int. J. Food Microbiol.*, 136, 3, 326–331, 2010.
157. Alizadeh, Z., Yousefi, S., Ahari, H., Optimization of bioactive preservative coatings of starch nanocrystal and ultrasonic extract of sour lemon peel on chicken fillets. *Int. J. Food Microbiol.*, 300, 31–42, 2019.
158. Alotaibi, S. and Tahergorabi, R., Development of a sweet potato starch-based coating and its effect on quality attributes of shrimp during refrigerated storage. *LWT*, 88, 203–209, 2018.
159. Raeisi, M. *et al.*, Effect of sodium alginate coating incorporated with nisin, Cinnamomum zeylanicum, and rosemary essential oils on microbial quality of chicken meat and fate of Listeria monocytogenes during refrigeration. *Int. J. Food Microbiol.*, 238, 139–145, 2016.
160. Song, Y. *et al.*, Effect of sodium alginate-based edible coating containing different anti-oxidants on quality and shelf life of refrigerated bream (Megalobrama amblycephala). *Food Control*, 22, 3–4, 608–615, 2011.
161. Lu, F. *et al.*, Cinnamon and nisin in alginate–calcium coating maintain quality of fresh northern snakehead fish fillets. *LWT-Food Sci. Technol.*, 43, 9, 1331–1335, 2010.
162. Vital, A.C.P. *et al.*, Effect of edible and active coating (with rosemary and oregano essential oils) on beef characteristics and consumer acceptability. *PLoS One*, 11, 8, e0160535, 2016.
163. Alamatian, S. *et al.*, Modeling of osmotic treatment of ostrich meat coated by tragacanth and salep. *Meat Sci.*, 156, 231–239, 2019.
164. Gadang, V. *et al.*, Evaluation of antibacterial activity of whey protein isolate coating incorporated with nisin, grape seed extract, malic acid, and EDTA on a turkey frankfurter system. *J. Food Sci.*, 73, 8, M389–M394, 2008.
165. Min, S., Rumsey, T.R., Krochta, J.M., Diffusion of the antimicrobial lysozyme from a whey protein coating on smoked salmon. *J. Food Eng.*, 84, 1, 39–47, 2008.
166. Min, S. *et al.*, Listeria monocytogenes inhibition by whey protein films and coatings incorporating lysozyme. *J. Food Prot.*, 68, 11, 2317–2325, 2005.
167. Sathivel, S., Chitosan and protein coatings affect yield, moisture loss, and lipid oxidation of pink salmon (*Oncorhynchus gorbuscha*) fillets during frozen storage. *J. Food Sci.*, 70, 8, e455–e459, 2005.
168. Farajzadeh, F. *et al.*, The effect of chitosan-gelatin coating on the quality of shrimp (*Litopenaeus vannamei*) under refrigerated condition. *Food Control*, 67, 163–170, 2016.

169. Alparslan, Y. *et al.*, Quality assessment of shrimps preserved with orange leaf essential oil incorporated gelatin. *LWT-Food Sci. Technol.*, 72, 457–466, 2016.
170. Noori, S., Zeynali, F., Almasi, H., Antimicrobial and antioxidant efficiency of nanoemulsion-based edible coating containing ginger (Zingiber officinale) essential oil and its effect on safety and quality attributes of chicken breast fillets. *Food Control*, 84, 312–320, 2018.
171. Lungu, B. and Johnson, M., Fate of Listeria monocytogenes inoculated onto the surface of model turkey frankfurter pieces treated with zein coatings containing nisin, sodium diacetate, and sodium lactate at 4 C. *J. Food Prot.*, 68, 4, 855–859, 2005.
172. Bolívar-Monsalve, J. *et al.*, Mechanisms of action of novel ingredients used in edible films to preserve microbial quality and oxidative stability in sausages - A review. *Trends Food Sci. Technol.*, 89, 100–109, 2019.
173. Duan, J. *et al.*, Effect of combined chitosan-krill oil coating and modified atmosphere packaging on the storability of cold-stored lingcod (Ophiodon elongates) fillets. *Food Chem.*, 122, 4, 1035–1042, 2010.
174. Galus, S. and Kadzińska, J., Food applications of emulsion-based edible films and coatings. *Trends Food Sci. Technol.*, 45, 2, 273–283, 2015.
175. Jasour, M.S. *et al.*, Chitosan coating incorporated with the lactoperoxidase system: An active edible coating for fish preservation. *J. Sci. Food Agric.*, 95, 6, 1373–1378, 2015.
176. Oliveira, M.M.M.d., Brugnera, D.F., Piccoli, R.H., Essential oils of thyme and Rosemary in the control of Listeria monocytogenes in raw beef. *Braz. J. Microbiol.*, 44, 4, 1181–1188, 2013.
177. Gharibzahedi, S.M.T. and Mohammadnabi, S., Effect of novel bioactive edible coatings based on jujube gum and nettle oil-loaded nanoemulsions on the shelf life of Beluga sturgeon fillets. *Int. J. Biol. Macromol.*, 95, 769–777, 2017.
178. Bartkiene, E. *et al.*, A concept of mould spoilage prevention and acrylamide reduction in wheat bread: Application of lactobacilli in combination with a cranberry coating. *Food Control*, 91, 284–293, 2018.
179. Saraiva, L.E.F. *et al.*, Development and application of edible film of active potato starch to extend mini panettone shelf life. *LWT*, 73, 311–319, 2016.
180. Berti, S. *et al.*, Edible coatings on Gouda cheese as a barrier against external contamination during ripening. *Food Biosci.*, 31, 100447, 2019.
181. Ortiz de Elguea-Culebras, G. *et al.*, Optimization of a chitosan solution as potential carrier for the incorporation of Santolina chamaecyparissus L. solid by-product in an edible vegetal coating on 'Manchego' cheese. *Food Hydrocolloids*, 89, 272–282, 2019.
182. Costa, M.J. *et al.*, Use of edible films and coatings in cheese preservation: Opportunities and challenges. *Food Research Int.*, 107, 84–92, 2018.
183. Pena-Serna, C., Penna, A.L.B., Lopes Filho, J.F., Zein-based blend coatings: Impact on the quality of a model cheese of short ripening period. *J. Food Eng.*, 171, 208–213, 2016.

184. Ramos, Ó.L. *et al.*, Evaluation of antimicrobial edible coatings from a whey protein isolate base to improve the shelf life of cheese. *J. Dairy Sci.*, 95, 11, 6282–6292, 2012.
185. Youssef, A.M. *et al.*, Development of bionanocomposite materials and its use in coating of Ras cheese. *Food Chem.*, 270, 467–475, 2019.
186. Lucera, A. *et al.*, Effect of active coating on microbiological and sensory properties of fresh mozzarella cheese. *Food Packag. Shelf life*, 1, 1, 25–29, 2014.
187. Guitián, M.V. *et al.*, Anti-Listeria monocytogenes effect of bacteriocin-incorporated agar edible coatings applied on cheese. *Int. Dairy J.*, 97, 92–98, 2019.
188. Lin, S.Y. and Krochta, J., Fluidized-bed system for whey protein film coating of peanuts. *J. Food Process Eng.*, 29, 5, 532–546, 2006.
189. Maghsoudlou, A. *et al.*, Evaluation of anti-fungal activity of chitosan and its effect on the moisture absorption and organoleptic characteristics of pistachio nuts. *Int. J. Adv. Sci. Eng. Inf. Technol.*, 2, 4, 336–340, 2012.
190. Vanaei, M. *et al.*, Novel edible coating based on Aloe Vera Gel to Maintain pistachio Quality. *Int. J. Sci. Eng. Technol.*, 3, 8, 1016–1019, 2014.
191. Azimzadeh, B. and Jahadi, M., Effect of chitosan edible coating with Laurus nobilis extract on shelf life of cashew. *Food Sci. Nutr.*, 6, 4, 871–877, 2018.
192. Chinma, C., Ariahu, C., Abu, J., Shelf life extension of toasted groundnuts through the application of cassava starch and soy protein-based edible coating. *Niger. Food J.*, 32, 1, 133–138, 2014.
193. Davoodi, F. and Naji, M.H., Study of the effect of sodium alginate coating containing pomegranate peel extract on chemical, sensory and microbial quality of walnut kernel. *Environ. Health Eng. Manag. J.*, 5, 4, 249–257, 2018.
194. Lin, S.Y. and Krochta, J., Whey protein coating efficiency on mechanically roughened hydrophobic peanut surfaces. *J. Food Sci.*, 71, 6, E270–E275, 2006.
195. Van Immerseel, F., Nys, Y., Bain, M., *Improving the safety and quality of eggs and egg products: egg safety and nutritional quality*, Elsevier, New York, US, 2011.
196. Torrico, D.D. *et al.*, Mineral oil–chitosan emulsion coatings affect quality and shelf life of coated eggs during refrigerated and room temperature storage. *J. Food Sci.*, 76, 4, S262–S268, 2011.
197. Caner, C. and Cansiz, O., Effectiveness of chitosan-based coating in improving shelf life of eggs. *J. Sci. Food Agric.*, 87, 2, 227–232, 2007.
198. Caner, C. and Cansiz, Ö., Chitosan coating minimises eggshell breakage and improves egg quality. *J. Sci. Food Agric.*, 88, 1, 56–61, 2008.
199. Xie, L. *et al.*, Edible film coating to minimize eggshell breakage and reduce post-wash bacterial contamination measured by dye penetration in eggs. *J. Food Sci.*, 67, 1, 280–284, 2002.
200. Suppakul, P., Jutakorn, K., Bangchokedee, Y., Efficacy of cellulose-based coating on enhancing the shelf life of fresh eggs. *J. Food Eng.*, 98, 2, 207–213, 2010.

201. Alleoni, A.C.C. and Antunes, A., Albumen foam stability and s-ovalbumin contents in eggs coated with whey protein concentrate. *Braz. J. Poultry Sci.*, 6, 2, 105–110, 2004.
202. Musa, T.N., Ulaiwi, W.S., Al-Hajo, N.N., The effect of shellac as coating material on the internal quality of chicken eggs. *Int. J. Poult. Sci.*, 10, 1, 38–41, 2011.
203. Wardy, W. *et al.*, Chitosan-soybean oil emulsion coating affects physico-functional and sensory quality of eggs during storage. *LWT-Food Sci. Technol.*, 44, 10, 2349–2355, 2011.
204. Izadi, S. *et al.*, Production of low-fat shrimps by using hydrocolloid coatings. *J. Food Sci. Technol.*, 52, 9, 6037–6042, 2015.
205. Lalam, S. *et al.*, Experimental study on transport mechanisms during deep fat frying of chicken nuggets. *LWT-Food Sci. Technol.*, 50, 1, 110–119, 2013.
206. Garmakhany, A.D. *et al.*, Study of oil uptake and some quality attributes of potato chips affected by hydrocolloids. *Eur. J. Lipid Sci. Technol.*, 110, 11, 1045–1049, 2008.
207. Kurek, M., Ščetar, M., Galić, K., Edible coatings minimize fat uptake in deep fat fried products: A review. *Food Hydrocolloids*, 71, 225–235, 2017.
208. Nedovic, V. *et al.*, An overview of encapsulation technologies for food applications. *Procedia Food Sci.*, 1, 0, 1806–1815, 2011.
209. Tanja, P. *et al.*, Protection of probiotic microorganisms by microencapsulation. *Chem. Ind. Chem. Eng. Q.*, 13, 3, 169–174, 2007.
210. McClements, D.J., Requirements for food ingredient and nutraceutical delivery systems, in: *Encapsulation Technologies and Delivery Systems for Food Ingredients and Nutraceuticals*, N. Garti and D.J. McClements (Eds.), pp. 3–18, Woodhead Publishing, Cambridge, 2012.
211. Zuidam, N.J. and Shimoni, E., Overview of microencapsulates for use in food products or processes and methods to make them, in: *Encapsulation Technologies for Active Food Ingredients and Food Processing*, N.J. Zuidam and V.A. Nedović (Eds.), pp. 3–29, Springer, New York, USA, 2009.
212. Anu Bhushani, J. and Anandharamakrishnan, C., Electrospinning and electrospraying techniques: Potential food based applications. *Trends Food Sci. Technol.*, 38, 1, 21–33, 2014.
213. Gbassi, G.K. and Vandamme, T., Probiotic encapsulation technology: From microencapsulation to release into the gut. *Pharmaceutics*, 4, 1, 149–163, 2012.
214. Zaeim, D. *et al.*, Electrospray-assisted drying of live probiotics in acacia gum microparticles matrix. *Carbohydr. Polym.*, 183, 183–191, 2018.
215. de Vos, P. *et al.*, Encapsulation for preservation of functionality and targeted delivery of bioactive food components. *Int. Dairy J.*, 20, 4, 292–302, 2010.
216. Martín, M.J. *et al.*, Microencapsulation of bacteria: A review of different technologies and their impact on the probiotic effects. *Innovative Food Sci. Emerg. Technol.*, 27, Supplement C, 15–25, 2015.
217. Pedroso, D.d.L. *et al.*, Protection of Bifidobacterium lactis and Lactobacillus acidophilus by microencapsulation using spray-chilling. *Int. Dairy J.*, 26, 2, 127–132, 2012.

218. Mandal, S. and Hati, S., Microencapsulation of Bacterial Cells by Emulsion Technique for Probiotic Application. *Methods Mol. Biol.*, 1479, 273–279, 2017.
219. Kailasapathy, K., Microencapsulation of probiotic bacteria: technology and potential applications. *Curr. Issues Intest. Microbiol.*, 3, 2, 39–48, 2002.
220. Krasaekoopt, W., Bhandari, B., Deeth, H., Evaluation of encapsulation techniques of probiotics for yoghurt. *Int. Dairy J.*, 13, 1, 3–13, 2003.
221. Zaeim, D. *et al.*, Electrospray assisted fabrication of hydrogel microcapsules by single- and double-stage procedures for encapsulation of probiotics. *Food Bioprod. Process.*, 102, 250–259, 2017.
222. Zaeim, D. *et al.*, Double layer co-encapsulation of probiotics and prebiotics by electro-hydrodynamic atomization. *LWT*, 110, 102–109, 2019.
223. Jaworek, A. and Sobczyk, A.T., Electrospraying route to nanotechnology: An overview. *J. Electrostat.*, 66, 3–4, 197–219, 2008.
224. Bock, N., Dargaville, T.R., Woodruff, M.A., Electrospraying of polymers with therapeutic molecules: State of the art. *Prog. Polym. Sci.*, 37, 11, 1510–1551, 2012.
225. Teo, W.E. and Ramakrishna, S., A review on electrospinning design and nanofibre assemblies. *Nanotechnology*, 17, 14, R89–R106, 2006.
226. Mellema, M., Mechanism and reduction of fat uptake in deep-fat fried foods. *Trends Food Sci. Technol.*, 14, 9, 364–373, 2003.
227. de Oliveira Alves Sena, E. *et al.*, Calcium application via hydrocooling and edible coating for the conservation and quality of cashew apples. *Sci. Hortic.*, 256, 1–8, 2019.
228. Yan, J. *et al.*, The effect of the layer-by-layer (LBL) edible coating on strawberry quality and metabolites during storage. *Postharvest Biol. Technol.*, 147, 29–38, 2019.
229. Salas-Méndez, E.d.J. *et al.*, Application of edible nanolaminate coatings with antimicrobial extract of Flourensia cernua to extend the shelf life of tomato (*Solanum lycopersicum* L.) fruit. *Postharvest Biol. Technol.*, 150, 19–27, 2019.
230. Nawab, A., Alam, F., Hasnain, A., Mango kernel starch as a novel edible coating for enhancing shelf- life of tomato (*Solanum lycopersicum*) fruit. *Int. J. Biol. Macromol.*, 103, 581–586, 2017.
231. Wu, S., Lu, M., Wang, S., Effect of oligosaccharides derived from Laminaria japonica-incorporated pullulan coatings on preservation of cherry tomatoes. *Food Chem.*, 199, 296–300, 2016.
232. Li, J. *et al.*, Improvement of banana postharvest quality using a novel soybean protein isolate/cinnamaldehyde/zinc oxide bionanocomposite coating strategy. *Sci. Hortic.*, 258, 108786, 2019.

11

Biopolymers as Edible Coating for Food: Recent Trends

Ravichandran Santhosh, Abhinav Tiwari and Ashish Rawson*

Department of Food Safety and Quality Testing, Indian Institute of Food Processing Technology, Thanjavur, India

Abstract
Edible coating, a primary packaging layer that consists of biopolymers and can be consumed along with the foods, is an all time in demand. Polysaccharides, proteins, and lipids are the major edible biopolymers used in edible coating. The primary function of edible coating is controlling gas and water permeations in food surfaces with its environment. The secondary function includes preventing the products from microbial spoilage and mechanical damage. In addition, the edible coating improves consumer acceptance through enriched organoleptic and sensory attributes of the coated foods. The application of edible coating includes coating of whole and fresh-cut fruits and vegetables, seafood, meat, and its recent application in bakery products. Edible coating is also an active preservation technique for fresh and processed foods with shelf life extension by inhibiting nutritional losses and quality deterioration. These biopolymers act as the carrier material for antioxidant, antimicrobial, and antiviral agents to enhance safety of the food products. Edible coatings are the promising solution to post-harvest losses of perishables, food insecurity, and increasing environmental concerns.

Keywords: Biopolymers, edible coating, food, polysaccharides, proteins, lipids

11.1 Introduction

Biopolymer is the chain of monomer residues derived from biological systems such as plants and animals. These polymers consist of repeating units

**Corresponding author*: ashishrawson@gmail.com; ashish.rawson@iifpt.edu.in

Inamuddin, Rajender Boddula, Mohd Imran Ahamed and Abdullah M. Asiri (eds.) Polymer Coatings: Technology and Applications, (233–274) © 2020 Scrivener Publishing LLC

of sugars or amino acids or aliphatic polyesters from the biological sources. Polysaccharides and proteins are the two major biopolymers used in food industries. The interaction of biopolymers with food molecules assists in freezing, texture modification, gel and network formation that aids in developing edible films and coatings [1].

Biopolymers are also known as green polymers that have potential to degrade itself without any further processes or foreign agents. Most of the biopolymers are hydrocolloids, which possess various functional properties including formation of intermolecular interactions between the compounds that resulted into a film matrix and coating network. Biopolymers possess the potential to form coatings and films that extends the shelf life of fresh and processed foods [2]. Network forming ability, nutritious composition, ability to hold antioxidant and antimicrobial agents, renewable origin, low cost, ready availability, and no negative impact on the environment are the reasons for biopolymers in edible coating. Biopolymers are called as IV generation of packaging [3].

Edible coating consists of biopolymers that are edible and coated directly on the food surface either by immersion of foods into biopolymer solution or by spraying the biopolymer solution over the food surface. It is a primary packaging material for most of the perishable and semi-perishable foods. Edibility and biodegradability are the two major advantages of edible coating. Other advantages include improved aesthetic appearance and enhanced sensory characteristics which lead to high consumer acceptance rates of edible-coated foods.

The promising features of edible coating/films over synthetic as stated by Robertson [4] are as follows:

- It prevents waste occurrence by consuming the packaging material together with food.
- Even if they are not consumed, still will not possess any harm to the nature.
- It can be produced from various agricultural and food industries as a source of byproduct utilization.
- It can enhance the sensory characteristics of the coated product by incorporating flavors, colors, sweeteners, etc.
- It can enhance the nutritional value of the product.
- It prevents migration between other compounds in heterogeneous foods such as pizza, pies, and sweets.
- It carries active agents—antimicrobial and antioxidant.

11.2 Need for Edible Coatings

Food packaging plays the major role in food preservation throughout the distribution chain. The packaging process must balance the food protection and energy utilized for the packaging material production and disposal. The non-degradable nature of plastic packaging materials possesses severe challenges to the increasing landfills and municipal solid wastes. The increasing environmental concerns resulted in implementing strict regulations, which directly reduces production of such polymers [5].

Edible coating replaces the usage of synthetic plastics in the food packaging application. Edible coatings are generally prepared from non-toxic, non-polluting, and low-cost biopolymers that are derived from various agricultural commodities and by-products of food and agriculture processing industries [6]. In economic and ecological perspective, edible coating utilizes the composite biopolymers obtained from cake, meal, flour and straw—the agroindustry by-products [7].

Post-harvest losses are one of the greatest challenges for the fruit and vegetable producers. To overcome these losses, conventional treatments including chemical waxes and fungicides are practiced. However, continuous exposure to these chemicals causes hazardous to health and environment. The increasing consumer awareness among recommended consumption rate of fruits and vegetables and to eliminate the excessive packaging waste generation and landfill [8] paves the way for minimally processing/preserving methods such as edible coating. In edible-coated fruits and vegetables, the physiological changes and microbial growth that causes biochemical changes and quality loss are retarded. Highly perishable fruits are the most appropriate for edible coating, because of their liable nutrient content and short shelf life [9].

The addition of antimicrobial and antioxidant agents in food matrix has been employed to prevent food contamination. However, direct application of these agents leads to neutralized or evaporated or reduced effective concentration on the food surface due to rapid and unstable release mechanisms [10]. To overcome these drawbacks, edible coating has the potential to carry and release the active agents into food in a controlled manner [11].

The presence of antioxidants such as vitamin, polyphenol, etc., in fresh-cut fruits and vegetables having links with the prevention of cancer and degenerative diseases, that increases consumer demand. The major drawback associated with the products of fresh-cut fruits and vegetables are its processing, which causes quality deterioration associated with tissue breakdown that results in metabolic, physicochemical, and textural changes [12]. Edible coating that carries antioxidant and anti-browning agents is the promising method to preserve fresh-cut produces without any nutritional loss.

Dough freezing, a practice in bakery industries, has been followed for economic advantages [13]. Ice crystal formed during freezing reduced the dough strength and CO_2 retention, prolonged fermentation, reduction in viability, and yeast activity. That reduces the volume and texture of the bakery product [14]. Edible coating provides the necessary gas barrier to the carbon dioxide release from dough, prevents ice crystal formation during freezing, and helps to maintain structure and quality of the final product [15].

11.3 Functions of Edible Coating

The primary functions of edible coating are preventing gas (carbon dioxide, oxygen, and volatiles) transfer and creating a modified atmosphere. Such atmosphere delays fruits and vegetables ripening and extends shelf life. Controlling the oxygen intake leading to retardation of lipid oxidation in meat and also depleted oxygen levels affects aerobic microorganisms and reduces the number of spoilage microorganisms present in the meat and its products [16]. Similar to meat, spoilage of cheese is due to fat oxidation and microorganisms' growth that resulted unacceptable flavor, odor, and appearance changes. These changes can be controlled by applying a low oxygen permeability packaging to avoid oxidation of fats and growth of undesirable microorganisms, which cause modifications in flavor, odor, and appearance [17]. Even in the case of moisture retention in the meat products, biopolymers have turned out as an effective alternative [16].

Softening of fruit and vegetable tissues influences the sensory score and affects consumer acceptance. Softening causes cell wall damage that composed of cellulose, hemicellulose, and pectin. The enzymes, which are responsible for this softening, were pectinmethylesterase (PME) and polygalacturanase (PG). Texture enhancers are added to edible coatings to minimize tissue damages during fruits and vegetables storage [18].

In addition, edible coating acts as an active carrier of preservatives including antioxidants, antimicrobials, and anti-browning agents. Natural sources of these agents are spices, organic acids, fatty acid esters, and essential oils, that can be directly incorporated into the biopolymer solutions [19]. Essential oils have high antioxidant and antimicrobial activities. The efficacies of these essential oils are based on the inactivating mechanism such as degrading cell membrane, inhibiting enzyme activity, influencing bacterial genetics [20].

Changes in color or browning during post-harvest operations are due to enzyme-catalyzed reactions. Polyphenol oxidase (PPO) is the major browning enzyme that is breaking the phenolic and polyphenolic compounds and reducing the antioxidant activity of fruits [21]. Ascorbic acid, oxalic acid, citric acid, phenolic acid, and resorcinol are some of the anti-browning agents that are used in edible coating [18, 22]. Extending shelf life by employing these agents, hence, edible coating can be considered as one of the food preservation methods.

Edible coating exhibits a diverse function, i.e., edible coatings reduces the oil uptake during frying process, the oils in the fried foods had direct role in diabetic and coronary diseases [23]. The hydrocolloid coating reduces the oil content in the fried foods due to its fat-barrier properties [24].

11.4 Materials Used for Making Edible Coating

The biopolymers used for edible coating are divided based on their origin—plant and animal sources. This can be justified by the fact that a particular coating can only sustain until it is suitable for a specific food commodity.

11.4.1 Plant Source

11.4.1.1 Starch

Starch is the most employed agricultural raw material for edible coatings. Plants such as corn, potato, wheat, oat, barley, and soybean are some of the major sources of starch [25]. Such sources determine the physical and chemical status of the starch granules [26]. Starch, a polymeric carbohydrate, consists of two polysaccharides—amylose and amylopectin. Amylose is a linear polymer chain with α-1,4 anhydroglucose units and it is responsible for the cross-linking nature of starches [27]. Amylopectin is a branched chain with short α-1,4 units binded through α-1,6 glycosidic bonding [28].

Starches are generally semi-crystalline in the range of 15%–45% crystallinity. The differences in structure and molecular weight of these two polysaccharides cause variation in coating network formation. The insoluble nature of starch molecules in cold water is because of hydrogen bonding. While heating, crystalline structure disrupts and water interacts with hydroxyl groups of starch granules. Therefore, formation of amorphous region forms and assists the coating network formation [29].

Garcia, Martino [9] studied the effect of medium amylose—corn and potato starches (250 g/kg starch) and high amylose—amylomaize VI (modified) starch (650 g/kg) on the strawberry coating. The results suggested the high amylose coating forms the better network than medium amylose coating, with reduced weight loss, retained firmness, and extended maturation period. Above all, the shelf life of strawberries increased to 28 days (coated) from 14 days (control).

Chickpea is a legume that contains starch high in amylose having the network-forming ability. The chickpea starch-coated fresh-cut papaya showed negligible weight loss, i.e., 53.65% to 51% in day 12 of storage. Moreover, the water loss barrier created by the coating caused reduction in transpiration. In addition, shelf life is extended to 12 days in coated fruit, whereas the uncoated fruit is spoiled at the 8th day [30].

Figure 11.1 shows the scanning micrograph of uncoated and coated banana peel. The rice starch-coated banana shows uniform surface without any cracks or cleavage on the epidermal cells thus slows down the ripening process. Uncoated banana shows the presence of cleavage and non-uniform surface that indicates the accelerated mass transfer through banana surfaces caused higher respiration rate with other biochemical changes [31]. Similar results were observed for the plum fruit with the same formulation. Here, a significant reduction in respiration rate at the 3rd week of coated plum was observed compared to control samples during storage, which indicates the marketability of uncoated fruit ended by the second week, whereas the coated fruit exhibited extended shelf life for 7 more days [32].

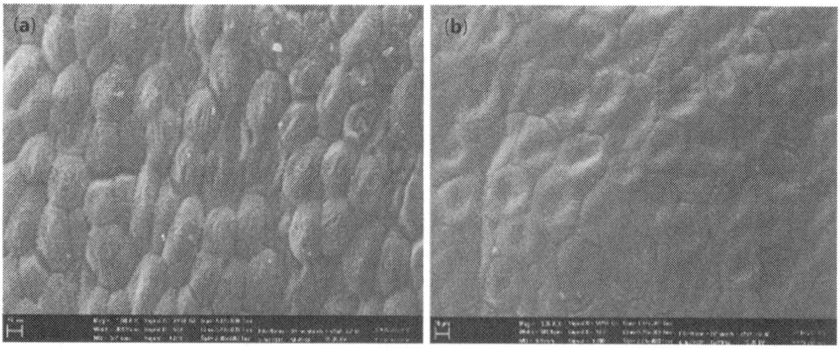

Figure 11.1 SEM micrographs of banana peel control (a) and coated (b) at 1,000× (Adapted from [31]).

Other major polysaccharide that is widely used in edible coating are alginates. Alginate is a polysaccharide derived from algae. Alginate represents alginic acid salts and its derivatives. The colloidal properties of alginates are ability to form gel and insoluble polymer when reacting with multivalent metal cations such as calcium [33]. Sodium alginate coated mango fruit shows higher antioxidant capacity than control fruit. Polyphenoloxidase activity in the coated mango remained lower than the control (uncoated mango) where it was observed to continuously increase [34]. The combination of sodium alginate with ascorbic and citric acids produced lesser browning on fresh-cut apples than pectin with same agents [35].

Chickpea starch, cassava starch, tapioca starch, corn starch, etc., are some of the other starches used for edible coating. The preservation of frozen dough and reduction in retrogradation rate are some emerging applications of the starch based edible coating [15, 36]. Different starches that are used for edible coating and its effect on the food quality are discussed in Table 11.1.

11.4.1.2 Cellulose Derivatives

Cellulose, the most abundant biopolymer that is present in all renewable sources such as wood, cotton, and plant-based materials, consists of a β (1-4) linked glucose residues [43]. Poor solubility and high crystalline nature of cellulose polymers make them difficult to use in packaging application. Poor water solubility of cellulose increases when soaked with alkali, swelling occurs, and reacting with chloroacetic acid, methyl chloride, or propylene oxide produces carboxymethyl cellulose, methyl cellulose, hydoxypropyl methylcellulose [44]. Therefore, the derivatives of cellulose including cellulose acetate, methylcellulose, hydroxypropyl methylcellulose, taurine cellulose are environmental friendly biopolymers, [45] filters, absorbents, and carrier material [1].

Methylcellulose and hydroxypropyl methylcellulose provide excellent barrier against gas transfer, but high moisture permeability [46]. These derivatives possess ability to form thermally stable coating, which is used for deep-fat frying products [47]. The principle of the thermally stable coatings was due to the methylcellulose and hydroxypropyl methylcellulose compounds exhibits thermo-gelation behavior [23]. This behavior occurs when suspensions are heated leading to formation of gel that returns below its gelation temperature, and viscosity is recovered [48].

Potato strips coated with methylcellulose reduces the oil absorption in the fried product around 40% without modifying the oil properties [49]. Similar results has been reported for reduction in oil uptake for fried potato chips [50], snacks [51], and chicken nuggets [52]. Hydroxypropyl

Table 11.1 Starch-based coating application on horticulture produce.

Biopolymers	Product	Inference	References
Rice starch, l-carrageenan, fatty acid esters of sucrose	Banana	Reduced weight loss, improved firmness, reduced starch degradation, and shelf life prolonged to 12 days from 6 days	[31]
Sodium alginate	Mango	Delay in weight loss, maintained higher concentrations of total phenolic compounds, and exhibited higher antioxidant enzyme activity	[34]
Cassava starch	Prickly pear	Delayed weight loss, softening and wilting of the fruit	[37]
Chickpea starch, stearic acid	Fresh-cut papaya	No weight loss and reduced microbial growth	[30]
Tapioca starch, natamycin, nisin	Gouda cheese	Protected against *Saccharamyces cerevisiae* and *Listeria innocua.* Maintained physicochemical properties	[38]
Rice starch, l-carrageenan, fatty acid esters of sucrose	Plum	Delayed respiration rate and retards the ethylene production	[32]

(Continued)

Table 11.1 Starch-based coating application on horticulture produce. (*Continued*)

Biopolymers	**Product**	**Inference**	**References**
Corn starch modified with ascorbic acid, tomato powder	Frozen bread dough	Bread with greater specific volume and less hardness	[15]
Waxy corn starch, gellan gum, sodium alginate	Rice cakes	Reduced retrogradation rate and hardness	[36]
Mango kernel starch, glycerol and/or sorbitol	Tomato	Retention of weight loss and firmness	[39]
Sodium alginate, pectin, ascorbic acid, citric acid, calcium chloride	Fresh-cut apple slices	Reduced microbial spoilage and wounding stress	[35]
Sodium alginate and/or pectin	Blueberry	Glossiness increased with decrease in lightness and increased blue hue color. Significant reduction in yeast and mesophilic aerobic bacteria.	[40]
Cassava starch, propolis extract	Strawberry	Maintained the higher free radical scavenging percentage and retained Vitamin-C content	[41]
Sodium alginate, ascorbic and citric acids (anti-browning agents)	Mango cubes	Ascorbic acid added coating shows color retention with enhanced antioxidant activity	[42]

methylcellulose coated chicken balls showed reduced fat content (up to 33.7%) after frying [53]. Methylcellulose coatings were more efficient in oil reduction than hydroxypropyl methylcellulose coatings [54]. During frying, coating dehydrates and deposits at the food surface, which inhibits oil absorption and creates a compact network [23].

Hydroxypropyl methylcellulose is a cellulose derivative in drug release applications due to its high hydrophilicity and non-toxicity. Major application includes controlled release, under hydration, a gelatinous layer forms through which active compound releases [55]. It is difficult to use as a stand-alone coating matrix. To overcome this drawback, additives mostly lipid components, namely, waxes, triglycerides, and fatty acids, are incorporated into the biopolymer matrix. This addition acts as barrier to moisture migration and decreases the water affinity of hydroxypropyl methylcellulose [56].

Addition of beeswax into hydroxypropyl methylcellulose matrix was one of the most significant approaches in edible coating application. Beeswax tightly packed the available crystalline structure and makes the hydroxypropyl methylcellulose more hydrophobic [57]. The hydroxypropyl methylcellulose-beeswax (20%) coated guava fruit extended the conservation time of the fruit to six days more than uncoated control fruit, with reduced respiration rate and transpiration processes. The coating acts effectively in maintaining the color and firmness of the fruit [58]. Plums coated with hydroxypropyl methylcellulose-beeswax formulation show reduced weight loss, softening, and bleeding. Also increase in beeswax ratio in the coating reduced film flexibility and increased plum bleeding [59]. Therefore, it is necessary to optimize additives ratio in coating matrix. This formulation produces similar result for cherry tomato [60], "Ortanique" mandarins [61], "Valencia" oranges [62], and plums [63].

11.4.1.3 Gum

Gums are a class of natural polysaccharides or carbohydrate polymers derived from renewable sources that have the ability to form gel or stabilizing emulsion system. Gum coating provides a semipermeable barrier to the food products, which cause reduced respiration rate, weight loss, and maintain the nutritional value. It can act as a carrier of plant extracts, essential oils, phenolic compounds, and vitamins to improve the mechanical, antioxidant, and antimicrobial activities [64].

Gellan gum is a water-soluble anionic polysaccharide extracted from *Pseudomonas elodea* through fermentation process. It consists of repeating units of [β-1,3-D-glucose, β-1,4-D-glucoronic acid, β-1,3-D-glucose, α-1,4-L-rhamnose], and two acyl units, acetate and glycerate linked to the glucose residue at the side of glucoronic acid [65]. It forms gels even at low concentrations when hot solutions are cooled in the presence of gel promoting cations [66]. This property helps to reduce the setting time of starch confectionary and prevents the sticky nature of cadies when exposed to warm environment [67]. Fish pieces that are deep-fried with gellan gum coating reduced the fat absorption by 63% [68]. Similar reduction in fat absorption was observed for low fat pastry mix [69].

Guar gum is a water-soluble polysaccharide extracted from the legume plant, *Cyamopsis tetragonolbus.* It provides high viscosities in aqueous solution even at low concentrations [70]. It is a galactomannan consisting of mannose [(1→4)-linked β-D-mannopyranose] backbone with galactose side groups [(1→6)-linked α-D-galactopyranose] [71]. It is a type of soluble dietary fiber, which is effective in reducing glycemic index of various products especially bread made with guar gum shows similar sensory and lower insulin production than control bread [72].

Xanthan gum is an exopolysaccharide synthesized by *Xanthomonas campestries,* which forms high viscous in cold and hot water at low concentration with excellent stability over a wide range of pH and temperature. It also shows resistance to enzymatic degradation [66]. The structure of xanthan gum consists of pentasaccharide units composed by glucose, mannose, glucoronic acid in the ratio of 2.8:2:2 [49]. The effect of xanthan gum coating enriched with citric acid and ascorbic acid on grapes explains the chelating effect exerted by citric acid in the coating, depletes the copper ion availability to polyphenoloxidase, and suppresses the enzyme activity [73].

Gum Arabic also known as acacia gum is the tree gum exudates of *Acacia senegal* and *Acacia seyal,* consists 1,3-linked β-D-galactopyranosyl units linked through 1,6-linkages [74]. It is the most used gum in fruits and vegetable coating [64]. Other gums that are used for edible coatings are karaya gum (*Sterculia urens*), tara gum (tara tree), peach gum (peach tree), cashew gum (*Anacardium occidentale*), flaxseed gum (*Linum usitatissimum*), tragacanth gum (*Astragalus gummifer*), and almond gum (*Amygdalus communis*). The effect of gum based edible coatings on different fruits and vegetables are tabulated in Table 11.2.

11.4.1.4 Protein

Proteins are hetero biopolymers, composed of 20 α-amino acids in addition to the amino and carboxyl groups involved in the formation of peptide bonds that contain side chains. These can be electrically charged or uncharged and either hydrophobic or polar, and each amino acid exhibits distinctive characteristics [91]. The primary structure of the protein

Table 11.2 Gum-based edible coatings of foods.

Biopolymers	Product	Inference	References
Xanthan gum, ascorbic and citric acids, glycerol	Grapes	Reduced polyphenoloxidase, ascorbic acid oxidase, and pectinmethyesterase enzyme activities.	[73]
Gum Arabic	Strawberry	Inhibited polyphenoloxidase activity, reduced decay percent, and delayed physicochemical changes.	[75]
Tragacanth gum, ascorbic acid	Fresh-cut apple slices	Inhibited polyphenoloxidase activity, retained ascorbic acid, and total phenol contents.	[76]
Cajanus cajan seed gums	Strawberry	Reduced citric acid consumption and physicochemical changes during storage.	[77]
Xanthan gum	Guava	Maintained ascorbic acid and polyphenol contents and reduced respiration rate.	[78]
Gum Arabic, tulsi extract	Guava	Reduced ripening and senescence. Inhibited mold growth.	[79]

(*Continued*)

Table 11.2 Gum-based edible coatings of foods. (*Continued*)

Biopolymers	Product	Inference	References
Gum Arabic, *Origanum vulgare* and *Rosemarinus officinalis*	Plum	Protected the soft rot incidence and phenolic compounds were retained	[80]
Gum Arabic	Raw and blanched tomatoes	Water activity, browning index and rate of increase in dry matter were reduced	[81]
Trangacanth, almond, quince seed gums	Banana slices	Reduced shrinkage and inhibited browning index	[82]
Flaxseed gum, lemongrass oil	Pomegranate	Reduced microbial growth and delayed physicochemical changes during storage	[83]
Peach gum	Cherry tomato	Retarded changes in sugar, ascorbic acid and acidity and reduced respiration rate	[84]
Gellan gum	Mango bars	Preserved color and volatiles with improved firmness and appearance	[85]
Gum Arabic, ginger oil/ginger extract	Papaya	Served as biofungicide for controlling anthracnose	[86]
Guar gum, cinnamon oil, potassium sorbate	Cucumber	Improved antioxidant and antibacterial activities	[87]

(*Continued*)

Table 11.2 Gum-based edible coatings of foods. (*Continued*)

Biopolymers	Product	Inference	References
Guar gum	Fried potato chips	Reduction in oil adsorption and crack formation	[88]
Almond gum	Fried potato chips	Improved color, texture, and sensorial characteristics	[89]
Xanthan gum, cinnamic acid	Fresh - cut pears	Inhibited polyphenoloxidase and peroxidase enzyme activities and retards oxidative browning	[90]

depends on its specific polypeptide chain and the spatial orientation of these chains is the secondary structure. The tertiary structure indicates the three-dimensional organization of the polypeptide chains to form globular, fibrous, or randomly coiled conformations. The quaternary structure depicts the result of interaction between different polypeptide chains that determines the final protein structure [92].

Proteins are classified into two types—globular and fibrous. The first one is soluble in water or aqueous solutions of acids, bases, or salts. The folded spherical structures of globular proteins are held by a combination of hydrogen, ionic, hydrophobic, and disulphide bonds. Wheat gluten, soy protein, corn zein, and whey protein are some of the globular proteins used for edible coating. The latter one is insoluble in water and structural element for the animal tissues. Collagen is a major fibrous protein used in edible coating. However, plant proteins are grouped into four categories—albumin (water-soluble), globulin (soluble in saline solutions), prolamine (soluble in aqueous alcohol), and glutelins (extractable in alkali) [93].

The denaturation of the protein structure is the necessary step to unfold the chains for network formation. The chain-to-chain interactions of protein molecules made the polymer matrix stronger and less permeable to gases and vapors [94]. Proteins have multiple sites for chemical interaction that permits property improvement and tailoring [95]. The intrinsic properties of the coating components and extrinsic processing factors affect the coating performance [96]. Intrinsic properties of proteins include amino acid composition, crystallinity, hydrophobicity/hydrophilicity, surface charge, Iso-electric point, molecular size, and three-dimensional shape. Processing temperature, drying conditions, pH, ionic strength, salt-type, relative humidity during processing and storage, shear and pressure are the extrinsic factors affecting protein based edible coating [91].

Zein is also known as the storage prolamine protein of corn. It consists of high amount of leucine, alanine, and other non-polar amino acids, which makes them water-resistant and thermoplastic [91, 97]. Zein is insoluble in water. The two major fractions are alpha-zein and beta-zein, which dissolves in 90% and 60% ethanol [97]. The formation of hydrophobic, hydrogen, and limited disulphide between the zein chains makes them an excellent biopolymer [98].

Gluten is the wheat protein consists of two insoluble portions—gliadins and glutenins. It consists of a large amount of non-polar amino acids and small portion of charged amino acids such as lysine, histidine and arginine that enhances the protein aggregates through hydrophobic interactions [99]. The disulphide bonds present in glutenin and gliadin determine the strength of protein matrix [100].

Soy protein is the globulin protein fraction of soybean. This protein consists of both polar and non-polar side chains. Soy protein exhibits strong intra- and inter-molecular interactions, such as hydrogen bonding, dipole-dipole, charge-charge, and hydrophobic interactions [101]. The two main globular proteins are glycinin (11S-globulin) and beta-conglycinin (7S-glubulin) [102]. The denaturation of glycinin forms intermolecular disulphide bonds that affects the mechanical strength of protein matrix [103].

Moreover, the cross-linking between protein chains enhances the protein network formation and improves the stability of the matrix. Thermal treatment [104], irradiation [105], enzymatic and chemical cross-linking [106, 107] induce various bond formations and improve the tensile strength and barrier properties of the coatings. The recent literature survey depicts a high number of publications related to the application of soy protein, in one or the other form, as the basis of the coating (Table 11.3).

11.4.1.5 Waxes

Wax represents a series of naturally or synthetically produced non polar substances. Wax consists of zero or little polar compounds that are hydrophilic, so that these compounds cannot interact with water molecule. This nature possesses difficulty for the wax compounds to spread and forms thin layer over food product. Wax coatings have been reported to exhibit strong resistance to moisture migration. Though, irregular thickness,

Table 11.3 Recent studies with application of plant proteins as potential coatings.

Biopolymers	Product	Inference	Reference
Soy protein isolate, honey (dip)	Fresh-cut pineapple	Shelf life extended (2 weeks), retention of phenolic compounds and color	[108]
Soy protein isolate, honey (dip)	Fresh-cut kajari melon	Reduced microbial growth with better sensory scores	[109]
Zein, tannic acid	Guava	Maintained fruit quality attributes, extended stability, and reduced gas exchanges	[110]
Soy protein isolate, chitosan	Apricot	Prevented weight loss, softening, and pectin degradation	[111]
Soy protein isolate, ferulic acid	Fresh-cut apple	Controlled weight loss and firmness	[112]
Soy protein isolate	Beef patties	Delayed lipid oxidation, maintained color deterioration and texture	[113]
Soy protein isolate	Potato pellet chips	Reduction in oil uptake and improved sensory characteristics	[114]
Zein, tannic acid	Guava	Maintained fruit quality attributes, extended stability, and reduced gas exchanges	[110]
Zein, hazelnut skin extract, α-tocopherol	Hazelnut	Reduced oxidative rancidity in the roasted product without any quality changes	[115]
Zein, polymeric chelator	Fish balls	Retards the bacteria growth and extended shelf life	[116]

greasy surfaces, and unacceptable taste of fat are the limitations to the application of wax coating [117].

Carnauba wax is a natural plant wax extracted from Brazillian palm tree (*Copernica cernifera*) [118]. The carnauba wax and carnauba wax with glycerol monolaurate coated Indian jujube shown in Figure 11.2 explain that the wax coating inhibits the chlorophyll degradation and slows down the respiration rate in the day 8 and 12 of storages when compared with control (uncoated) fruit [119]. The increase in redness during storage indicates the jujube ripening, senescence, and chlorophyll degradation in uncoated fruit.

In a study conducted, it was observed that the preservation of fruits through antagonistic microorganism such as fungi, yeast, and bacteria with inactivating mechanisms of competition of nutrients and niches, antibiosis, lytic enzymes, volatile inhibitory metabolites, pH decrease, and parasitism were possible [120]. The strawberries coated with candelilla wax and Bacillus subtilis HFC 103 biocontrol agent showed significant severity index against *Rhizopus stolonifera* and completely restricted the fruit decay till 6 days of storage [121].

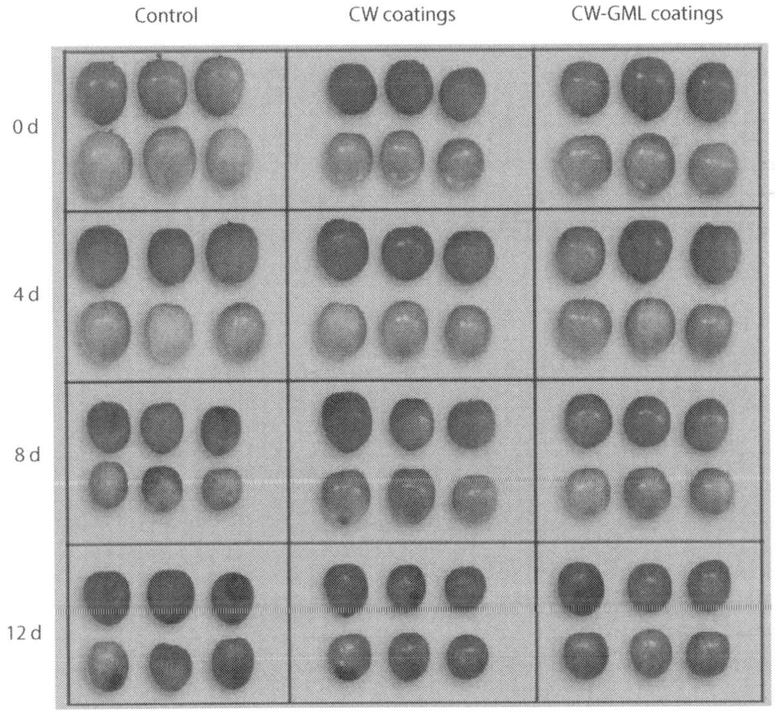

Figure 11.2 Control, carnauba wax (CW) coated, carnauba wax-glycerol monlaurate (CW-GML) coated Indian jujube fruits at 0, 4, 8, and 12 day of storage [119].

Lacquer wax is a fatty acid resource present in the mesocarp and seed of the berries from the Chinese tree *Toxicodendron vernicifluum*. It is consisting of hexadecadienoic acid, eicosenoic acid, and dodecadenioic acid. Therefore, lacquer wax was more elastic and softer than other waxes [122]. It is cheap, abundant, natural, edible, and easy to prepare. The lacquer wax coated kiwifruit slows down the respiration rate and retards the activity of pectinmethylesterase and polygalacturanose activities [123]. The different wax coated fruits and vegetables are depicted in Table 11.4.

Table 11.4 Recent application of wax as an edible coating.

Biopolymers	Product	Inference	References
Carnauba wax, glycerol monlaurate	Indian jujube	Glossiness maintained, reduced chlorophyll degradation	[119]
Lacquer wax	Kiwifruit	Suppressed the respiration rate, ethylene production and fruit softness	[123]
Carnauba wax, galactomannan	Guava	Reduced chlorophyll degradation and prevented from chilling injury	[124]
Candelilla wax, *Bacillus subtilis* HFC 103 (biocontrol agent)	Strawberry	Reduced fruit decay, protection against *Rhizopus stolonifer*	[121]
Candelilla wax, pectin, aloe mucilage	Avocado	Inhibited internal grey pulp development	[125]
Carnauba wax	Eggplant	Reduced physiological weight loss, enhanced antioxidant activity of eggplants	[126]
Candelilla wax, fermented extract of tarbush	Apple	Maintained firmness, retained sensory scores	[127]

11.4.2 Animal Source

11.4.2.1 Chitosan

Over the past decade, the focus of the marine processing industry has been directed towards the utilization of its waste in a feasibly better and creative way to cater the generated carbon footprints during the processing of the waste. One of these was the utilization of crustacean shells, which accounts for a high volume of waste that is being developed during the in shellfish processing [128]. This waste cannot be consumed directly for food purposes, but with deliberate use of processing can prove vital in several applications, one of which is developing it as an edible coating.

Chitin, the abundantly found biopolymer in the exoskeletons of these in-shell crustaceans, is structurally identical to cellulose. Chitosan, which turns out to be a copolymer of β-(1-4)-2-amino-D-glucose and β-(1-4)-2-acetamido-D-glucose, is the deacetylated form of chitin, which is known for the antimicrobial properties related to its cationic and film-forming properties [129]. Several *in vitro* trials on various food matrices have proved its ability to act as a natural food preservative. But the fact that chitosan coatings are highly permeable to water vapor limits, its use since control over the moisture is the most desirable characteristic of the coatings being used for foods stored in moist environment [130]. Hence, still extensive studies are needed to cater the limitations of chitosan to be used as a coating alone.

In the aforementioned context related to drawbacks of chitosan, investigations have been carried out depicting the increase in hydrophobicity by the inclusion of neutral lipids, waxes, and certain clays [131], whereas some chemical and physical alterations (methylated or sulfated polylactic acid) are also being utilized for increasing the mechanical properties by the means of ultrasound, irradiation, and crosslinking agents.

The outbreak of microbes in the processed food processing industry is the major concern luring over the food processors to store and monitor the frozen fruits and vegetables and even the fresh-cut meat and poultry flesh [132–134]. The antimicrobial activity of chitosan leads to a leakage of intercellular constituents and provokes the osmotic imbalance, resulting in the growth inhibition of the particular microbe. It's fascinating to observe a high number of publications (over 2,038) related to chitosan and its characteristic activity which is due to the physicochemical and biological attributes, out of which several supports the use of chitosan as edible coatings and even as films. There's no study till date which can describe the exact mechanism of microbial inhibition due to the antimicrobial activity of chitosan but many proposals are in line to depict the same. Some of the applications for the shelf life prolongation or preservation of different food matrices are compiled in Table 11.5.

Table 11.5 Recent studies on chitosan-based edible coatings.

Effect	Products	Added value	References
Antioxidant/Antimicrobial Activity	Blueberry	Extended shelf life of 5 days	[135]
	Cherry Tomato	Control over mycelial growth and fungi	[136]
	Cucumber	Control of *Phytophthora Drechsleri* with extended shelf life of 21 days	[137]
	Table Grape	Delayed mold growth (*A. niger*, *P. expansum*, and *R. tolonifera*)	[138]
Sensory Enhancement	Mango	Increased flesh firmness	[139]
	Star Fruit	50% extended glossiness with shelf life of 40 days	[140]
	Bamboo Shoot	Inhibited activity of PPO and peroxidase	[141]
	Litchi	Lower hydrogen peroxide and MDA content	[142]
	Pumpkin	Control over PPO and POD release	[143]
Preservation (Cheese)	Ricotta	Retained sensory attributes with enhanced shelf life up to 30 days and control over majorly affecting micro-organism (Listeria innocua, Escherichia coli, and mesophilic aerobic bacteria)	[144]
	Mozzarella		[145]
	Saloio		[146]
	Cheddar		[147]
	Soft white		[148]

(Continued)

Table 11.5 Recent studies on chitosan-based edible coatings. (*Continued*)

Effect	Products	Added value	References
Preservation (Meat and other products)	Shrimps	17 days shelf life with control over Pseudomonas	[149]
	Chicken Breast	Postponed oxidative deterioration and enhanced shelf life of 10 to 15 days	[150]
	Beef Fillets	30 days prolonged shelf life	[151]
	Dry fermented sausages	Better odor and flavor characteristics with notable increase in stability of the sausage core	[152]
	Beef steaks	21 days shelf life with reduced drip loss	[153]
	Rainbow trout fillets	22 days shelf life with control over coliform count	[154]

11.4.2.2 Animal Protein

11.4.2.2.1 Gelatin

Coming on to the meat processing industry, gelatin is one of its derivatives from the collagen which is extensively being used as a biopolymer for its impressive functional properties. As it comes under the section of protein-based biopolymer, it possesses 3-D structures governed by the several interactions, among them, the major are non-covalent forces [155]. The use of such polymer as a coating for highly perishable commodities like meat and fish cuts depends on the cost, availability, mechanical capacity, and its optical characteristics.

During the 1930s, pigskin was among the first materials been used to manufacture the gelatin and it still holds a large percentage of production under its use in the large-scale food production; else gelatin is also obtained from cattle bones but it comes out to be a more complex and costly method when compared with the former [156]. Apart from porcine and bovine gelatine which comes under the category of the meat derived gelatin, the utilization of fish gelatin is also on a high during the recent times, but only accounts to a mere 1.5% of the total production [157].

The attributes of the gelatin being manufactured are primarily based on the characteristics of initial collagen and on the extraction process, up to some extent. The process onsets from the warm water extraction, destabilization of the hydrogen, and the covalent bonds and terminates with the subsequent conversion into its solubilized form. The structural skeleton of gelatin consists of a polymeric α-chain, two covalently cross-linked α-chains which eventually give rise to a β-chain, and γ-chains. Gelatin's amino acid structure follows the trail as: Ala-Gly-Pro-Arg-Gy-Gly-4Hyp-Gly-Pro- (Figure 11.3). The main property because of which the use of gelatin has gained importance is the gel strength (or known as bloom value) it develops that also enables us to use it in the powder and granulated form [158].

Gelatin has a wide application with film-forming capacity as one of the well-known attributes. It has been studied extensively for its ability to protect the food from getting dried and from the exhibition to light and exposure to oxygen. But alongside substances like cross-linkers, plasticizers, oils, and other additives, gelatin coatings have overcome all its drawbacks and provided better stability to the food products [159]. It can even be seen in some literature that the use of gelatin alongside other biopolymers like whey proteins, starch, pectin, and chitosan have proved to be a better strategy to refine its mechanical and water resistance properties [160]. But when it comes to the utilization, major reported publications are for the

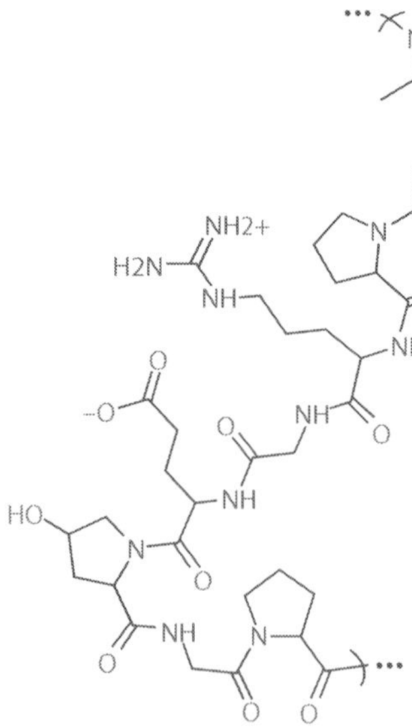

Figure 11.3 Amino Acid Skeleton of Gelatin.

casting of films and use of gelatin for the purpose of coating lies still in its making [161].

Unlike chitosan, Shakila, Jeevithan [162] claim the enhancement of antimicrobial properties and shelf stable product for fish steaks by the use of grouper bone gelatin when used with the clove and pepper essential oil as additives. This depicts the strength and the effect of the additives in the formation of the coating. Similarly, Wu, Liu [163] suggest the application of fish gelatin for the purpose of encapsulation for the monitored delivery of the active compounds. Here, also the formulation was incorporated with essential oil (cinnamon) to enhance its properties. Food grade gelatin has also been reported to be used for coating shrimps, incorporated with orange peel essential oil for an edge with the increased antioxidant activity [164]. Different matrices of gelatin are used as coatings for the various meat and fishery and horticultural produce in recent years (Table 11.6).

Table 11.6 Recent literatures with the application of edible gelatin-coating.

Product	Coating	Method	Reference
Nile tilapia fillets	Fish scale gelatin + Chitosan	Dipping in formulation matrix	[165]
Beef meat	Fish gelatin + Pectin		[166]
Hake medallions	Fish gelatin + Carvacrol		[167]
Strawberry	Gelatin + Mentha oil		[168]
Pastirma	Gelatin + Alginate		[169]
Tuna fillet	Gelatin + Chitosan + Clove oil		[170]
Beef steaks	Gelatin + Pectin + Sodium alginate		[171]
Shrimps	Gelatin + glycerol + Tween 20 + sorbitol + orange leaf extracts		[164]
Raw Beef	Gelatin + Tween 80 + Thyme & Rosemary oil		[172]
Pork loin	Pork gelatin		[173]
Kabanosy dry sausage	Porcine gelatin + Kappa-carrageenan + Glycerol		[174]

11.4.2.3 Milk Protein: Whey and Casein

Milk protein contains various essential components that make it relevant to be used as a biopolymer. Casein is found abundantly in the ruminant's milk proteins while the cow's milk gives around 18% of the whey protein.

The micelle structure of the casein yields to form spherical complexes which are colloids of calcium phosphate. This paves way for the utilization of the film-forming property of the casein protein in order for edible coating over the perishable food commodities. Here, the method of extraction governs the solubility of the film being produced. Coatings with essential oils from the Zataria multiflora and ginger rhizomes have been successfully deployed for the shelf life extension of certain food commodities. Now, coming on to the whey protein, which is developed during the cheese production, the principal compound present deals with the properties which are the key to make the why protein cater as a coating material.

Under the classification, the Whey Protein Isolate (WPI) and Whey Protein Concentrates (WPC) are the most widely used as coating materials. Literatures reported for the application of coating onto the roasted peanut kernels [175], cut apples [176], pike-perch fillets [177], and even on carrots claims the efficacy of WPI as an effective coating material. WPI also depicted wonderful results when used under modified atmospheres and vacuum packaging for tuna [178]. Blended formulations also have proved beneficial in the past as Catarino, Alves-Silva [179] reports the use of cinnamon and rosemary essential oils applied on the sausages for its manufacturing. This was supervised for a period of four months and even proved beneficial in delaying microbial spoilage and preventing discoloration.

11.4.2.4 Shellac

Shellac is a natural biopolymer being prepared by the *Kerriar lacca* insect's secretions, which is popular in some of the South Asian countries including Thailand, India, and China. It mainly consists of a combination of hard and soft resins and single ester with polyhydroxy-polybasic acids, *viz.*, laccijalaric, jalaric, and aleuritic acid [180]. This is gaining interest in recent times due to its excellent protection against the moisture and because of its exceptional glossy finish films and its properties that are produced on the surfaces [181]. This may be used as a singly or can be formulated for combinations also. The sole purpose of this intervention in the shellac's formulation is to control the permeability of the coating for delaying the senescence process of the agricultural produce [182]. This primarily affects the oxidizing enzymes that govern the ethylene production, moisture loss, and the firmness of the commodity. But, shellac is also prone to the instability caused because of polymerization thus reducing its mechanical strength; therefore, the need of modification arises while using the shellac. Thus, research on the best possible formulation can be done with the efficiency of coating in mind and can be further optimized to ensure high marketability [183].

The coating which was being used until the introduction of shellac had many limitations which direct emphasis on the quality which is reflected on the stability of the product. Polysaccharides, as well as the proteins, could not handle the situations as the issue of low moisture protection persisted. The lipid coatings, due to their low polarity, gave high moisture barrier but their low cohesiveness laid down the difficulty in the formation of independent polymer [156]. Thus, this formed a gap to be filled by a hydrophilic polymer to improve the weakness against the moisture, and thus, the use of shellac began with the polysaccharides and protein-polymer [180].

Saberi, Chockchaisawasdee [184] depicted the utilization of shellac while formulating homogenous layer of the pea starch/guar gum along with Tween 20. Other studies published regarding the combination of carnauba wax and shellac alongside lemongrass oil for coating apples also had a high impact. The coated apples had lower count of yeast and molds with even emergence of bacteria was reduced to an extent [185]. Later, Soradech, Nunthanid [186] investigated the delay in ripening characteristics of banana on using the shellac-based composite coating. This ended up increasing the shelf life with the quality parameters retained which were monitored on the color profile of the surface of the banana (Figure 11.4).

11.5 Composite Coatings

Composite coatings consist of two or more biopolymers that form the network together to overcome the limitations of individual component. Composite coating exhibits the advantage of each biopolymer and produces unique characteristic coating. Polysaccharide-polysaccharide, protein-protein, polysaccharide-protein, polysaccharide-lipid, and protein-lipid are the different blends that are used for composite edible coatings.

Potassium sorbate, organic salt of sorbic acid, holds scope as effective fungistatic agent in perishables. Kowalczyk, Kordowska-Wiater [187] developed a composite coating for fungus infected apricots, coating

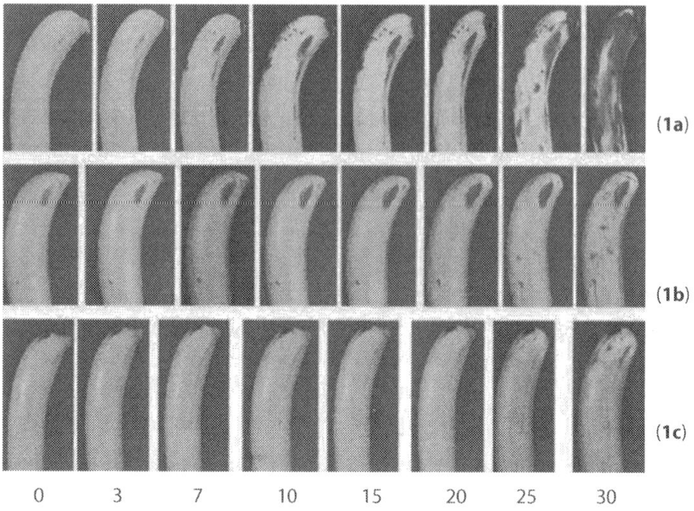

Figure 11.4 Color profile of banana without coating (1a), with shellac coating (1b) and with the formulated composite film (1c). (Adapted from [186]).

consists of carboxymethycellulose and candelilla wax blend with and without potassium sorbate. *Botrytis cinerea* and *Monilinia fructigena* were the most sensitive to the antifungal treatment. The significant reduction in oxygen consumption was observed for the coated apricots, which in turn relates with the controlled enzyme activity and softness of the fruit.

Nano polymeric materials can have an edge in the delivery and retaining the bioactive in the food matrix. They are generally introduced into a polysaccharide or protein matrix to form a mixture of particles for the purpose of coating with at least one of the component in the nanoscale. This is also known as "temporal distribution system" that may release the active component at the time of storage. These can be used in the form of nanoemulsion, polymeric nanoparticles, solid lipid nanoparticles (SLNs), lipid nanocarriers, inorganic/organic nanocomposites, and even as nanotubes and nanofibers with direct application on the food surface (Figure 11.5).

The micro-fibrillated carrot was coated with starch as biopolymer in order to have an extended shelf life. The bioactive acted as a medium to strengthen the coating's mechanical property and its functionality and compatibility [189]. Similarly, chitosan coating reinforced with the cellulose nanofibers (CNF) (manufactured in the name of Avicel®) improved the glass transition temperature and its vitreous nature [190]. Wang, Liu [191] reported that CNF was used along with the fish gelatin and palmitic acid

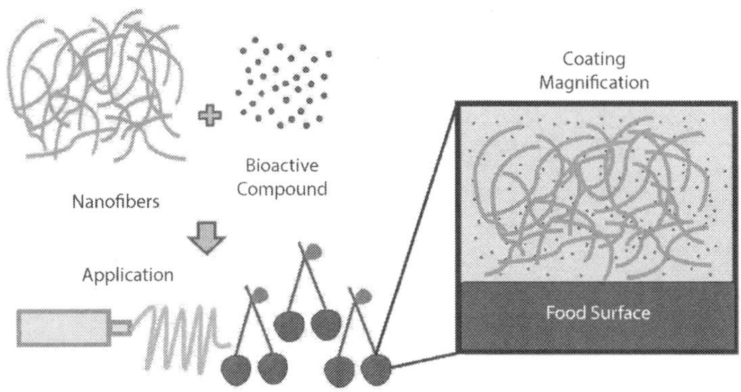

Figure 11.5 Schematic representation of nanofiber application as coating (Adapted from [188]).

formulation to increase resistance against the gases as well as the strength up to some extent. In another reported study, the bioactive compound from the turmeric rhizomes, curcumin, encapsulated in zein nanofiber was used to coat apples for improving its shelf stability. The surface of these apples was inoculated with *Penicillium expansum* and *Botrytis cineral* to check the efficacy of the composite coating which was electrospun over the surface of apples. The use of nanocomposite coating successfully inhibited the microorganisms for a fortnight's time [192]. Cui, Yuan [193] developed a composite with mounted bacteriophage, encapsulated in lysosomes for a controlled release, to counter the microbial attack on stored beef. This composite assembly was mounted on the chitosan and the product came out to be a very efficient antibacterial coating without affecting the sensorial characteristics of beef (Figure 11.6).

11.6 Current Trends

Bionanocomposites consist of reinforced matrices with organic or inorganic (nanoparticles) with a 1-D nano structure (1–100 nm). Conventional microscopic fillers present poor interaction at the interface, while nanofillers present better interaction between both phases due to high surface/volume ratio [194]. These materials exhibit improved thermal, mechanical, and barrier properties due to the high aspect ratio and the surface area of the particles formulated. The properties of these composites are judged by the range of the dispersion of nanoparticles in the biopolymeric structure and by the interaction within them [195].

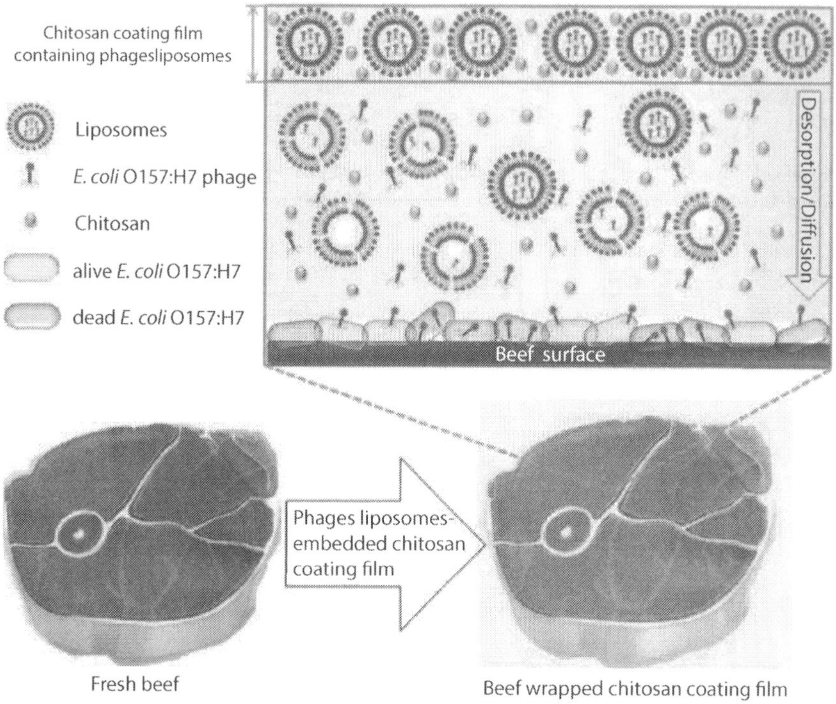

Figure 11.6 Schematic representation of antibacterial composite coating of chitosan on beef (Adapted from [193]).

Edible coatings, with time, has changed the use and perception associated with the biopolymers for food coatings. With inclusion of nanomaterials, the reach of supplying the materials that are essentially needed has increased with a tremendous rate. The bioactive which earlier were considered to be ingested only by its natural source can now be coated in the biopolymer and delivered to a pre-known targeted site [196]. But, evolution follows the invention, and thus, there remains a need for improvement in a particular entity and its methods to be used, always.

In the case of coatings, such changes are being brought up by making the coatings self-able, so that it can detect any deliberate change occurring during its presence over the particular food commodity. This calls for the development of smart or active coatings. This is a novel approach towards monitoring the horticultural produce. It makes use of immobilized nanoparticles (antimicrobial, at times) and has the capability to monitor the internal gas composition and to cease the ethanol fermentation. These coatings can also be as self-adhesive coatings which are able to communicate with the end consumer about its state and condition at the time

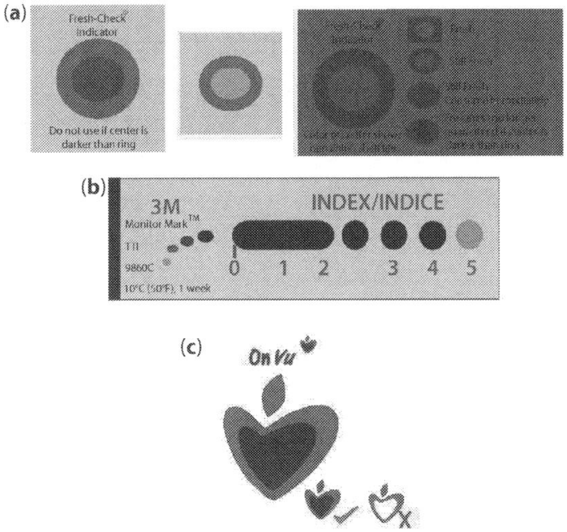

Figure 11.7 Commercially available indicators; (a). Fresh-check® label; (b). Monitor Mark™ label; (c). OnVu™ label (Adapted from [198]).

of ingestion. This also turns out to be a helping technique which can contribute to improving the HACCP and QACCP systems [197]. These sensors can be used as indicators and data carriers wherever it fits the purpose and indicates the possible attribute which is being monitored to check the quality. Figure 11.7 illustrates one of the fine examples of such techniques commercially being used nowadays.

Secondly, the next most talked about technique in recent time is the layer-by-layer or L-B-L coating technique which uses different biopolymers themselves to overcome the drawback of the others [199]. This limits the use of synthetic additives and makes our aim towards green manufacturing up to some extent. It consists of formation of groups of sequentially adsorbed complementary molecules which yields the formation of multilayer coatings. It is better than the earlier used Langmuir-Blodgett and self-assembled monolayer techniques which needed expensive machinery and amphiphilic materials [200]. Its applications span from its use in sensors, drug delivery systems, catalysis and to even in the energy storage and conversion [201].

11.7 Conclusion

In accordance with the industrial and substantial viewpoints, biopolymer coatings are the future materials in the processing industry. Several naturally occurring biopolymers are being used to produce shelf stable and self-able composite for the purpose of coating of the food commodities. However, it is critical to realize the reason that biopolymer coatings can become a mandate luring with time due to its attributes. The general need of novel fabrication and the nanomaterials can help in the betterment of the known barrier (to water, oxygen, and UV/light) properties of such coatings. Notably, this can implement unique features with add-ons in biopolymer films, opening a medium of novel applications for food industry. However, the perniciousness, Geno-toxicity, and neurotoxic effects are not exonerated till date. So, it becomes important to focus the studies towards the nanoparticle migration into meat, fish, and even beverages with its risks accompanying furthermore possible intake, before the nanoparticle application in industrial packaging. Moreover, government assistance in the emerging countries and reducing a huge cost leap in comparison of biopolymers to that of the plastic polymers are the necessary factors to be considered, to increase the biodegradable packaging market.

References

1. Prameela, K., Mohan, C.M., Ramakrishna, C., Biopolymers for Food Design: Consumer-Friendly Natural Ingredients, in: *Biopolymers for Food Design*, A.M. Grumezescu and A.M. Holban (Eds.), pp. 1–32, Academic Press, Cambridge, UK, 2018.
2. Baldwin, E.A., Edible coatings for fresh fruits and vegetables: Past, present, and future, in: *Edible coatings and films to improve food quality*, J.M. Krochta, E.A. Baldwin, M. Nisperos-Carriedo (Eds.), pp. 25–64, CRC Press, Lancaster, PA, 1994.
3. Lazic, V. and Gvozdenovic, J., Biopolimers [i.e Biopolymers] as packaging materials. *Mleko i mlecni proizvodi*, 18, 19–22, 2007.
4. Robertson, G., Edible, biobased and biodegradable food packaging materials, in: *Food packaging-Principles and practice*, pp. 49–90, CRC Press, Boca Raton, FL, 2013.
5. Song, J. *et al.*, Biodegradable and compostable alternatives to conventional plastics. *Philos. Trans. R. Soc. B: Biol. Sci.*, 364, 1526, 2127–2139, 2009.
6. Vásconez, M.B. *et al.*, Antimicrobial activity and physical properties of chitosan–tapioca starch based edible films and coatings. *Food. Res. Int.*, 42, 7, 762–769, 2009.
7. Su, J.-F. *et al.*, Structure and properties of carboxymethyl cellulose/soy protein isolate blend edible films crosslinked by Maillard reactions. *Carbohydr. Polym.*, 79, 1, 145–153, 2010.
8. Ruban, S.W., Biobased packaging-application in meat industry. *Vet. World*, 2, 2, 79–82, 2009.

9. Garcia, M.A., Martino, M.N., Zaritzky, N.E., Starch-based coatings: Effect on refrigerated strawberry (*Fragaria ananassa*) quality. *J. Sci. Food Agric.*, 76, 3, 411–420, 1998.
10. Pranoto, Y., Rakshit, S., Salokhe, V., Enhancing antimicrobial activity of chitosan films by incorporating garlic oil, potassium sorbate and nisin. *LWT-Food Sci. Technol.*, 38, 8, 859–865, 2005.
11. Chollet, E. *et al.*, Monitoring nisin desorption from a multi-layer polyethylene-based film coated with nisin loaded HPMC film and diffusion in agarose gel by an immunoassay (ELISA) method and a numerical modeling. *Innov. Food Sci. Emerg. Technol.*, 10, 2, 208–214, 2009.
12. García-Betanzos, C.I. *et al.*, Physicochemical, total phenols and pectin methylesterase changes on quality maintenance on guava fruit (*Psidium guajava L.*) coated with candeuba wax solid lipid nanoparticles-xanthan gum. *Food Res. Int.*, 101, 218–227, 2017.
13. Halagarda, M., Effects of trehalose and dough additives incorporating enzymes on physical characteristics and sensory properties of frozen savory Danish dough. *LWT-Food Sci. Technol.*, 86, 603–610, 2017.
14. Steffolani, M.E. *et al.*, Use of enzymes to minimize dough freezing damage. *Food Bioprocess Tech.*, 5, 6, 2242–2255, 2012.
15. Galvão, A.M.M.T. *et al.*, Coating development with modified starch and tomato powder for application in frozen dough. *Food Packaging Shelf*, 16, 194–203, 2018.
16. Song, Y. *et al.*, Effect of sodium alginate-based edible coating containing different anti-oxidants on quality and shelf life of refrigerated bream (Megalobrama amblycephala). *Food Control*, 22, 3–4, 608–615, 2011.
17. Popović, S.Z. *et al.*, Biopolymer Packaging Materials for Food Shelf-Life Prolongation, in: *Biopolymers for Food Design*, A.M. Grumezescu and A.M. Holban (Eds.), pp. 223–277, Academic Press, Cambridge, UK, 2018.
18. Oms-Oliu, G. *et al.*, Recent approaches using chemical treatments to preserve quality of fresh-cut fruit: A review. *Postharvest Biol. Tec.*, 57, 3, 139–148, 2010.
19. Han, J.H., Antimicrobial food packaging, in: *Novel Food Packaging Techniques*, vol. 8, R. Ahvenainen (Ed.), pp. 50–70, Woodhead Publishing Ltd, Cambridge, UK, 2003.
20. Atarés, L. and Chiralt, A., Essential oils as additives in biodegradable films and coatings for active food packaging. *Trends Food Sci. Technol.*, 48, 51–62, 2016.
21. Kerch, G., Chitosan films and coatings prevent losses of fresh fruit nutritional quality: A review. *Trends Food Sci. Technol.*, 46, 2, 159–166, 2015.
22. Rojas-Graü, M.A., Soliva-Fortuny, R., Martín-Belloso, O., Edible coatings to incorporate active ingredients to fresh cut fruits: A review. *Trends Food Sci. Technol.*, 20, 10, 438–447, 2009.
23. Suárez, R.B. *et al.*, Comparison of the deep frying process in coated and uncoated dough systems. *J. Food Eng.*, 84, 3, 383–393, 2008.
24. Rayner, M. *et al.*, Development and application of soy-protein films to reduce fat intake in deep-fried foods. *J. Sci. Food Agric.*, 80, 6, 777–782, 2000.

25. Jabeen, N., Majid, I., Nayik, G.A., Bioplastics and food packaging: A review. *Cogent Food Agric.*, 1, 1, 1117749, 1-6, 2015.
26. Molavi, H. *et al.*, A review on biodegradable starch based film. *J. Microbiol. Biotechnol. Food Sci.*, 4, 5, 456–461, 2015.
27. Bertuzzi, M.A. *et al.*, Water vapor permeability of edible starch based films. *J. Food Eng.*, 80, 3, 972–978, 2007.
28. Peressini, D. *et al.*, Starch–methylcellulose based edible films: Rheological properties of film-forming dispersions. *J. Food Eng.*, 59, 1, 25–32, 2003.
29. Cazón, P. *et al.*, Polysaccharide-based films and coatings for food packaging: A review. *Food Hydrocolloids*, 68, 136–148, 2017.
30. Kathiresan, S. and Lasekan, O., Effects of glycerol and stearic acid on the performance of chickpea starch-based coatings applied to fresh-cut papaya. *CYTA-J. Food*, 17, 1, 365–374, 2019.
31. Thakur, R. *et al.*, A starch edible surface coating delays banana fruit ripening. *LWT-Food Sci. Technol.*, 100, 341–347, 2019.
32. Thakur, R. *et al.*, Development and application of rice starch based edible coating to improve the postharvest storage potential and quality of plum fruit (*Prunus salicina*). *Sci. Hortic.-Amsterdam*, 237, 59–66, 2018.
33. Montero-Calderón, M., Rojas-Graü, M.A., Martín-Belloso, O., Effect of packaging conditions on quality and shelf-life of fresh-cut pineapple (*Ananas comosus*). *Postharvest Biol. Tec.*, 50, 2–3, 182–189, 2008.
34. Rastegar, S., Hassanzadeh Khankahdani, H., Rahimzadeh, M., Effectiveness of alginate coating on antioxidant enzymes and biochemical changes during storage of mango fruit. *J. Food Biochem.*, e12990, 43, 11, 1–10, 2019.
35. Guerreiro, A.C. *et al.*, The effect of edible coatings on the nutritional quality of 'Bravo de Esmolfe' fresh-cut apple through shelf-life. *LWT-Food Sci. Technol.*, 75, 210–219, 2017.
36. Eom, H. *et al.*, Development of a starch/gum-based edible coating for rice cakes to retard retrogradation during storage. *LWT-Food Sci. Technol.*, 97, 516–522, 2018.
37. Nunes, V.X. *et al.*, Effect of Cassava Starch Coating on the Quality and Shelf Life of Prickly Pear in Refrigerated Storage. *J. Exp. Agric. Int.*, 37, 6, 1–11, 2019.
38. Berti, S. *et al.*, Edible coatings on Gouda cheese as a barrier against external contamination during ripening. *Food Biosci.*, 100447, 31, 1–8, 2019.
39. Nawab, A., Alam, F., Hasnain, A., Mango kernel starch as a novel edible coating for enhancing shelf-life of tomato (*Solanum lycopersicum*) fruit. *Int. J. Biol. Macromol.*, 103, 581–586, 2017.
40. Mannozzi, C. *et al.*, Study on the efficacy of edible coatings on quality of blueberry fruits during shelf-life. *LWT-Food Sci. Technol.*, 85, 440–444, 2017.
41. Thomas, A.B. *et al.*, Cassava starch edible coating incorporated with propolis on bioactive compounds in strawberries. *Ciênc. Agrotec.*, 40, 1, 87–96, 2016.
42. Robles-Sánchez, R.M. *et al.*, Influence of alginate-based edible coating as carrier of antibrowning agents on bioactive compounds and antioxidant activity in fresh-cut Kent mangoes. *LWT-Food Sci. Technol.*, 50, 1, 240–246, 2013.

43. Wang, S., Lu, A., Zhang, L., Recent advances in regenerated cellulose materials. *Prog. Polym. Sci.*, 53, 169–206, 2016.
44. Bourtoom, T., Edible films and coatings: Characteristics and properties. *Int. Food Res. J.*, 15, 3, 237–248, 2008.
45. Rydz, J. *et al.*, Present and future of biodegradable polymers for food packaging applications, in: *Biopolymers for Food Design*, A.M. Grumezescu and A.M. Holban (Eds.), pp. 431–467, Academic Press, Cambridge, UK, 2018.
46. Villalobos, R. *et al.*, Gloss and transparency of hydroxypropyl methylcellulose films containing surfactants as affected by their microstructure. *Food Hydrocolloids*, 19, 1, 53–61, 2005.
47. Varela, P. and Fiszman, S., Hydrocolloids in fried foods. A review. *Food Hydrocolloids*, 25, 8, 1801–1812, 2011.
48. Mallikarjunan, P. *et al.*, Edible coatings for deep-fat frying of starchy products. *LWT-Food Sci. Technol.*, 30, 7, 709–714, 1997.
49. Garcia, M. *et al.*, Methylcellulose coatings applied to reduce oil uptake in fried products. *Food Sci. Technol. Int.*, 10, 5, 339–346, 2004.
50. Tavera-Quiroz, M.J. *et al.*, Plasticized methylcellulose coating for reducing oil uptake in potato chips. *J. Sci. Food Agric.*, 92, 7, 1346–1353, 2012.
51. Primo-Martín, C. *et al.*, Performance of cellulose derivatives in deep-fried battered snacks: Oil barrier and crispy properties. *Food Hydrocolloids*, 24, 8, 702–708, 2010.
52. Martelli, M.d.R. *et al.*, Reduction of oil uptake in deep fat fried chicken nuggets using edible coatings based on cassava starch and methylcellulose. *Ital. J. Food Sci.*, 20, 1, 111–117, 2008.
53. Balasubramaniam, V. *et al.*, The effect of edible film on oil uptake and moisture retention of a deep-fat fried poultry product. *J. Food Process Eng.*, 20, 1, 17–29, 1997.
54. Garcıa, M. *et al.*, Edible coatings from cellulose derivatives to reduce oil uptake in fried products. *Innov. Food Sci. Emerg. Technol.*, 3, 4, 391–397, 2002.
55. Ford, J.L., Thermal analysis of hydroxypropylmethylcellulose and methylcellulose: Powders, gels and matrix tablets. *Int. J. Pharm.*, 179, 2, 209–228, 1999.
56. Fahs, A. *et al.*, Hydroxypropyl methylcellulose (HPMC) formulated films: Relevance to adhesion and friction surface properties. *Carbohydr. Polym.*, 80, 1, 105–114, 2010.
57. Jafari, S.M. *et al.*, Hydrophobicity, thermal and micro-structural properties of whey protein concentrate–pullulan–beeswax films. *Int. J. Biol. Macromol.*, 80, 506–511, 2015.
58. Formiga, A.S. *et al.*, Use of edible coatings based on hydroxypropyl methylcellulose and beeswax in the conservation of red guava 'Pedro Sato'. *Food Chem.*, 290, 144–151, 2019.
59. Navarro-Tarazaga, M.L., Sothornvit, R., Pérez-Gago, M.a.B., Effect of plasticizer type and amount on hydroxypropyl methylcellulose–beeswax edible film properties and postharvest quality of coated plums (cv. Angeleno). *J. Agric. Food Chem.*, 56, 20, 9502–9509, 2008.

60. Fagundes, C. *et al.*, Effect of antifungal hydroxypropyl methylcellulose-beeswax edible coatings on gray mold development and quality attributes of cold-stored cherry tomato fruit. *Postharvest Biol. Tec.*, 92, 1–8, 2014.
61. Navarro-Tarazaga, M.L. *et al.*, Fatty Acid Effect on Hydroxypropyl Methylcellulose- Beeswax Edible Film Properties and Postharvest Quality of Coated 'Ortanique' Mandarins. *J. Agric. Food Chem.*, 56, 22, 10689–10696, 2008.
62. Valencia-Chamorro, S.A. *et al.*, Effect of antifungal hydroxypropyl methylcellulose (HPMC)–lipid edible composite coatings on postharvest decay development and quality attributes of cold-stored 'Valencia' oranges. *Postharvest Biol. Tec.*, 54, 2, 72–79, 2009.
63. Gunaydin, S. *et al.*, Effect of hydroxypropyl methylcellulose-beeswax composite edible coatings formulated with or without antifungal agents on physicochemical properties of plums during cold storage. *J. Food Qual.*, 2017, 1–9, 2017.
64. Tahir, H.E. *et al.*, Recent developments in gum edible coating applications for fruits and vegetables preservation: A review. *Carbohydr. Polym.*, 224, 115-141, 2019.
65. Giavasis, I., Harvey, L.M., McNeil, B., Gellan gum. *Crit. Rev. Biotechnol.*, 20, 3, 177–211, 2000.
66. Sworn, G., Gellan gum, in: *Handbook of hydrocolloids*, G.O. Phillips and P.A. Williams (Eds.), pp. 204–227, Woodhead Publishing Ltd, Cambridge, UK, 2009.
67. Saha, D. and Bhattacharya, S., Hydrocolloids as thickening and gelling agents in food: A critical review. *J. Food Sci. Technol.*, 47, 6, 587–597, 2010.
68. Nussinovitch, A., *Hydrocolloid applications: Gum technology in the food and other industries*, Springer, London, UK, 1997.
69. Williams, R. and Mittal, G., Low-fat fried foods with edible coatings: Modeling and simulation. *J. Food Sci.*, 64, 2, 317–322, 1999.
70. Miyazawa, T. and Funazukuri, T., Noncatalytic hydrolysis of guar gum under hydrothermal conditions. *Carbohydr. Res.*, 341, 7, 870–877, 2006.
71. Cunha, P.L. *et al.*, Low viscosity hydrogel of guar gum: Preparation and physicochemical characterization. *Int. J. Biol. Macromol.*, 37, 1–2, 99–104, 2005.
72. Ekström, L.M., Björck, I.M., Östman, E.M., On the possibility to affect the course of glycaemia, insulinaemia, and perceived hunger/satiety to bread meals in healthy volunteers. *Food Funct.*, 4, 4, 522–529, 2013.
73. Golly, M.K. *et al.*, Shelf-life extension of grape (Pinot noir) by xanthan gum enriched with ascorbic and citric acid during cold temperature storage. *J. Food Sci. Technol.*, 1–12, 4867–4878, 2019.
74. Lopez-Torrez, L. *et al.*, Acacia senegal vs. Acacia seyal gums–Part 1: Composition and structure of hyperbranched plant exudates. *Food Hydrocolloids*, 51, 41–53, 2015.
75. Tahir, H.E. *et al.*, Quality and postharvest-shelf life of cold-stored strawberry fruit as affected by gum arabic (Acacia senegal) edible coating. *J. Food Biochem.*, 42, 3, e12527, 1–9, 2018.

76. Jafari, S., Hojjati, M., Noshad, M., Influence of soluble soybean polysaccharide and tragacanth gum based edible coating to improve the quality of fresh-cut apple slices. *J. Food Process. Preserv.*, 42, 6, e13638, 1–8, 2018.
77. del C Robles-Flores, G. *et al.*, Development and Evaluation of a Film and Edible Coating Obtained from the Cajanus cajan Seed Applied to Fresh Strawberry Fruit. *Food Bioprocess Tech.*, 11, 12, 2172–2181, 2018.
78. Kurt, A., Toker, O.S., Tornuk, F., Effect of xanthan and locust bean gum synergistic interaction on characteristics of biodegradable edible film. *Int. J. Biol. Macromol.*, 102, 1035–1044, 2017.
79. Murmu, S.B. and Mishra, H.N., Optimization of the arabic gum based edible coating formulations with sodium caseinate and tulsi extract for guava. *LWT-Food Sci. Technol.*, 80, 271–279, 2017.
80. Andrade, S.C. *et al.*, Control of Rhizopus soft rot and quality responses in plums (*Prunus domestica L.*) coated with gum arabic, oregano and rosemary essential oils. *J. Food Process. Preserv.*, 41, 6, e13251, 1–14, 2017.
81. Adiamo, O.Q., Eltoum, Y.A., Babiker, E.E., Effects of Gum Arabic Edible Coatings and Sun-Drying on the Storage Life and Quality of Raw and Blanched Tomato Slices. *J. Culinary Sci. Technol.*, 17, 1, 45–58, 2019.
82. Farahmandfar, R., Mohseni, M., Asnaashari, M., Effects of quince seed, almond, and tragacanth gum coating on the banana slices properties during the process of hot air drying. *Food Sci. Nutr.*, 5, 6, 1057–1064, 2017.
83. Yousuf, B. and Srivastava, A.K., Flaxseed gum in combination with lemongrass essential oil as an effective edible coating for ready-to-eat pomegranate arils. *Int. J. Biol. Macromol.*, 104, 1030–1038, 2017.
84. Li, C., Tao, J., Zhang, H., Peach gum polysaccharides-based edible coatings extend shelf life of cherry tomatoes. *Biotech.*, 7, 3, 168, 2017.
85. Danalache, F. *et al.*, Optimisation of gellan gum edible coating for ready-to-eat mango (Mangifera indica L.) bars. *Int. J. Biol. Macromol.*, 84, 43–53, 2016.
86. Ali, A., Hei, G.K., Keat, Y.W., Efficacy of ginger oil and extract combined with gum Arabic on anthracnose and quality of papaya fruit during cold storage. *J. Food Sci. Technol.*, 53, 3, 1435–1444, 2016.
87. Saha, A. *et al.*, Natural gums of plant origin as edible coatings for food industry applications. *Crit. Rev. Biotechnol.*, 37, 8, 959–973, 2017.
88. Yu, L. *et al.*, Effect of guar gum with glycerol coating on the properties and oil absorption of fried potato chips. *Food Hydrocolloids*, 54, 211–219, 2016.
89. Bouaziz, F. *et al.*, Feasibility of using almond gum as coating agent to improve the quality of fried potato chips: Evaluation of sensorial properties. *LWT-Food Sci. Technol.*, 65, 800–807, 2016.
90. Sharma, S. and Rao, T.R., Xanthan gum based edible coating enriched with cinnamic acid prevents browning and extends the shelf-life of fresh-cut pears. *LWT-Food Sci. Technol.*, 62, 1, 791–800, 2015.
91. Damodaran, S., Amino acids, peptides and proteins, in: *Fennema's food chemistry*, vol. 4, pp. 217–329, CRC Press, Boca Raton, FL, 2008.

92. Kannan, A., Hettiarachchy, N., Marshall, M., Food Proteins–Peptides: Chemistry and Structure, in: *Food Proteins and Peptides*, pp. 18–41, CRC Press, Boca Raton, FL, 2012.
93. Osborne, T., The vegetable proteins Longmans Green. *J. Soc. Chem. Ind.*, 43, 17, 413–452, 1924.
94. Wittaya, T., Protein-based edible films: Characteristics and improvement of properties, in: *Structure and Function of Food Engineering*, A.A. Eissa (Ed.), pp. 43–70, InTechOpen, London, UK, 2012.
95. Barone, J.R. and Schmidt, W.F., Nonfood applications of proteinaceous renewable materials. *J. Chem. Educ.*, 83, 7, 1003, 2006.
96. Panyam, D. and Kilara, A., Enhancing the functionality of food proteins by enzymatic modification. *Trends Food Sci. Technol.*, 7, 4, 120–125, 1996.
97. Shukla, R. and Cheryan, M., Zein: The industrial protein from corn. *Ind. Crops Prod.*, 13, 3, 171–192, 2001.
98. Gennadios, A. *et al.*, Edible coating and film based on proteins, in: *Edible coatings and films to improve food quality*, J.M. Krochta, E.A. Baldwin, M. Nisperos-Carriedo (Eds.), pp. 201–303, CRC Press, Lancaster, PA, 1994.
99. Chism, G. and Haard, N., Characteristics of edible plant tissues. *Food Chem.*, pp. 943–1011, 1996.
100. Kinsella, J., Relationships between structure and functional properties of food proteins, in: *Food proteins*, vol. 1, P.F. Fox and J.J. Condon (Eds.), pp. 51–103, Applied Science Publishers, London, UK, 1982.
101. Zhang, J., Mungara, P., Jane, J.-L., Mechanical and thermal properties of extruded soy protein sheets. *Polymer*, 42, 6, 2569–2578, 2001.
102. Kunte, L. *et al.*, Cast films from soy protein isolates and fractions. *Cereal Chem.*, 74, 2, 115–118, 1997.
103. Petruccelli, S. and Añón, M., Relationship between the method of obtention and the structural and functional properties of soy proteins isolates. 1. Structural and hydration properties. *J. Agric. Food Chem.*, 42, 10, 2161–2169, 1994.
104. Kinsella, J.E. and Morr, C.V., Milk proteins: Physicochemical and functional properties. *Crit. Rev. Food Sci. Nutr.*, 21, 3, 197–262, 1984.
105. Lacroix, M. *et al.*, Use of γ-irradiation to produce films from whey, casein and soya proteins: Structure and functionals characteristics. *Radiat. Phys. Chem.*, 63, 3–6, 827–832, 2002.
106. Lim, L.-T., Mine, Y., Tung, M.A., Transglutaminase cross-linked egg white protein films: Tensile properties and oxygen permeability. *J. Agric. Food Chem.*, 46, 10, 4022–4029, 1998.
107. Bigi, A. *et al.*, Stabilization of gelatin films by crosslinking with genipin. *Biomaterials*, 23, 24, 4827–4832, 2002.
108. Yousuf, B. and Srivastava, A.K., Impact of honey treatments and soy protein isolate-based coating on fresh-cut pineapple during storage at 4° C. *Food Packaging Shelf*, 21, 100361, 1–9, 2019.
109. Yousuf, B. and Srivastava, A.K., A novel approach for quality maintenance and shelf life extension of fresh-cut Kajari melon: Effect of treatments

with honey and soy protein isolate. *LWT-Food Sci. Technol.*, 79, 568–578, 2017.

110. Santos, T.M. *et al.*, Enhancing storage stability of guava with tannic acid-crosslinked zein coatings. *Food Chem.*, 257, 252–258, 2018.
111. Zhang, L. *et al.*, Impact of soybean protein isolate-chitosan edible coating on the softening of apricot fruit during storage. *LWT-Food Sci. Technol.*, 96, 604–611, 2018.
112. Alves, M.M., Gonçalves, M., Rocha, C.M., Effect of ferulic acid on the performance of soy protein isolate-based edible coatings applied to fresh-cut apples. *LWT-Food Sci. Technol.*, 80, 409–415, 2017.
113. Guerrero, P. *et al.*, Application of soy protein coatings and their effect on the quality and shelf-life stability of beef patties. *RSC Adv.*, 5, 11, 8182–8189, 2015.
114. Angor, M.M., Reducing fat content of fried potato pellet chips using carboxymethyl cellulose and soy protein isolate solutions as coating films. *J. Agr. Sci.*, 8, 3, 162–168, 2016.
115. Yıldırım, E. and Barutçu Mazı, I., Effect of zein coating enriched by addition of functional constituents on the lipid oxidation of roasted hazelnuts. *J. Food Process Eng.*, 40, 4, e12515, 1–9, 2017.
116. Chen, B.-J. *et al.*, Edible antimicrobial coating incorporating a polymeric iron chelator and its application in the preservation of surimi product. *Food Bioprocess Tech.*, 9, 6, 1031–1039, 2016.
117. Šuput, D.Z. *et al.*, Edible films and coatings: Sources, properties and application. *Food Feed Res.*, 42, 1, 11–22, 2015.
118. Barman, K., Asrey, R., Pal, R., Putrescine and carnauba wax pretreatments alleviate chilling injury, enhance shelf life and preserve pomegranate fruit quality during cold storage. *Sci. Hortic.-Amsterdam*, 130, 4, 795–800, 2011.
119. Chen, H., Sun, Z., Yang, H., Effect of carnauba wax-based coating containing glycerol monolaurate on the quality maintenance and shelf-life of Indian jujube (*Zizyphus mauritiana Lamk.*) fruit during storage. *Sci. Hortic.-Amsterdam*, 244, 157–164, 2019.
120. Couëdelo, L. *et al.*, Lymphatic absorption of α-linolenic acid in rats fed flaxseed oil-based emulsion. *Br. J. Nutr.*, 105, 7, 1026–1035, 2011.
121. Oregel-Zamudio, E. *et al.*, Effect of candelilla wax edible coatings combined with biocontrol bacteria on strawberry quality during the shelf-life. *Sci. Hortic.-Amsterdam*, 214, 273–279, 2017.
122. Chen, H. *et al.*, Synthesis and properties of a lacquer wax-based quarternary ammonium gemini surfactant. *Molecules*, 19, 3, 3596–3606, 2014.
123. Hu, H., Zhou, H., Li, P., Lacquer wax coating improves the sensory and quality attributes of kiwifruit during ambient storage. *Sci. Hortic.-Amsterdam*, 244, 31–41, 2019.
124. Germano, T.A. *et al.*, Galactomannan-carnauba wax coating improves the antioxidant status and reduces chilling injury of 'Paluma' guava. *Postharvest Biol. Tec.*, 149, 9–17, 2019.

125. Aguirre-Joya, J.A. *et al.*, Effects of a natural bioactive coating on the quality and shelf life prolongation at different storage conditions of avocado (Persea americana Mill.) cv. Hass. *Food Packaging Shelf*, 14, 102–107, 2017.
126. Singh, S. *et al.*, Carnauba wax-based edible coating enhances shelf-life and retain quality of eggplant (*Solanum melongena*) fruits. *LWT-Food Sci. Technol.*, 74, 420–426, 2016.
127. De León-Zapata, M.A. *et al.*, Edible candelilla wax coating with fermented extract of tarbush improves the shelf life and quality of apples. *Food Packaging Shelf*, 3, 70–75, 2015.
128. Muxika, A. *et al.*, Chitosan as a bioactive polymer: Processing, properties and applications. *Int. J. Biol. Macromol.*, 105, 1358–1368, 2017.
129. Yadav, M. *et al.*, Seafood waste: A source for preparation of commercially employable chitin/chitosan materials. *Bioresour. Bioprocess.*, 6, 1, 8, 1–20, 2019.
130. Bautista-Baños, S., Romanazzi, G., Jiménez-Aparicio, A., *Chitosan in the preservation of agricultural commodities*, Academic Press, Cambridge, MA, 2016.
131. Elsabee, M.Z. and Abdou, E.S., Chitosan based edible films and coatings: A review. *Mater. Sci. Eng. C*, 33, 4, 1819–1841, 2013.
132. Dehghani, S., Hosseini, S.V., Regenstein, J.M., Edible films and coatings in seafood preservation: A review. *Food Chem.*, 240, 505–513, 2018.
133. Nunes, E.M. *et al.*, Edible Coatings, in: *Edible Food Packaging: Materials and Processing Technologies*, p. 413, CRC Press, Boca Raton, FL, 2017.
134. Sun, D.-W., *Handbook of frozen food processing and packaging*, CRC Press, Boca Raton, FL, 2016.
135. Vieira, J.M. *et al.*, Effect of chitosan–Aloe vera coating on postharvest quality of blueberry (*Vaccinium corymbosum*) fruit. *Postharvest Biol. Tec.*, 116, 88–97, 2016.
136. Guerra, I.C.D. *et al.*, Coatings comprising chitosan and Mentha piperita L. or Mentha × villosa Huds essential oils to prevent common postharvest mold infections and maintain the quality of cherry tomato fruit. *Int. J. Food Microbiol.*, 214, 168–178, 2015.
137. Mohammadi, A., Hashemi, M., Hosseini, S.M., Chitosan nanoparticles loaded with Cinnamomum zeylanicum essential oil enhance the shelf life of cucumber during cold storage. *Postharvest Biol. Tec.*, 110, 203–213, 2015.
138. Guerra, I.C.D. *et al.*, The effects of composite coatings containing chitosan and Mentha (piperita L. or x villosa Huds) essential oil on postharvest mold occurrence and quality of table grape cv. Isabella. *Innov. Food Sci. Emerg. Technol.*, 34, 112–121, 2016.
139. Cissé, M. *et al.*, Preservation of mango quality by using functional chitosan-lactoperoxidase systems coatings. *Postharvest Biol. Tec.*, 101, 10–14, 2015.
140. Zaki, N.H.M., Som, H.Z.M., Haiyee, Z.A., Application of palm stearin-chitosan edible coating on star fruits (*Aver-rhoa carambola L.*). *Malaysian J. Anal. Sci.*, 16, 3, 325–334, 2012.

141. Yang, H. *et al.*, Effects of combined aqueous chlorine dioxide and chitosan coatings on microbial growth and quality maintenance of fresh-cut bamboo shoots (*Phyllostachys praecox f. prevernalis.*) during storage. *Food Bioprocess Tech.*, 8, 5, 1011–1019, 2015.
142. Sun, D. *et al.*, Improved preservation effects of litchi fruit by combining chitosan coating with ascorbic acid treatment during postharvest storage. *Afr. J. Biotechnol.*, 9, 22, 3272–3279, 2010.
143. Ponce, A.G. *et al.*, Antimicrobial and antioxidant activities of edible coatings enriched with natural plant extracts: *In vitro* and *in vivo* studies. *Postharvest Biol. Tec.*, 49, 2, 294–300, 2008.
144. Di Pierro, P. *et al.*, Chitosan/whey protein film as active coating to extend Ricotta cheese shelf-life. *LWT-Food Sci. Technol.*, 44, 10, 2324–2327, 2011.
145. Duan, J. *et al.*, Antimicrobial chitosan-lysozyme (CL) films and coatings for enhancing microbial safety of mozzarella cheese. *J. Food Sci.*, 72, 9, M355–M362, 2007.
146. Fajardo, P. *et al.*, Evaluation of a chitosan-based edible film as carrier of natamycin to improve the storability of Saloio cheese. *J. Food Eng.*, 101, 4, 349–356, 2010.
147. Moreira, M.d.R. *et al.*, Antimicrobial effectiveness of bioactive packaging materials from edible chitosan and casein polymers: Assessment on carrot, cheese, and salami. *J. Food Sci.*, 76, 1, M54–M63, 2011.
148. Youssef, A.M. *et al.*, Evaluation of bionanocomposites as packaging material on properties of soft white cheese during storage period. *Carbohydr. Polym.*, 132, 274–285, 2015.
149. Arancibia, M. *et al.*, Chitosan coatings enriched with active shrimp waste for shrimp preservation. *Food Control*, 54, 259–266, 2015.
150. Bazargani-Gilani, B., Aliakbarlu, J., Tajik, H., Effect of pomegranate juice dipping and chitosan coating enriched with Zataria multiflora Boiss essential oil on the shelf-life of chicken meat during refrigerated storage. *Innov. Food Sci. Emerg. Technol.*, 29, 280–287, 2015.
151. Chen, H. *et al.*, Development of active packaging film containing bioactive components encapsulated in β-cyclodextrin and its application. *Food Hydrocolloids*, 90, 360–366, 2019.
152. Šuput, B.Š.L.P.D. *et al.*, Application of chitosan coating with oregano essential oil on dry fermented sausage. *J. Food Nutr. Res.*, 51, 1, 60–68, 2012.
153. Gudjónsdóttir, M. *et al.*, Effects of electrospun chitosan wrapping for dry-ageing of beef, as studied by microbiological, physicochemical and low-field nuclear magnetic resonance analysis. *Food Chem.*, 184, 167–175, 2015.
154. Albertos, I. *et al.*, Effect of edible chitosan/clove oil films and high-pressure processing on the microbiological shelf life of trout fillets. *J. Sci. Food Agric.*, 95, 14, 2858–2865, 2015.
155. Calva-Estrada, S.J., Jiménez-Fernández, M., Lugo-Cervantes, E., Protein-based films: Advances in the development of biomaterials applicable to food packaging. *Food Eng. Rev.*, 11, 2, 78–92, 2019.

156. Ramos, M. *et al.*, Gelatin-based films and coatings for food packaging applications. *Coatings*, 6, 4, 41, 1–20, 2016.
157. Sultana, S., Ali, M.E., Ahamad, M.N.U., Gelatin and collagen, in: *Preparation and Processing of Religious and Cultural Foods*, Md. Eaqub Ali and Nina Naquiah Ahmad Nizar (Eds.), p. 215, Woodhead Publishing Ltd, Duxford, UK, 2018.
158. Dille, M.J., Hattrem, M.N., Draget, K.I., Bioactively filled gelatin gels; challenges and opportunities. *Food Hydrocolloids*, 76, 17–29, 2018.
159. Luangapai, F., Peanparkdee, M., Iwamoto, S., Biopolymer films for food industries: Properties, applications, and future aspects based on chitosan. *Rev. Agric. Sci.*, 7, 59–67, 2019.
160. Ansorena, M.R., Pereda, M., Marcovich, N.E., Edible Films, in: *Polymers for Food Applications*, T.J. Gutierrez (Ed.), pp. 5–24, Springer International Publishing, Switzerland, AG, 2018.
161. Hanani, Z.N., Roos, Y.H., Kerry, J., Use and application of gelatin as potential biodegradable packaging materials for food products. *Int. J. Biol. Macromol.*, 71, 94–102, 2014.
162. Shakila, R.J. *et al.*, Suitability of antimicrobial grouper bone gelatin films as edible coatings for vacuum-packaged fish steaks. *J. Aquat. Food Prod. T.*, 25, 5, 724–734, 2016.
163. Wu, J. *et al.*, The preparation, characterization, antimicrobial stability and *in vitro* release evaluation of fish gelatin films incorporated with cinnamon essential oil nanoliposomes. *Food Hydrocolloids*, 43, 427–435, 2015.
164. Alparslan, Y. *et al.*, Quality assessment of shrimps preserved with orange leaf essential oil incorporated gelatin. *LWT-Food Sci. Technol.*, 72, 457–466, 2016.
165. Alcântara, L.O. *et al.*, Wettability of edible coatings on Nile tilapia fillets *(Oreochromis niloticus)*. *J. Food Eng.*, 247, 152–159, 2019.
166. Bermúdez-Oria, A. *et al.*, Effect of edible pectin-fish gelatin films containing the olive antioxidants hydroxytyrosol and 3, 4-dihydroxyphenylglycol on beef meat during refrigerated storage. *Meat Sci.*, 148, 213–218, 2019.
167. Neira, L.M. *et al.*, Shelf life extension of refrigerated breaded hake medallions packed into active edible fish gelatin films. *Packag. Technol. Sci.*, 32, 1–10, 2019.
168. Aitboulahsen, M. *et al.*, Gelatin-based edible coating combined with mentha pulegium essential oil as bioactive packaging for strawberries. *J. Food Qual.*, 2018, 1–8, 2018.
169. Abdallah, M.R. *et al.*, Application of alginate and gelatin-based edible coating materials as alternatives to traditional coating for improving the quality of pastirma. *Food Sci. Biotechnol.*, 27, 6, 1589–1597, 2018.
170. Kumar, K.S. *et al.*, Shelf life extension of tuna fillets by gelatin and chitosan based edible coating incorporated with clove oil. *Fishery Technol.*, 55, 104–113, 2018.
171. Cardoso, G.P. *et al.*, Selection of a chitosan gelatin-based edible coating for color preservation of beef in retail display. *Meat Sci.*, 114, 85–94, 2016.

172. Oliveira, M.M.M.d., Brugnera, D.F., Piccoli, R.H., Essential oils of thyme and Rosemary in the control of *Listeria monocytogenes* in raw beef. *Braz. J. Microbiol.*, 44, 4, 1181–1188, 2013.
173. Herring, J.L. *et al.*, Oxidative stability of gelatin coated pork at refrigerated storage. *Meat Sci.*, 85, 4, 651–656, 2010.
174. Tyburcy, A. and Kozyra, D., Effects of composite surface coating and pre-drying on the properties of kabanosy dry sausage. *Meat Sci.*, 86, 2, 405–410, 2010.
175. Riveros, C.G. *et al.*, Preserving sensory attributes of roasted peanuts using edible coatings. *Int. J. Food Sci. Technol.*, 48, 4, 850–859, 2013.
176. Marquez, G.R. *et al.*, Fresh-cut fruit and vegetable coatings by transglutaminase-crosslinked whey protein/pectin edible films. *LWT-Food Sci. Technol.*, 75, 124–130, 2017.
177. Shokri, S. and Ehsani, A., Efficacy of whey protein coating incorporated with lactoperoxidase and α-tocopherol in shelf life extension of Pike-Perch fillets during refrigeration. *LWT-Food Sci. Technol.*, 85, 225–231, 2017.
178. Xie, J. *et al.*, Effects of whey protein films on the quality of thawed bigeye tuna (*Thunnus obesus*) chunks under modified atmosphere packaging and vacuum packaging conditions. *Food Sci. Biotechnol.*, 26, 4, 937–945, 2017.
179. Catarino, M.D. *et al.*, Development and performance of whey protein active coatings with Origanum virens essential oils in the quality and shelf life improvement of processed meat products. *Food Control*, 80, 273–280, 2017.
180. Luangtana-anan, M. and Limmatvapirat, S., Shellac-Based Coating Polymer for Agricultural Applications, in: *Polymers for Agri-Food Applications*, T.J. Gutierrez (Ed.), pp. 487–524, Springer International Publishing, Switzerland, AG, 2019.
181. Patel, A.R., Shellac-Based Oleogels, in: *Edible Oleogels*, A. Marangoni and N. Garti (Eds.), pp. 173–192, Academic Press and AOCS Press, Urbana, IL, 2018.
182. Álvarez, K., Famá, L., Gutiérrez, T.J., Physicochemical, antimicrobial and mechanical properties of thermoplastic materials based on biopolymers with application in the food industry, in: *Advances in physicochemical properties of biopolymers: Part-1*, M. Masuelli and D. Renard (Eds.), pp. 358–400, Bentham Books, Sharjah, UAE, 2017.
183. Gutiérrez, T.J. and Álvarez, K., Biopolymers as microencapsulation materials in the food industry, in: *Advances in physicochemical properties of biopolymers: Part-2*, M. Masuelli and D. Renard (Eds.), pp. 296–322, Bentham Books, Sharjah, UAE, 2017.
184. Saberi, B. *et al.*, Development of biocomposite films incorporated with different amounts of shellac, emulsifier, and surfactant. *Food Hydrocolloids*, 72, 174–184, 2017.
185. Jo, W.-S. *et al.*, Quality and microbial safety of 'Fuji' apples coated with carnauba-shellac wax containing lemongrass oil. *LWT-Food Sci. Technol.*, 55, 2, 490–497, 2014.

186. Soradech, S. *et al.*, Utilization of shellac and gelatin composite film for coating to extend the shelf life of banana. *Food Control*, 73, 1310–1317, 2017.
187. Kowalczyk, D. *et al.*, Antifungal resistance and physicochemical attributes of apricots coated with potassium sorbate-added carboxymethyl cellulose-based emulsion. *Int. J. Food Sci. Technol.*, 53, 3, 728–734, 2018.
188. Zambrano-Zaragoza, M. *et al.*, Nanosystems in edible coatings: A novel strategy for food preservation. *Int. J. Mol. Sci.*, 19, 3, 705, 2018.
189. Guimarães, I.C. *et al.*, Cellulose microfibrillated suspension of carrots obtained by mechanical defibrillation and their application in edible starch films. *Ind. Crops Prod.*, 89, 285–294, 2016.
190. Deng, Z. *et al.*, Cellulose nanocrystal reinforced chitosan coatings for improving the storability of postharvest pears under both ambient and cold storages. *J. Food Sci.*, 82, 2, 453–462, 2017.
191. Wang, W. *et al.*, Effects of cellulose nanofibers filling and palmitic acid emulsions coating on the physical properties of fish gelatin films. *Food Biophys.*, 12, 1, 23–32, 2017.
192. Aytac, Z. *et al.*, Antibacterial electrospun zein nanofibrous web encapsulating thymol/cyclodextrin-inclusion complex for food packaging. *Food Chem.*, 233, 117–124, 2017.
193. Cui, H., Yuan, L., Lin, L., Novel chitosan film embedded with liposome-encapsulated phage for biocontrol of *Escherichia coli O157: H7* in beef. *Carbohydr. Polym.*, 177, 156–164, 2017.
194. Avella, M. *et al.*, Biodegradable starch/clay nanocomposite films for food packaging applications. *Food Chem.*, 93, 3, 467–474, 2005.
195. Ray, S.S. and Bousmina, M., Biodegradable polymers and their layered silicate nanocomposites: In greening the 21st century materials world. *Prog. Mater. Sci.*, 50, 8, 962–1079, 2005.
196. Lv, M. *et al.*, Engineering nanomaterials-based biosensors for food safety detection. *Biosens. Bioelectron.*, 106, 122–128, 2018.
197. Schaefer, D. and Cheung, W.M., Smart Packaging: Opportunities and Challenges. *Procedia CIRP*, 72, 1022–1027, 2018.
198. Ghaani, M. *et al.*, An overview of the intelligent packaging technologies in the food sector. *Trends Food Sci. Technol.*, 51, 1–11, 2016.
199. Shutava, T.G., Livanovich, K.S., Sharamet, A.A., Layer-by-layer films of polysaccharides modified with polyethylene glycol and dextran. *Colloids Surf. B*, 173, 412–420, 2019.
200. Kim, J.-H., Hong, W.-s., Oh, S.-W., Effect of layer-by-layer antimicrobial edible coating of alginate and chitosan with grapefruit seed extract for shelf-life extension of shrimp (Litopenaeus vannamei) stored at 4 C. *Int. J. Biol. Macromol.*, 120, 1468–1473, 2018.
201. Ban, Z. *et al.*, Effects of postharvest application of chitosan-based layer-by-layer assemblies on regulation of ribosomal and defense proteins in strawberry fruit *(Fragaria × ananassa). Sci. Hortic.-Amsterdam*, 240, 293–302, 2018.

12

Polymer Coatings for Pharmaceutical Applications

Muhammad Harris Shoaib*, Rabia Ismail Yousuf, Farrukh Rafiq Ahmed, Fatima Ramzan Ali, Faaiza Qazi, Kamran Ahmed and Farya Zafar

Department of Pharmaceutics, Faculty of Pharmacy and Pharmaceutical Sciences, University of Karachi, Karachi, Pakistan

Abstract

Polymeric coating of pharmaceutical dosage form is one of the effective tool to improve the patient compliance, shelf life and therapeutic efficacy of drugs by providing protection against physical degradation, chemical degradation and unpleasant taste. Added advantage offers potential opportunities to modify the drug release characteristics. This chapter outlines the historical evolution of concepts and practices of coating pharmaceuticals for providing better drug delivery systems. The types of coatings for pharmaceutical products, primarily solvent-based coatings, solvent-less coatings and polymer coatings for micro/nano particulate drug delivery systems, as well as mechanism of drug release through coating system are discussed. Pharmaceutical products utilizing such approaches and polymers range from simple coated tablets to those employing more complex technologies for nano particulate drug delivery systems. Technologies including self-organized/assembled coating, coating by coprecipitation, coating by adsorption and coating through layer by layer assembly are described.

Keywords: Polymer, pharmaceutical coating, solvent-based coating, solvent-less coating, nano particle coating, drug release mechanism

12.1 Introduction

The coating of pharmaceutical solid dosage forms holds multiple benefits such as masking of the unpleasant taste and odor, protection of the drug

**Corresponding author*: mhshoaib@uok.edu.pk; harrisshoaib2000@yahoo.com

Inamuddin, Rajender Boddula, Mohd Imran Ahamed and Abdullah M. Asiri (eds.) Polymer Coatings: Technology and Applications, (275–318) © 2020 Scrivener Publishing LLC

substances from oxidation, and degradation due to photo and pH sensitivity and thereby enhancing the stability of the drug substance. Moreover, the coating of solid drug products with polymers also helps in modifying drug release characteristics. Thin film coating also enhances the product appearance and helps in identification of the product. Furthermore, this also reduces the friction in high speed packaging machine for increasing the production output. The coating applications on solid dosage forms also facilitates its swallowing and may help reduce the interaction between two or more drug substances by forming polymeric coating layers in dosage forms [1–3].

12.2 Polymers for Coating Pharmaceuticals, A Historical Perspective

Coating of pharmaceutical products is more than a thousand-year-old practice. The first coating procedure was reported by Muhammad ibn Zakariya al-Razi (*Rhazes*) around 900 A.D. where he used mucilage of *Psyllium* seed for coating pills. Later on, Ibn e Sina (*Avicenna*) introduced gold and silver coating in pharmacy [4, 5]. In nineteenth century, pharmacists frequently used gelatin and sugar as a coating material for pills to mask the bitter taste. In 1884, German Physician, Dr. Paul Unna introduced keratin coated pills having resistance in a gastric environment with disintegration and dissolution characteristics in the intestinal medium [6]. Sugar coating was a dominant method till mid of the twentieth century, but due various reasons such as higher time consumption, less automation, higher batch-to-batch variation, the requirement of specialized skills, and a considerable increase in the final tablet weight resulted in a shift of practice to use polymer-based film coating instead in 1950s. As a result, the production run time did not only reduce from days to hours but also led to enhanced benefits such as improved tablet strength, aesthetic appearance, and easier swallowing with a higher level of patient compliance [7–9].

Modified release coating polymers such as cellulose acetate phthalate (CAP) was also introduced in 1951 for enteric coated drug delivery systems [10]. After CAP, hydroxypropyl cellulose phthalate (HP50 and HP55), hydroxypropyl methylcellulose (HPMC), and ethyl cellulose (EC) were introduced. These polymers and their mixtures have variable solubility characteristics in different organic solvents and pH. Disintegration time ranges from a few minutes to several hours also because of different

EC and HPMC proportions [11]. In 1953, Swintosky of Smith Kline and French used hydrogenated castor oil-EC combination for tablets and pellets coating. In the same year, Rohm and Hass marketed Eudragit®, copolymer acrylic acids, and methacrylic acid and their esters. Different Eudragit type L and S polymers were introduced with different ratios of acrylic and methacrylic acid at that time. Till date, at regular intervals, different other Eudragit type and grades are available from Evonik Industries [12]. In the 1970s, the efforts of Higuchi and coworkers at ALZA Corporation led to the discovery of osmotic drug delivery system whereby the drug is released from tablets or capsule through an orifice due to osmotic pressure. The cellulose acetate coated semi-permeable polymeric membraneon the surface allows water to move inside that causes the dissolution of the drug in anosmogen containing core [13–16]. The first tablet product based on ALZA "elementary osmotic pump" (EOP) Acutrium® was marketed by Ciba Geigy in the early 80s for delivering phenylpropanolamine [17].

Concept of coating hard and soft gelatin capsule was also introduced in the 1980s and was subsequently patented. The finished capsule shell was used to be subcoated with HPMC suspension. This subcoating, later on, was used for improved enteric coating on capsule shell. Murthy *et al.* in 1986 studied and compared coating Eudragit® L 30 D, Aquateric®, and Coateric® an aqueous enteric polymerson hard gelatin capsules shell. Eudragit® L 30 D-55 coating on soft gelatin capsule was also reported by Felton and her colleagues [18–20].

Dry coating, a type of solventless coating using hydroxypropyl methylcellulose acetate succinate (HPMCAS) as a novel enteric coating method, was developed by Obara *et al.* in 1999. Method reduced not only the processing time but also suitable for water sensitive materials [21]. Other solventless coating methods such as compression coating, electrostatic coating, hot-melt coating, supercritical fluid spray coating, and photocurable coating were also introduced for developing different controlled release products during this period and in early 2000 [22–30].

Nanoparticle polymeric coating for therapeutic purpose was another major development when nanoparticles coated with block copolymer, poly(oxyethylene)-poly(oxypropylene), substantially increased the time of blood circulation of delivery vehicles by reducing the liver and spleen uptake [31, 32]. These nanoparticles if uncoated were found to be rapidly opsonized and removed by the reticuloendothelial system of liver, spleen lungs, and bone marrow [33–35].

12.3 Types of Coatings Used on Pharmaceutical Drug Products

There are two major categories of various procedures used to coat polymeric substances on the drug dosage forms and delivery systems. The first and most popular category comprises of solvent (aqueous/organic)-based coating procedures while the other is rest of the methods which doesn't use any solvent to disperse or dissolve the polymeric material.

12.3.1 Solvent-Based Coatings

Solvent-based polymeric coatings on pharmaceuticals are applied to various dosage forms available commercially such as tablets, capsules, and multiparticulate systems, e.g., pellets, etc., in solvent-based coating procedures usually a single or multiple coats of a solution or dispersion of polymeric material, in an either aqueous or organic solvent, is applied through spray atomization technique. Moreover, in lesser and less popular cases, coatings are also applied by pouring the same solutions or dispersions which is also called Ladle process [36, 37].

12.3.1.1 Sugar Coating

It is one of the oldest methods of coating and which is still used. Although the major ingredient is sugar (sucrose) for coating which is applied through various steps and coats, the process, however, also needs an initial coating of a hydrophobic material such as shellac, a natural polymeric resin for the sealing phase. Shellac protects the tablets from physical and chemical degradation due to the subsequent coating steps containing aqueous-based sugar solutions. Other materials such as Zein and Cellulose acetate phthalate are used for the same purpose. Major phases in the coating after sealing are subcoating, smoothing, coloring, polishing, and imprinting. The overall method is performed in conventional coating pans by Ladle process; however, the spray coating technique has also gained popularity in the last two decades. Excipients used have wide acceptance with less regulatory issues. The disadvantages attributed to it are more labor skills required as the process is more art than science. During the process, the tablet weight is increased considerably and affects the disintegration time [37].

12.3.1.2 *Film Coating*

Film coating is a procedure whereby a thin film of polymeric substance is applied over the surfaces of dosage forms and drug delivery systems. It is usually accomplished through atomized spraying of solution or dispersion of polymers in an either aqueous or organic solvent. Depending upon the soluble and permeable properties rendered by the coated film onto the surface of the dosage form which is characteristic of the chemical nature and physical properties of the polymer, the film coatings can be broadly classified as soluble, insoluble, gastro-resistant (pH-dependent solubility), and semi-permeable membranes.

12.3.1.3 *Soluble Film Coating*

Soluble film coating is usually applied as a thin film over the dosage form to protect the product from atmospheric oxidation, friability, and render taste-masking properties to the product as well. The thin film coat ensures that the shape of the tablet along with embossing is intact and visible. Various polymeric substances used to coat soluble films are discussed below.

12.3.1.3.1 Cellulose Ethers

Cellulose ethers are produced by substituting hydrogen atoms of hydroxyl groups with alkyl groups in anhydrous units of cellulose. These are high molecular weight compounds and their properties depend on molecular weight, chemical structure, substitution group, its degree, and distribution. These properties that are significant for coating include viscosity, solubility, surface activity, thermoplasticity of film, and stability against heat, oxidation, hydrolysis, and biodegradation. Viscosity is directly related to the molecular mass. Most commonly used cellulose ethers (as mentioned in Tables 12.1) for soluble film coatings include hydroxypropylmethylcellulose (HPMC), methylcellulose (MC), hydroxyethylcellulose (HEC), and hydroxypropyl cellulose (HPC) [38].

12.3.1.3.1.1 Methylcellulose (MC)

Methyl ether of cellulose is known as methylcellulose (MC). It is produced by the reaction of methyl chloride and alkaline cellulose. It is made from natural cellulose and is white or milky white in color. It is used as a film-former, emulsifier, adhesive, and thickening agent. However, its use in

Table 12.1 Physical attributes of some commercial soluble and non-soluble coating polymers.

Coating polymer	Other names	Grades	Solubility	Molecular weight	Glass transition temperature (Tg)	Viscosity	Pharmaceutical application	Manufacturer	Reference
Hydroxypropyl methyl cellulose	Hypromellose	K4M, K15MP, K100MP, E10MP CR E4MP, K100LVP, E3PREM.LV, E5PREM.LV, E6PREM.LV, E15PREM.LV, E50PREM.LV, K3PREM.LV	Practically insoluble in ethanol (95%), ether and chloroform. Soluble in water (cold), mixture of dichloromethane with methanol or ethanol and mixture of alcohol with water.	10,000–1500,000	180°C	24–120,000 cP	In oral pharmaceutical formulations like tablets and capsules different viscosity grades are widely used to retard drug release at 10-80% w/w levels.	Hebei Derek Chemical Limited, Kima Chemical (Zibo) Co., Ltd, Sidley Chemical Co. Ltd, etc.	[40, 45]

(Continued)

Table 12.1 Physical attributes of some commercial soluble and non-soluble coating polymers. (*Continued*)

Coating polymer	Other names	Grades	Solubility	Molecular weight	Glass transition temperature (Tg)	Viscosity	Pharmaceutical application	Manufacturer	Reference
Methylcellulose	Methocel	A4MP, A15-LV, A15CP, A4CP	Practically insoluble in water (hot), ethanol, toluene, acetone, saturated salt solutions, chloroform, ether and methanol. Soluble in equal volumes of chloroform and ethanol and glacial acetic acid. It forms a clear to opalescent viscous colloidal dispersion as methylcellulose swells and slowly disperses.	10,000–220,000 Da	184°C –197°C	15–4,000 cP	In tablet formulations may be added to produce sustained release of drug	SE Tylose GmbH & Co. KG, The Dow Chemical Company, etc.	[40]

(*Continued*)

Table 12.1 Physical attributes of some commercial soluble and non-soluble coating polymers. (*Continued*)

Coating polymer	Other names	Grades	Solubility	Molecular weight	Glass transition temperature (Tg)	Viscosity	Pharmaceutical application	Manufacturer	Reference
Ethylcellulose	Ethocel	EthocelStd 4 Premium N-7, EthocelStd 7FP Premium, EthocelStd 7 Premium N-10, EthocelStd 10FP Premium, EthocelStd 10P Premium N-14, EthocelStd 20P Premium N-22, EthocelStd 45P Premium N-50, N-100 EthocelStd 100FP Premium, EthocelStd 100P	Practically insoluble in water, propylene glycol and glycerin. Freely soluble in mixture of ethanol and hydrocarbons, chloroform, tetrahydrofuran and methyl acetate if ethoxyl group of ethyl cellulose is less than 46.5% whereas greater than 46.5% ethoxyl group of ethyl cellulose show solubility in methanol, toluene, chloroform, ethyl acetate and ethanol (95%).	454.5 g/mol	129°C –133°C	3–110 cP	In oral formulations ethylcellulose is used to coat granules and tables to modify the drug release.	Sidley Chemical Co. Ltd etc.	[40, 45].

(*Continued*)

Table 12.1 Physical attributes of some commercial soluble and non-soluble coating polymers. (*Continued*)

Coating polymer	Other names	Grades	Solubility	Molecular weight	Glass transition temp. (Tg)	Viscosity	Pharmaceutical application	Manufacturer	Reference
Hydroxyethyl cellulose	Cellosize WP (02), WP and QP (09, 3, 40, 300, 4400), QP (10,000, 15,000, 30,000, 52,000, 100 M) Natrosol HHR, H4R, HR, MHR, MR, KR, GR, ER, JR, LR.	HHR, H4R, HR, MHR, MR, KR, GR, ER, JR, LR.	Hydroxyethylcellulose forms uniform and clear solution in both cold and hot water. Partially solubilize or swell in some polar organic solvents like glycols. Practically insoluble in most organic solvents, toluene, acetone, ether and ethanol.	806.9 g/mol	120°C	Cellosize WP 7–20 cP, WP and QP 10–5,200 cP, QP150–21,000 cP. Natrosol HHR, H4R, HR, MHR 800–5,000 cP, MR, KR, GR, ER 25-6500 cPs, JR, LR 750–400 cP.	Hydroxyethylcellulose is a nonionic, water soluble polymer that is used to form film coats on tablets.	Kima Chemical (Zibo) Co., Ltd, Sidley Chemical Co. Ltd, etc.	[40]

(*Continued*)

Table 12.1 Physical attributes of some commercial soluble and non-soluble coating polymers. (*Continued*)

Coating polymer	Other names	Grades	Solubility	Molecular weight	Glass transition temp. (Tg)	Viscosity	Pharmaceutical application	Manufacturer	Reference
Hydroxypropyl cellulose	Hyprolose, Klucel, Methocel	Klucel HF, Klucel MF, Klucel GF, Klucel JF, Klucel LF, Klucel EF	Soluble in water (1 in 2 parts), methanol (1 in 2 parts), ethanol (1 in 2.5 parts), dichloromethane (1 in 10 parts), propylene glycol (1 in 5 parts), propan-2-ol (1 in 5 parts), Practically insoluble in oils, hydrocarbons, carbon tetrachloride and aromatic hydrocarbons.	50,000–1,250,000	105°C	Klucel HF-EF750–4000 cP.	It is used as a film forming and extended release matrix former agent. It is also used as a binder in tablet dosage form.	Hebei Derek Chemical Limited, Kima Chemical (Zibo) Co., Ltd,	[40, 45]

(*Continued*)

Table 12.1 Physical attributes of some commercial soluble and non-soluble coating polymers. (*Continued*)

Coating polymer	Other names	Grades	Solubility	Molecular weight	Glass transition temp. (Tg)	Viscosity	Pharmaceutical application	Manufacturer	Reference
Polyvinyl-pyrrolidone	Kollidon, Plasdone-polyvidone	Povidone K-12, Povidone K-15, Povidone K-17, Povidone K-25, Povidone K-30, Povidone K-60, Povidone K-90, Povidone K-120	Practically soluble in different organic solvents and water.	2,500–3,000,000	168°C	10–20 cP	Povidone solutions may be used as coating agents.	BASF	[40, 45]
Sodium Carboxymethyl Cellulose	Carmellose sodium	Akucell AF (0305,2785,3085)	Practically insoluble in toluene, ethanol, acetone, and ether.	90,000–700,000	135°C	150–12,000 cP	In oral pharmaceutical formulations widely used as a coating agent.	Shanghai Honest ChemCo., Ltd	[40, 46]

film coating is rare, due to unavailability of low viscosity material, as per compendial requirement [37].

A clear solution is formed upon the dissolution of methylcellulose in cold water. However, upon heating, gel is formed. The gelling process is based on flocculation, where molecules lose their water of hydration, and as a result, molecules pack together more closely. These films stretch when pulled; this property is also observed in other cellulose derivatives. A 3% solution of MC produces a film of ~1,000 μm thickness and have enough strength to peel off in one piece [39].

It is slightly hygroscopic, and the bulk must be stored in a cool dry place in a well-closed container. At room temperature, its solution is stable in alkalis and dilute acids having pH 3–11. Methylcellulose powder is stable, although slightly hygroscopic. At pH lower than 3, it is susceptible to acid-catalyzed hydrolysis. It is also prone to microbial spoilage, and therefore, preservatives should be added to its solution [40].

12.3.1.3.1.2 Hydroxypropylmethyl Cellulose (HPMC)

Hydroxypropyl methylcellulose (HPMC), also known as hypromellose, is a soluble methylcellulose ether. It is used as a binder, thickening agent, film former, and hydrophilic matrix material. Different viscosity grades of HPMC are available for fabricating hydrophilic matrix systems that range from 4,000–100,000 mPa.s as shown in Table 12.2. The use of HPMC for film coating dates back to the initial days of coating [41].

It is mixed alkyl and hydroxyalkyl cellulose ether consisting of methoxy and hydroxypropyl groups. It is synthesized by reacting alkali treated cellulose with methyl chloride in order to add methoxy groups and then with propylene oxide to add propylene glycol ether groups. HPMC is most widely used cellulose ether due to its wide availability, ease of use, enhanced film forming capability, biocompatibility, and biodegradability. It is soluble in both organic solvents and aqueous media used for film coating. Due to its water solubility, the HPMC film can be colored by the use of pigments to get the clear film. Its non-tacky nature makes processing relatively easier. An aqueous solution of 10%–15% w/w low viscosity polymer can be sprayed for forming a film. The film produced by HPMC is strong, but not stronger than methylcellulose [37, 42].

It has linear structure and nonionic nature, which is ideal for micelle formation or for development of the film. The larger hydroxypropyl groups tend to keep polymer chains apart due to steric hindrance. When a 3% HPMC solution is cast, a film of 500 μm thickness is produced. This film is strong and can be peeled in one piece [42].

Table 12.2 Characterization of some film coating polymers related to commercial viscosity grade available.

Coating polymer	Grades	Viscosity
Hydroxypropyl methyl cellulose	K4M, K15MP, K100MP, E4MP, E10MP CR, K100LVP, E3 PREM.LV, E5 PREM.LV, E6 PREM.LV, E15 PREM.LV, E50 PREM.LV, K3 PREM.LV	K100LVP (80–120 cP), K4M (3,000–5,600 cP), K15MP (12,000–21,000 cP), K100MP (80,000–120,000 cP), E4MP (3,500–5,600 cP), E10MP CR (8,000-13,000 cP), E3 PREM.LV (2.4–3.6 cP), E5 PREM.LV (4–6 cP), E6 PREM.LV (5–7 cP), E15 PREM.LV (12–18 cP), E50 PREM.LV (40–60 cP), K3 PREM.LV (2.4–3.6)
Methylcellulose	A4MP, A15-LV, A15CP, A4CP	A4MP (4,000 cP), A15-LV (15 cP), A15CP (1,500 cP), A4CP (400 cP).
Ethylcellulose	Ethocel Std 4 Premium N-7, Ethocel Std 7FP Premium, Ethocel Std 7 Premium N-10, Ethocel Std 10FP Premium, Ethocel Std 10P Premium N-14, Ethocel Std 20P Premium N-22, Ethocel Std 45P Premium N-50, N-100 Ethocel Std 100FP Premium, Ethocel Std 100P	EthocelStd 4 Premium (3–5.5 cP) N-7 (5.6-8 cP), EthocelStd 7FP Premium (6–8 cP), EthocelStd 7 Premium (6–8 cP), N-10 (8–11 cP), EthocelStd 10 FP Premium (9–11 cP), EthocelStd 10 P Premium (9–11 cP), N-14 (12–16 cP), EthocelStd 20P Premium (18–22 cP), N-22 (18–24 cP), EthocelStd 45P Premium (41–49 cP), N-50 (40–52 cP), N-100 (80–105 cP), EthocelStd 100FP Premium (90–110 cP), EthocelStd 100P (90–110 cP)

(Continued)

Table 12.2 Characterization of some film coating polymers related to commercial viscosity grade available. (*Continued*)

Coating polymer	Grades	Viscosity
Hydroxyethylcellulose	Cellosize (Types and Grades) WP (02), WP and QP (09, 3, 40, 300, 4400), QP (10,000, 15,000, 30,000, 52,000, 100 M) Natrosol HHR, H4R, HR, MHR, MR, KR, GR, ER, JR, LR.	Cellosize WP (02; 5% w/v) 7-20cP , WP and QP (09 - 3; 5% w/v) 60–350 cP, WP and QP (40–4400; 2% w/v) 70-5200 cP, QP(10,000–15,000;2% w/v) 5,700–21,000 cP, (30,000–100 M; 1% w/v) 950–3,000 cP. Natrosol HHR (1% w/v) 3,400–5,000 cP, H4R (1% w/v) 2,600–3,300 cP, HR (1% w/v) 1,500–2,500 cP, MHR (1% w/v) 800–1,500 cP, MR (2% w/v) 4,500–6,500 cP, KR (2% w/v) 1,500–2,500 cP, GR(2% w/v) 150–400 cP, ER (2% w/v)25–105 cP, JR (5%) 150–400 cP, LR (5%) 75–150 cP.

(*Continued*)

Table 12.2 Characterization of some film coating polymers related to commercial viscosity grade available. (*Continued*)

Coating polymer	Grades	Viscosity
Hydroxypropyl cellulose	Klucel HF, Klucel MF, Klucel GF, Klucel JF, Klucel LF, Klucel EF	Klucel HF (1%) 1,500–3,000 cP, Klucel MF (2%) 4,000–6,500 cP, Klucel GF (2%) 150–400 cP, Klucel JF (5%) 150–400 cP, ,Klucel LF (5%) 75–150 cP, Klucel EF (10%) 200–600 cP.
Polyvinylpyrrolidone	Povidone K-12, Povidone K-15, Povidone K-17, Povidone K-25, Povidone K-30, Povidone K-60, Povidone K-90, Povidone K-120	Aqueous viscosity of polyvinylpyrrolidone is expressed as a K-value that ranges from 10 to 20.
Sodium Carboxymethyl Cellulose	Akucell AF (0305; low), (2785; medium), (3085; high)	Akucell AF (0305) 10–15 cP, Akucell AF (2785) 1,500–2,500 cP, (3085) 8,000–12,000 cP.

HPMC solution is stable at pH 3–11. It forms a viscous solution upon dissolution in water with improved enzyme-resistance and higher stability during long-term storage. However, these solutions are susceptible to microbial spoilage and should be preserved with a preservative. The powder form of the material should be stored and kept in a cool and dry place. The common brands available for coating applications of HPMC are Tylopur® (low viscosity Hypromellose), Pharmacoat®, and Tylose® MO (modified HPMC) [40].

12.3.1.3.1.3 Hydroxyethyl Cellulose (HEC)

HEC is water soluble cellulose ether and is insoluble in organic solvents. It is used as a film coating agent, binder, controlled release matrix former, viscosity modifier, suspending agent and thickening agent in topical and ophthalmic preparations. It is mentioned in US-NF and there is no specific requirement for the presence of hydroxyethyl group. The pharmacopeia also allows the presence of additives to enhance powder dispersion in water and to avoid caking on storage [37].

HEC is highly hygroscopic and therefore should be stored in a tightly closed container at cool and dry place. It is tasteless, odorless grayish-white, yellowish-white, or white color hygroscopic powder. The aqueous solutions of hydroxyethyl cellulose are stable in the range from pH 2–12 and viscosity are also unaffected [43]. Although below pH 5 there are chances of hydrolysis and at high pH, oxidation may occur. Increase in temperature may lead to a reduction in viscosity. However, it is restored on cooling. It is susceptible to enzymatic degradation by bacteria and fungi present in the atmosphere. Thus, antimicrobial agents must be added to its aqueous solution. Available brands for HEC for coating applications include Natrosol®, Tylose® H, Tylose® HS, and Tylose® CER [40].

12.3.1.3.1.4 Hydroxypropyl Cellulose (HPC)

Hydroxypropyl cellulose (HPC) is a water-soluble, non-ionic, hygroscopic polymer that is used for film coatings, binder, and controlled release matrix. Furthermore, it has also been used as a viscosity enhancer, suspending agent, stabilizing agent, thickening agent, and emulsifying agent. It is prepared by partial or complete substitution of the hydroxyl group on the glucose unit of cellulose by hydroxypropyl groups. The viscosity of polymer depends upon the degree of polymerization [44]. It is soluble in the both aqueous and alcoholic medium. Its film appears to be tacky, that makes the film weak and resists rapid coating. In order to increase adhesion to the substrate, it is used with other polymers [37].

It is stable material but become hygroscopic after drying, thus for appropriate storage, it is required to be kept at an ambient conditions with low moisture content in a tightly closed container. Its aqueous solution is stable at pH 6–8. At low pH, the solution undergoes acid hydrolysis also leading to decrease in viscosity. At high pH, oxidation may occur, leading to polymer degradation and loss of viscosity [40]. Coating application brands of HPC are available as Klucel® (Ashland) and Celny®.

12.3.1.3.1.5 Carboxymethyl Cellulose (CMC and Na-CMC)

Carboxymethylcellulose (CMC) is a polyelectrolyte polymer with anionic character. It is also available in the form of sodium salt (Na-CMC). It is carboxy methylether derivative of cellulose achieved through the semi-synthetic process. Common names employed are Carmellose Sodium, Carmellosum Natricum, Cellulose gum, Cethylose, Carbose D. Various brands of Na-CMC are available including Akucell®, Aqualon®, Aquasorb®, Blanose®, Walocel® C, Finnfix®Tylose®CB; Tylose® MGA. Na-CMC is a tasteless, odorless, white granular powder that becomes hygroscopic after drying. It is soluble in water (hot and cold) and forms mucilage due to the presence of the polar carboxyl group. CMC is used as film forming agent, viscosity enhancer, binding, suspending, thickening, stabilizing agents, disintegrant, and water-absorbing agent [40, 43].

12.3.1.3.2 Polymethacrylates

Poly(methacrylic acid) was the first methacrylate polymer introduced in 1880. Poly(methyl methacrylate) (PMMA) was developed 48 years later. It was marketed in 1993 and is most widely used methacrylate polymer. The worldwide production of methacrylate polymers exceeds two million metric tons annually [38].

Chemically, polymethacrylates are ionic (aninonic and cationic) synthetic polymers of methacrylic acid, its esters (varying ratios), and dimethylamino ethyl methacrylates. They are non-irritant and non-toxic polymers and are mostly used for film coating applications on tablets [40].

Polymethacrylates can be classified as pH-independent and pH-dependent polymers. pH-independent polymers are used for extended-release formulations, while pH-dependent polymers are used for enteric coatings; these are a methacrylic acid copolymer, which is soluble at the pH range of proximal to distal intestine. The Eudragit E with amino functional groups tends to exhibit pH-independent solubility in the gastrointestinal pH range. The other grades of polymethacrylates marketed are discussed in the proceeding section of gastro-resistant coatings [40, 43].

12.3.1.3.3 Polyvinyl Derivatives

12.3.1.3.3.1 Polyvinyl Alcohol (PVA)

Polyvinyl alcohol is a hydrophilic polymeric material with high aqueous solubility. It has chemical empirical formula of $(C_2H_4O)n$, where *n* varies between 500–5,000. The commercially available grades of polyvinyl alcohol have molecular weight in the range of 20,000–200,000 $gmol^{-1}$. It is primarily a hydrolytic product of polyvinyl acetate and the physicochemical properties are thus dependent on the degree of hydrolysis and polymerization. It has also been used as a coating agent, viscosity increasing agent, stabilizing agent,lubricant, and for the preparation of sustained release formulations [40, 43].

12.3.1.3.3.2 Polyvinylpyrrolidone (PVP)

Polyvinylpyrrolidone, chemically expressed by formula $(C_6H_9NO)n$, is a non-ionic synthetic hydrophilic polymer having high hygroscopicity and excellent water solubility. It also has high solubility in chlorinated (e.g., dichloromethane and chloroform) and alcoholic (e.g., ethanol) solvents. Common names employed for PVP are polyvidone, povidonum, E1201, Povipharm, 1-vinyl-2-pyrrolidinone, and poly[1-(2-oxo-1-pyrrolidinyl) ethylene] polymer. The marketed brands for PVP are Plasdone®(Ashland) and Kollidon® (BASF). The molecular weights of commercially available brands range from 2,500 to 3,000,000 $gmol^{-1}$. Its aqueous viscosity (K-value) ranges from 10 to 120 as mentioned in Table 12.2. It is used as dissolution enhancer, binder, disintegrant, suspending agent. Since it has high Hygroscopicity, it is recommended to use it with other suitable insoluble polymeric agents for coating of granules and tablets [40, 43].

12.3.1.3.3.3 Polyvinylpyrrolidone-Vinyl Acetate Copolymer (Copovidone)

Polyvinylpyrrolidone-vinyl acetate copolymer is commonly known as Copolyvidone, PVP/VA copolymer, poly(1-vinylpyrrolidone-co-vinyl acetate), polymer with 1-vinyl-2-pyrrolidinone, acetic acid vinyl ester, acetic acid ethenyl ester, 2:3 ratio copolymer of vinyl acetate and 1-vinyl-2-pyrrolidone, and 2:3 copolymer of ethenyl acetate and 1-ethenylpyrrolidin-2-one. The brands available are Plasdone® S-630, Luviskol® VA (BASF), Kollidon® VA 64 (BASF). It is used as a film forming agent, binder, and granulation aid [40, 43].

12.3.1.4 *Insoluble Film Coating*

12.3.1.4.1 Ethylcellulose (EC)

Ethylcellulose is a water-insoluble, non-irritant, non-allergenic, and non-toxic polymer which is primarily used to modify drug release by applying hydrophobic coats. It is composed of β-anhydroglucose monomeric units with interlinking acetal bonds. Ethylcellulose powder is white to light tan-color, free flowing, and tasteless. This polymer is widely used as a tablet coating material and also sometimes used as tablet binder and viscosity enhancing agent while rarely used as a tablet filler and flavoring agent [43]. The available brands of ethylcellulose (EC) are Aquacoat® ECT (30% latex dispersion of EC) and Surelease® (30% colloidal dispersion in aqueous aluminum hydroxide). Both of these brands also contain additives in various proportions to support the elastic formation of the film on drug products [40].

12.3.1.5 *Gastro-Resistant Film Coating*

The gastro-resistant coating was first reported by Unna in 1884. It is the most established technology of all modified released drug delivery systems. Acid sensitive and gastric irritant drug products are commonly coated with gastro-resistant polymeric materials which ultimately prevent the release of such molecules in the gastric cavity. It is achieved by the coating of the delivery system with weakly acidic polymer, consisting of carboxylic acid group, which remain intact in the acidic environment while dissolves near neutral pH. Shellac, gelatin, and keratin were initially used to fulfill this purpose. Over time, various synthesized polymers (as mentioned in Table 12.3) were available in the market to play this role [47]. Commonly used polymers for enteric coating include cellulose acetate phthalate (CAP), polyvinyl acetate phthalate (PVAP), and hydroxypropyl methylcellulose phthalate (HPMCP) and poly-methacrylates [47].

12.3.1.5.1 Shellac

Since 1930, shellac has been frequently used alone or in combination with other materials as an enteric coating polymer. Wruble investigated the use of various solutions for enteric coating and concluded that the shellac ammonia solutions were the best among all. However, due to numerous reported disadvantages, it is no longer used for this purpose. Being of natural origin, the material may vary from batch to batch. Its tackiness also

Table 12.3 Physical attributes of some commercial gastro-resistant polymers.

Coating polymer	Other names	Type	Solubility	Molecular weight	Glass transition temp. (Tg)	Viscosity	Pharmaceutical application	Manufacturer	Reference
Hydroxy-propyl methyl-cellulose phthalate	Hypromellose Phthalate, Hydroxy-propylmethyl-cellulosephtalat	Organic (Cellulose ether)	Soluble at pH greater than 4.2 while at low pH show limited solubility	240.30 g/mol	129.6°C −130.7°C	320–480 cP	Enteric coating agent for tablet and granules: 5-10% soluble in solvent mixture of dichloromethane (50): ethanol (50) and water (20) : ethanol (80)	Sigma-Aldrich, Career henan chemical co, Henan DaKen Chemical Co., LTD. Etc.	[40, 42]
Cellulose acetate phthalate	Aquacoat CPD; CAP; Cellacefate; cellulose acetate benzene-1,2-dicarboxylate	Organic (Cellulose acetate)	Practically insoluble in water and alcohols. It is soluble in certain buffered aqueous solutions (pH ≥6.0)	3921.085 g/mol	160°C −170°C	500–900 cP	Enteric film coating, matrix binder in pharmaceuticals	Sigma-Aldrich, Henan DaKen Chemical CO.,LTD. Henan Tianfu Chemical Co.,Ltd.	[40, 42]

(Continued)

Table 12.3 Physical attributes of some commercial gastro-resistant polymers. (*Continued*)

Coating polymer	Other names	Type	Solubility	Molecular weight	Glass transition temp. (Tg)	Viscosity	Pharmaceutical application	Manufacturer	Reference
Polyvinyl acetate phthalate	Phthalavin; PVAP; Opaseal; Sureteric	Organic (vinyl acetate polymer)	Practically insoluble in water but soluble in methanol and ethanol. Sparingly soluble in propan-2-ol and acetone. Polyvinylacetate phthalate(200mg/L) shows pH dependent solubility in buffer solutions, i.e., below pH 5 insoluble and above pH 5 soluble. Ionic strength also influences solubility.	60,700 g/mol	23.5°C–42.5°C	75–110 cP	Used as enteric film coating agent in 9%–10% and sealant in tablet sugar coating in 28%–29%	Colorcon (Opadry-Enteric), Hubei Jusheng Technology Co., Ltd.S.Z. PhyStandard Bio-Tech. Co., Ltd.	[40, 42]

(*Continued*)

Table 12.3 Physical attributes of some commercial gastro-resistant polymers. (*Continued*)

Coating polymer	Other names	Type	Solubility	Molecular weight	Glass transition temp. (Tg)	Viscosity	Pharmaceutical application	Manufacturer	Reference
Polymethacrylates	Acryl-EZE; Eudragit (E, L, S, FS, RL, RS); Kollicoat MAE, Eastacryl 30D	Acrylic-based polymers	Acetone and Alcohol	150,000–800,000 g/mol	45°C –150°C	Eudragit E3–1.2 cP, Eudragit NE 30 D (450 cP), Eudragit L and S (50–200 cP), Eudragit FS 30 D (420 cP), Eudragit L 30 D-55 (415 cP), Eudragit L 100-55 (100–200 cP), Eudragit RL and RS (415 cP), Eudragit RL and RS 30 D (42,000 cP), Kollicoat MAE and Kollicoat MAE 100 P 30 DP (415 cP), Eastacryl 30D (145 cP)	Eudragit E (film-forming agent) and Eudragit L 30 D-55, S & FS, East acryl 30 D and Kollicoat MAE 30 DP (enteric coating agent)	Evonik (Eudragit E, L, S, FS, RL, RS); (Acryl-EZE®, Acryl-EZE® II, Opadry® Enteric, Sureteric®BASF (Kollicoat), etc.	[40, 59]

(*Continued*)

Table 12.3 Physical attributes of some commercial gastro-resistant polymers. (*Continued*)

Coating polymer	Other names	Type	Solubility	Molecular weight	Glass transition temp. (Tg)	Viscosity	Pharmaceutical application	Manufacturer	Reference
Hydroxypropyl methyl cellulose Phthalate	Hypromellose Phthalate, Cellulose phthalate hydroxypropyl methyl ether, HP-50, HP-55	Semi synthetic Organic (cellulose-Phthalate derivative)	Ethanol/Water mass ratio 80/20, methylene, chloride/ ethanolratio 60/40, mixture of dichloro-methane and methyl alcohol (1:1), ethyl or methyl alcohol and acetone (1:1)	80,000–130,000 g/mol	133°C–137 °C	HP-55: 320-480 cP, HP-50: 440–560 cP	12%–15% as enteric coating agent.	Colorcon (Opadry Enteric), Shin-Etsu Chemical Co. Ltd;	[40, 60–62]

makes it difficult to apply as a coating material. Delivery systems coated with shellac, which is stored for 12 months failed to disintegrate at intestinal pH. This occurs due to the polymerization of shellac. Since most of the enteric coated formulations tend to dissolve between pH 3.8 to 6.9, the failure of shellac to dissolve below pH 7 is a major drawback [48].

Shellac resin comprises esters of shellolic acids and aleuritic, which impart the enteric coating characteristics of the material. The precise composition of shellac may differ depending on the source of the polymeric substance and/or manufacturing process. According to NF, four types of shellac are listed based on the method of preparation [49]. It is usually applied as an alcoholic solution or aqueous dispersion. Although shellac possesses various disadvantages and is not preferred to be used as an enteric coating polymer, its use with slight modification in terms of its functionality and to increase its stability has been extensively reported in the literature [50].

12.3.1.5.2 Cellulose Acetate Phthalate (CAP)

Cellulose acetate phthalate is one of the most popular and oldest synthetic polymer for gastro-resistant coating introduced and patented in 1940 by Eastman Kodak. It is a mixed cellulose ester, wherein esterification of one carboxyl group of phthalic acid occurs due to reaction with the hydroxyl group of cellulose. It is a white powder with high flowability and a slightly acetic acid odor. As per the requirements of JP, the acetyl and phthalyl content must be 17%–22% and 30%–40%, respectively, while the USNF requirements are 21.5%–26% and 30%–36%, respectively. As it is prone to hydrolysis, the pharmacopeias specify the maximum limits of loss on drying and quantity of free acid [51, 52].

It is a polymeric acid and ionizes and dissolves in the high pH region of the gastrointestinal tract. However, literature also suggests that dissolution of CAP is attributed to the hydrolytic action of intestinal esterases rather than ionization. It is also permeable to water vapor, ionic substances, and gastric fluid. The permeability depends on numerous factors such as type and quantity of plasticizer and the presence or absence of other solid materials [48].

CAP stability is not affected if it is kept in a closed container at a cool and dry place. However, at high temperature and humidity, slow hydrolysis may occur leading to an increase in its viscosity, free acid content, and acetic acid odor [40].

Cellulose acetate phthalate is commonly used to coat oral enteric delivery systems such as tablets and granules. Beside various companies offering various grades of CAP, Aqua Coat® CPD is the most recognized product of CAP for coating applications.

12.3.1.5.3 Polyvinyl Acetate Phthlate (PVAP)

Polyvinyl acetate phthalate is another commonly used polymer for gastro-resistant coating applications that was introduced and patented by a Canadian company, Charles E. Frost, and later, it was further studied by Millar in 1957, who characterized the impact of the polymer's phthalyl content on the pH of tablet disintegration medium. The optimal content of phthalyl was found to be between 60% and 70%. Though, currently, this range has been revised for commercially available polymers, and has become part of the USP-NF monograph. It is prepared by random synthesis of polyvinyl alcohol with phthalic anhydride and acetic acid [51].

Like other polymers containing phthalate, its aqueous solubility is pH dependent, i.e., at low pH, it is insoluble though quickly dissolves at pH 5. It is less permeable to simulated gastric fluid and water vapors than CAP and other polymers of the same class. This property makes it a good enteric polymer. Like other phthalate containing polymers, it is prone to hydrolysis; however, it is more stable as compare to HPMCP and CAP [48].

12.3.1.5.4 Hydroxypropyl Methylcellulose Phthalate (HPMCP)

Hydroxypropyl methylcellulose phthalate (HPMCP) is a derivative of cellulose in which hydroxyl groups are substituted by 2-hydroxypropyl ethers, methyl ethers, or phthalyl esters. It was introduced in 1971 as an enteric coating polymer. Common names employed for HPMCP are hydroxypropyl methylcellulose benzene-1,2-dicarboxylate, cellulose phthalate hydroxy propyl methyl ether, methylhydroxypropylcellulose phthalate, 2-hydroxypropyl methyl cellulose phthalate, Hypromellose phthalate, HP-5, and HP-55 [53, 54].

It is a white powder or granular substances. USP and JP monographs describe two types of substitution, i.e., HPMCP 220824 and HPMCP 20073. The specification denotes to the percentage substituent of respective methoxyl, hydroxypropyl, and carboxybenzoyl groups [51].

HPMCP like other phthalate containing enteric coating polymers is insoluble in water and gastric medium but ionizes and dissolves in the upper small intestine where pH increases. The numerical descriptions that follow letter HP indicate the pH at which it dissolves. For, e.g., HP-55 hydrates and dissolves at pH 5.5. HPMCP disintegrates and releases more rapidly in the intestine as compared to other polymers. However, it could also release in the stomach of some individuals in the fed state [48].

It is physically and chemically stable for 3–4 years at ambient temperature and for 2–3 months at stress condition (40°C and 75% relative humidity). It is not susceptible to microbial degradation at ambient storage conditions and generally more stable than cellulose acetate phthalate. HPMCP is extensively used in the development of various conventional and advanced gastro-resistant drug delivery systems [55].

12.3.1.5.5 Polymethacrylates

Polymethacrylates are synthetic polymers with anionic and cationic characteristics. They are synthesized products of methacrylic acid, methacrylic acid esters, and dimethyl aminoethyl methacrylates in different ratios [56]. These polymers are marketed by Evonik Industries commonly denoted as Eudragit®. They can be classified as pH independent and pH dependent polymers as discussed earlier. Polymers such as Eudragit L 30 D-55 / L 100-55 and Eudragit L 100 have a high content of carboxylic acid groups and are soluble at low pH of proximal intestine. Hydrophobic polymers such as Eudragit S 100 and Eudragit FS 30 D dissolve at higher pH and are used for colonic drug delivery systems [57, 58].

The use of plasticizers is recommended along with the polymers. Due to the sticky nature of the polymers, it is also suggested to add pigments and opacifier. These polymers can bind to the large quantity of pigments, approximately 2–3 times of polymer used. In order to improve the gloss of the final product, polyethylene glycols are commonly used.

Eudragit S100 and L100 are commercially available in powder form and concentrates in an organic solvent, which can be further diluted for processing. These polymers comply with US-NF requirements [51].

Dry powders of polymethacrylates are stable for 3 years when stored below 30°C in a well-closed container. Above this temperature, clumps are formed, which can be easily broken. Although, it doesn't affect the quality of the product. Dispersions of polymethacrylates are stable for 18 months when stored between 5°C–25°C in a tightly closed container. Phase separation may occur in dispersions below 8°C [40].

12.3.1.6 Semi-Permeable Film Coating

Osmotic drug delivery devices offer a zero-order drug release pattern, which is almost independent of the pH, food, and other hydrodynamic conditions of gastro intestinal tract. Rose-Nelson osmotic pump is considered as the first drug delivery system working on the principle of osmosis [63]. In early 1970s, ALZA Corporation proposed several variations in Rose-Nelson pump and was first to design an oral osmotic drug delivery system that was called as OROS. In the current era, numerous medicinal products are designed by OROS technology [64].

The rate controlling membrane (film coated semi-permeable membrane) plays a significant role in the design of an osmotic system and impart selective permeability character for water molecules against the drugs and other ingredients existing in the device core like osmotic agents. These film coated membranes are required to possess adequate water permeability and enough strength to bear the pressure within the device and maintain its physical stability during the life span of the system. The thickness of membrane is typically kept in between 0.2 mm and 0.3 mm. The membrane should be inert, firm, and non-swelling [64–66]. The drug release rate and thickness of semi-permeable membrane have an inverse relation. As the thickness of the semi-permeable membrane increases the resistance to water diffusion also increases and hence drug release rate decreases. Generally, 1/1000 to 50/1000 portions of a plasticizer are used alone or with coating substances [65–67].

There are many polymers that are a constituent part of coating recipes that can be used to create semi-permeable membrane for osmotic DDS. Commonly utilized polymers for coating of tablet cores in osmotic systems are cellulose (C) esters such as cellulose acetate (mono, di, and tri) and cellulose acetate butyrate and cellulose propionate, etc. Moreover, cellulose ethers such as ethyl cellulose (EC) and certain products of Eudragits are also utilized for the purpose.

12.3.1.6.1 Cellulose Acetate (CA)

Cellulose acetate is an acetyl substituted derivative of cellulose polymer and is widely used for semi-permeable coating applications. It is hygroscopic, white free flowing powder with acetyl substitution in the range of 32% to 39.5%. Though other derivatives such as cellulose di and triacetate are also termed as cellulose acetates, they differ in hydroxyl group substitution range and thus imparting different solubility character. CA is generally used to create membranes that control the rate of water influx for osmotic devices. CA membranes are not soluble, yet semi-permeable for water molecules to penetrate them. Water permeability of CA film is adequately achievable and can be easily rectified by differentiating the degrees of acetylation from 32% to 38%. Water permeability characteristic of CA film declines as the acetyl content in CA rises. Absorption of water through the membrane is also augmented by incorporation of hydrophilic natured flux and permeation enhancers. It has been reported that the inclusion of plasticizers in cellulose acetate for coating application on tablet formulations usually reduces the glass transition temperature, increase in the polymer-chain movement augments the elasticity, and alters the permeation capability of the membrane [66].

12.3.1.6.2 Cellulose Acetate Butyrate (CAB)

Cellulose acetate butyrate also called as cellaburate is produced from reacting acetic anhydride, acetic acid, and cellulose fibers. It is also hygroscopic, white to yellowish-white free flowing powder. CAB is also a semipermeable polymer and has been used for controlling the release of drugs. It has been found as a commercial alternative to CA for thinner semipermeable coating applications with increased robustness and decreased drying times [68].

12.3.1.7 Mucoadhesive Coating Polymers

Mucoadhesive polymers are a class of natural and synthetic polymers which are coated on dosage forms with a target to adhere to gastric mucosa (mucin) and thereby increasing the residence time in the given part of the gastrointestinal tract. Moreover, they are also known to deliver larger payload of drug to be absorbed directly from the mucosal lining [69]. Most of the cellulose derivatives discussed above are used for this purpose especially for designing multi-particulate gastro-retentive drug delivery systems [70]. Beside cellulose esters and derivative, there are certain other polymeric materials which have gained enormous popularity for this particular role.

12.3.1.7.1 Chitosan

Chitosan is a deacetylated product of chitin, an inelastic mucopolysaccharide. It is creamy-white or white odorless flakes or powder, may look cottonlike as fiber formation occurs during precipitation. Chitosan polymers are commercially available with different molecular weights ranging from 50,000 to 2,000,000 $gmol^{-1}$. Moreover, grades with varying degrees of deacetylation (40%–98%) and viscosities are also available. It is used as a tablet film forming agent, disintegrant, mucoadhesive coating. It is also used as viscosity enhancing material and binder [40, 43].

12.3.1.7.2 Poly(Ethylene Oxide)

Polyethylene oxide (PEO), a non-ionic homopolymer, is prepared from polymerization of ethylene oxide using a catalyst. It is represented by a chemical formula $(CH_2CH_2O)n$ where n is the average number of oxyethylene groups. Synonyms for PEO are polyoxyethylene, polyoxirane, and polyoxiante that are available under the brands Polyox® (Colorcon). PEO is off-white to white powder with good flowability and have a slight ammonia like odor. It is soluble in both water and organic solvents like methylene chloride, acetonitrile, and chloroform. It is rapidly eliminated from the body thus showing low levels of toxicity. It is used as a coating agent, mucoadhesive, thickening agent, binder, and for the preparation of immediate and sustained release formulations depending upon the molecular weight [40, 43].

12.3.1.7.3 Poly(Ethylene Glycol)

Polyethylene glycol (PEG) is a non-toxic, non-immunogenic polyether diol, having a general formula $HOCH_2(CH_2O\ CH_2)_m$-CH_2OH where m is the average number of oxyethylene group. It is a polymerized product of ethylene oxide with common names used as Macrogols, Lipoxol, and Pluriol E. Common brands available for PEG are Carbowax® (Dow Chemical Company), Lutrol® E (BASF), and Carbowax Sentry® (polyethylene glycol and methoxy polyethylene glycol MPEGs). It is mainly used for the preparation of solid dosage form and parenteral due to its combined elimination from hepatic and renal pathways. Being an amphiphilic polymer, it is soluble in both water and organic solvents. The molecular weight ranges from 500 to 20,000 g mol^{-1}. In liquid formulations like oral solutions, solid dispersions, and injections, it increases the solubility of poorly soluble drugs. It prevents bacterial surface growth and deposition of proteinaceous material and thereby is widely used as "Stealth" polymer to ensure long circulation of nanoparticles inside the bloodstream [40, 43].

12.3.2 Solvent-Less Coating Procedures

12.3.2.1 Compression Coating

It is one of the oldest type of dry coating or press coating technique which is characterized as solvent-free method. It is a two-step method, where after compression of the tablet core, coating material is compressed around the core. Procedure needs proper placing of the core tablets on the outer layer forming material as variation in the coating is usually observed if the core is not properly placed at the center [71]. Drug release can be effectively controlled after applying the rate controlling hydrophilic membrane around the core surface. Cellulose-based derivative such as hydroxyethylcellulose (HEC), hydroxypropylmethylcellulose (HPMC), hydroxypropylcellulose (HPC), and carboxymethylcellulose (CMC) are commonly used because of its water solubility and gel forming nature [24]. HEC was used for developing delayed release system while non-cellulose polymers such as alginate-chitosan complex for time controlled drug delivery, polyethylene oxide (PEO) and polyethylene glycol (PEG) for controlling the lag time before release [72–75]. Compression coating was also used as a taste masking agent for clinical studies was also reported by Hadfield *et al.* in 1987 [76].

12.3.2.2 Hot Melt Coating

Hot melt coating procedure was used for taste masking, stability for controlling the drug release. The coating material is applied in the molten state which was then solidified on cooling [77]. Drug releases mechanism either diffusion or dialysis, enzymatic breakdown or pH dependent dissolution. The coating material beside waxes such as beeswax and carnuba wax or some oils, versatile polymers such as Gelucires with different melting points and polyethylene glycols are also used. Four-step process of coating that begins with the warming of equipment, preheating of the substrate, melting and spreading of the coating material and congealing [23, 77, 78].

12.3.2.3 Dry Powder Coating

Many polymers such as HPMCAS, Eudragit® RSPO, and Eudragit® RLPO were effectively used in dry powder coating where the film forming polymers and plasticizer are separately fed either in a tablet coating pan of fluid bed coater. Powder forms several layers of coating which may be stabilized by spraying a small amount of water during annealing. Refined dry coating process reported by Cerea *et al.* in 2004 where no water even in small quantity is required. Theophylline tablets were coated using Eudragit E PO for coating theophylline tablets [79–81].

12.3.2.4 Electrostatic Spray Powder Coating

This technique was first introduced in USA in the 1950s and now extensively used in different industries for coating due to its distinct advantages such as low operational cost, short processing time, and highly valued for energy savings. In this coating, dry powders are charged by an electrostatic spray gun and sprayed on the surface of the grounded substrate with electrostatic attractions and mechanical forces to form a deposited powder layer. The thickness of each coated layer is controlled by generating an electrical charge which hinders further layering. This deposited substrate is cured for film formation in an oven at high temperature for a certain period of time. This technique significantly enhances the coating efficiency by increasing the adhesion of coating powder and yields more uniform coating in terms of coating thickness and surface morphology by controlling the feeding of coating powder [82, 83].

12.3.2.5 *Supercritical Fluid-Based Coating*

The supercritical fluid-based coating is used to coat small particles uniformly under a supercritical condition. The physical and thermal properties of supercritical fluids fall in between pure liquids and gases because their temperature and pressure are greater than their critical temperature (Tc) and pressure (Pc). For pharmaceutical purposes, Carbon dioxide is an ideal supercritical fluid having low critical pressure (72 bar) and temperature (31.1°C). It is non-toxic, inexpensive, easily available, nonflammable, environment-friendly, and can be recycled easily but in the absence of cosolvents, its use is limited for coating. The limitation for this method is that the supercritical fluid dissolves coating material only while leaves the core undissolved. The rapid expansion of supercritical solutions (RESS) is a commonly used process in pharmaceutical industry. The active ingredients are dispersed in a high pressure vessel containing supercritical solution of the coating agent. The solvent power of supercritical liquid is reduced on the rapid expansion of suspension resulting in the precipitation of coating material on to the dispersed drug particles. Cosolvents like acetone and methanol are used to increase the solubility of the compound in supercritical fluid [83, 84].

12.3.2.6 *Photocurable Coating*

Photocuring coating is the only chemical technique for the formation of coating films but its application is limited to the photocurable materials. The cross-linked network is rapidly formed by the free-radical polymerization reaction of photocurable materials at room temperature or below. This method is not suitable for photosensitive drugs [83].

12.3.3 Polymer Coatings for Micro/Nano Particulate Drug Delivery Systems (DDS)

Nano/microparticulate drug delivery systems are a class of novel drug delivery platforms which are designed to achieve targeted drug delivery *in vivo* through various routes of administrations. Although the coating of polymers has been a long discussed topic of conventional dosage forms either to enhance the appearance and increasing the stability of the product or to render functional characteristics to such systems as discussed earlier. Coating of polymeric materials on smart novel drug delivery systems offer various constituent properties to systems and/or at various instances, offer functional characteristics to the particulate drug delivery platforms which may include "stealth coating" of nanoparticles for long circulation and plasma half-life, and conjugating targeting ligands (antibodies, receptor binders) for targeted delivery is the drug payload [85].

12.3.3.1 Types of Polymer Coating Systems for Specialized DDS

In case of micro/nanoparticulate drug delivery systems, coating of polymeric materials around the core are either achieved during the fabrication step of the particulate system or after fabrication stage. The major coating techniques on micro/nanoparticulate systems are briefly discussed here.

12.3.3.1.1 Self-Organized/Assembled Coating

Amphiphilic polymeric materials such as block co-polymers consisting of a hydrophilic block and a hydrophobic block are used to prepare a polymeric radial core-shell architecture called micelles. The core of the micelle substance is constituted of drug substance with a shell of radially arranged short amphiphilic polymeric chains. This is usually achieved by dispersing block polymeric substance in an organic drug solution followed by dialysis and then forming an oil/water emulsion where the drug is consolidated into the organic phase and the hydrophobic ends of the polymers face the drug. This is followed by solvent evaporation and freeze drying so that this ensures that aqueous component of the system stays till last and is removed in a solid state so that the hydrophilic ends are exposed towards the exterior. Commonly used polymeric substances for micelle formation are poly(ethylene-glycol)-*b*-poly(L-lysine) (PEG-*b*-PLL), poly(ethylene-oxide)-*b*-poly(propylene-oxide)-*b*-poly(ethylene-oxide) (PEO-*b*-PPO-*b*-PEO), poly(ethylene-glycol)-*b*-poly(aspartic-acid) (PEG-*b*-PAA), etc. Moreover, certain block copolymers such as poly(ethylene-glycol)-*b*-poly(ethylene-imine) (PEG-*b*-PEI) and poly(ethylene-glycol)-*b*-poly-(dimethylaminoethyl-methacrylate) (PEG-b-PAMA), etc., are also used for the synthesis of gold nanoparticles with micellar core-shell architecture [86–88].

12.3.3.1.2 Coating by Coprecipitation

Coprecipitation technique is a form of micro/nanoparticles synthetic procedure which yields particles with core-shell architecture. The shell, in this case, is composed of polymeric material and the core consists of the drug substance alone or dispersed in a matrix. The process involves the dispersion of a solution of film-forming polymer and drug substance in an organic phase into a non-solvent phase consisting of surfactants in a controlled fashion. This results in the precipitation of polymeric substance encapsulating drug substance as the organic phase is diffused out into the bulk. The three key stages in coprecipitation are (1) nucleation, (2) growth, and (3) aggregation. The process is controlled in a way so as to achieve a uniform sized particle by avoiding excessive aggregation of the particles. The commonly demonstrated polymers for coprecipitation include poly(caprolactone) (PCL), poly(lactide (PLA)), poly(lactide-co-glycolide) (PLGA), and poly(alkylcyanoacrylate) (PACA) [88, 89].

12.3.3.1.3 Coating by Adsorption

Polymeric materials can be coated on micro/nanoparticulate systems by applying coats of polymers through adsorption technique. This is achieved by the adsorption of dissolved or dispersed polymer on the surface of preformed particulate matter. The polymers, in this case, posses certain partial charges and are usually dispersed in the aqueous phase. Consequently, they are precipitated onto the surface of the particulate matter; finally, the dispersion medium is evaporated to further precipitate the polymer on the surfaces. The polymers which have been demonstrated for coating by this technique include poly(methacrylate) (PMA), poly(methyl methacrylate) (PMMA), and poly(caprolactone) (PCL) [89].

12.3.3.1.4 Coating through Layer-by-Layer (LbL) Assembly

Layer-by-layer assembly is a simple yet versatile technique used to coat multiple layers of differently charged polymers on the surfaces of micro/nanoparticulate systems. The method involves immersing core material of particulate matter consisting of macromolecules or macromolecular dispersion of drug substance in a solution of differently charged polymer followed by incubation and washing. Subsequently, the particulate matter is again immersed and incubated in another polymer with opposite charge. Thus, by repeating this procedure, multiple layers of polyelectrolyte polymers can be achieved. The cationic polymers which have been demonstrated for LbL assembly include chitosan, poly(ethylene-imine) (PEI) poly(allylamine hydrochloride) (PAH), poly(dimethyldiallylamide ammonium chloride) (PDDA), similarly anionic polymer used in conjunction include poly(styrene sulphonate) (PSS), carrageenan, poly(vinylpyrrolidone) (PVP), etc. [88, 90].

12.4 Mechanism of Drug Release through Coating Systems

12.4.1 Diffusion

In diffusion-controlled drug delivery system the polymer is employed as a barrier to drug transfer and thereby controls the rate of drug release. From different types of polymethacrylates, polyalkylacrylates have shown diffusion controlled drug delivery. Diffusion controlled systems can be classified into two types: reservoir devices (polymer in form of a membrane is coated over the drug) and matrix devices (dispersed drug in a polymer matrix).

In case of coated reservoir devices, the drug core is generally enclosed by hydrophobic polymeric material and through the mesh (the space between macromolecular chains) of these polymers the drug liberates out. The release rate of drug from a reservoir device is governed by the simplified form of Fick's first law of diffusion, as given below:

$$J = \frac{DK\Delta C}{d}$$

Equation 12.1. Equation for Fick's first law of diffusion

Where, "J" is the rate of drug transfer across a unit plane area, "D" is the diffusion coefficient of the drug in the polymer, "K" is the partition coefficient, "ΔC" is concentration gradient across the polymer membrane and "d" is the thickness of the diffusion layer [91].

12.4.2 Dissolution

Dissolution-controlled systems can be categorized as reservoir type and matrix systems. In case of reservoir systems, hydrophilic polymeric substances form a sheath or membranous layer around drug particles. The libration rate of drug can be controlled by type of the polymer used for the membrane and its thickness. The formulations containing polymers like carboxymethyl cellulose, gelatin, hydroxyl propylmethyl cellulose, starch, and polyvinylpyrrolidone have been found to show dissolution dependent drug release [92].

12.4.3 Erosion

The release of the drug from biodegradable polymers such as polyglycolic acid, poly lactic-co-glycolic acid, and poly anhydrides, etc., is usually dependent upon the rate of erosion of the polymeric substance that encases the drug particles. The drug is usually trapped in the polymeric matrix system by physical means or on the other hand, the drug could also be chemically bonded to the polymeric structure. In this case, the release of the drug would take place by the chemical bond degradation between polymer and the drug. The mass erosion pattern relay on the aqueous solubility of the polymer. If the polymer is hydrophobic, then the diffusion of water into the matrix will be limited, restricting the polymer erosion to the surface only. In the case of a hydrophilic polymer, erosion takes place throughout the polymer matrix (bulk erosion). The zero-order drug release can be rate limiting and primary factor only if the surface erosion occurs and the surface area almost remains constant [92, 93].

12.4.4 Osmosis

The rate at which drug exits from a delivery system can also be monitored by applying the process of osmosis in the design of a drug delivery system. One of the most important components in the design of an osmotic drug delivery system is the semi-permeable membrane [66].

The phenomena of osmosis exhibits migration of solvent molecules from the region of lower concentration to the region of higher concentration across a semi-permeable membrane. In this type of controlled drug delivery system, the drug (either in a solution or a solid form) and an osmotic agent are enclosed by a semi-permeable membrane having orifice for drug passage out of the system and thereby determines the rate of the drug release [94, 95]. The Liquid (solvent) influx monitors the rate of delivery of the drugvia the semi-permeable membrane from an elementary osmotic system, in turn, it carries the drug out of the system through an opening. The transfer rate of a drug (impermeable) through the membrane is mathematically expressed by the given equation [96].

$$\frac{dm}{dt} = \frac{A}{h} L_p(\sigma\Delta\pi - \Delta P).C$$

Equation 12.2. Modified equation for osmotic delivery of drugs through Rose-Nelson pump

Where, "*dm/dt*" is the rate of drug transport across a semi-permeable membrane, "*A*" is the surface area of the semi-permeable membrane, "*h*" is the thickness, "*Lp*" is the mechanical permeability of membrane, "σ" is the reflection coefficient, "$\Delta\pi$" is the osmotic pressure gradient across the membrane, "Δp" is the hydrostatic pressure difference across the membrane, and "*C*" is the concentration of drug in the distributed fluid.

12.5 Ideal Characteristics of Coating Polymers

12.5.1 Solubility

The solubility of coating polymer in a gastrointestinal fluid is a key factor for determining the application of polymer as immediate-release, delayed-release, and prolonged-release film coats. Furthermore, it also determines whether the coating polymer solution or suspension is sprayed or delivered as a pseudo latex or latex dispersion [97, 98].

12.5.2 Viscosity

The viscosity of coating polymers in solution is a limiting factor with regard to the ease of application and appearance of the finished dosage form. Highly viscous suspension having a viscosity greater than 400–500 mPas is difficult to spray. Lower viscosity suspensions having viscosity around 100 mPas or less are evenly sprayed and provide a smoother coat. To prevent settling of dispersed particles during coating, suspension must be continuously stirred [98, 99].

12.5.3 Permeability

Appropriate permeability is necessary for various functional properties of the polymer films. For barrier applications, moisture protection, and taste masking, low permeability is required while intrinsic permeability of the polymer films may be changed for some prolonged-release dosage forms by incorporating additional formulation components [98, 99].

12.5.4 Glass Transition Temperature

Glass Transition Temperature (Tg) expresses the molecular motion of the polymer which is dependent on many factors affecting polymer chain movement. Free volume, the empty space in between the ends of the polymer chain, facilitates the segmental motion in polymers. Generally, polymers with higher Tg values are desirable as they enhance processing and handling qualities, dissolution and bioavailability, and physical stability. Long polymer chains having high molecular weight and polymers with polar and bulky side groups possess higher Tg values due to smaller free volumes. In comparison to branched and flexible polymers, linear and rigid polymer chains have higher Tg values. Similarly cross-linked chains have a higher Tg value due to higher entropy [98, 99].

12.5.5 Mechanical Strength

Polymer films with a high tensile strength provide better protection and are more resistant to cracking and splitting. Young's modulus and tensile strength are in a linear relationship with the molecular weight of the polymer. Higher the molecular weight greater the tensile strength and Young's modulus [98, 99].

12.6 Conclusion

The successful design of coated pharmaceutical product requires knowledge of the polymers, technologies and release mechanisms to ensure that the most appropriate system is adopted for the targeted drug. This chapter presents a broad overview of types of polymers, covering its characteristics, applications and commercialized systems using both natural and synthetic polymers. The variability of the polymer, with respect to diversified physicochemical properties and stability also needs to be considered when deciding on a strategy for coated pharmaceutical product design. It is essential that the coating processes and coating formulations are carefully controlled and optimized to ensure batch to batch reproducibility of the product. Based on the valuable economic and clinical benefits that the coating polymers can provide, it is expected that the significance of polymers for coating pharmaceuticals will continue to evolve in the near future especially in the field of nanotechnology and target specific drug delivery system.

References

1. Cole, G. *et al.*, *Pharmaceutical Coating Technology*, G. *Cole* (Ed.), Taylor & Francis, London, 2002.
2. Savage, G.V. and Rhodes, C.T., The Sustained Release Coating of Solid Dosage Forms: A Historical Review. *Drug Dev. Ind. Pharm.*, 21, 1, 93–118, 1995.

3. Aulton, M.E. and Taylor, K.M.G., *Aulton's Pharmaceutics. The Design and Manufacture of Medicines*, Fifth Edition, Elsevier Ltd., China, 2018.
4. Fahr, A., *Voigt's Pharmaceutical Technology*, John Wiley & Sons, Stuttgart, Germany, 2018.
5. Kremers, E. and Sonnedecker, G., *Kremers and Urdang's history of pharmacy*, Amer. Inst. History of Pharmacy, J. B. Lippincott & Co, USA, 1986.
6. Unna, P.G., *Eine neue form medicamentoser einverleibung*, vol. 2, pp. 507–509, Fortschr. Med (Berlin), 1884.
7. Shen, E., Materials and equipment innovations have advanced tablet coating from sugar to copolymers and simplified pharma production. *Pharm. Technol.*, 41, 7, 34–36, 2017.
8. Loyd, V. and Allen, J., *Ansel's Pharmaceutical Dosage Forms and Drug Delivery Systems*, 11th ed., Wolters Kluwer (India) Pvt Ltd., New Delhi, 2018.
9. Lieberman, H., Lachman, L., Schwartz, J.B., *Pharmaceutical Dosage Forms: Tablets*, vol. 3, Informa Health Care, New York, 1989.
10. Hillery, A.M. and Park, K., *Drug Delivery: Fundamentals and Applications*, Second edition, CRC Press; Taylor & Francis Group, Florida, USA, 2017.
11. Rowe, R.C., Materials used in the film coating of oral dosage forms, in: *Materials used in Pharmaceutical Formulation, Critical Reports on Applied Chemistry*, A.T. Florence (Ed.), pp. 1–36, Blackwell Scientific Publications, Oxford, 1984.
12. Florence, A.T., A Short History of Controlled Drug Release and an Introduction, in: *Controlled Release in Oral Drug Delivery*, C.G. Wilson and P.J. Crowley (Eds.), pp. 1–26, Springer US, Boston, MA, 2011.
13. T. Higuchi and H.M. Leeper, *Osmotically driven fluid dispenser*, Alza Corp, US Patent No. 4,034,756, 1977.
14. M. Nakano, T. Higuchi, A., Hussain, A., *Osmotic dispenser*, Alza Corp, US Patent No. 3,995,632, 1976.
15. F. Theeuwes and T. Higuchi, *Osmotic dispensing device with maximum and minimum sizes for the passageway*, Alza Corp, US Patent No. 3,916,899, 1975.
16. Theeuwes, F., Elementary osmotic pump. *J. Pharm. Sci.*, 64, 12, 1987–1991, 1975.
17. Wen, H. and Park, K., *Oral controlled release formulation design and drug delivery, Theory to practice*, pp. 21–32, John Wiley & Sons, Stuttgart, Germany, 2010.
18. Felton, L.A. *et al.*, Physical and enteric properties of soft gelatin capsules coated with eudragit® L 30 D-55. *Int. J. Pharm.*, 113, 1, 17–24, 1995.
19. J.W. Matthews and G. Vergilio, Process for coaing gelatin capsules, U.S. Patent, United States, 1989.
20. Murthy, K.S. *et al.*, Comparative evaluation of aqueous enteric polymers in capsule coatings. *Pharm. Technol.*, 10, 36–46, 1986.
21. Obara, S. *et al.*, Dry coating: An innovative enteric coating method using a cellulose derivative. *Eur. J. Pharm. Biopharm.*, 47, 1, 51–59, 1999.

22. Barthelemy, P. *et al.*, Compritol 888 ATO: An innovative hot-melt coating agent for prolonged-release drug formulations. *Eur. J. Pharm. Biopharm.*, 47, 1, 87–90, 1999.
23. Achanta, A.S. *et al.*, Development of Hot Melt Coating Methods. *Drug Dev. Ind. Pharm.*, 23, 5, 441–449, 1997.
24. Ozeki, Y. *et al.*, Evaluation of novel one-step dry-coated tablets as a platform for delayed-release tablets. *J. Control. Release*, 95, 1, 51–60, 2004.
25. Ozeki, Y. *et al.*, Evaluation of the compression characteristics and physical properties of the newly invented one-step dry-coated tablets. *Int. J. Pharm.*, 267, 1–2, 69–78, 2003.
26. Thies, C. *et al.*, A supercritical fluid-based coating technology 1: Process considerations. *J. Microencapsulation*, 20, 1, 87–96, 2003.
27. Ribeiro Dos Santos, I. *et al.*, A supercritical fluid-based coating technology. 3: Preparation and characterization of bovine serum albumin particles coated with lipids. *J. Microencapsulation*, 20, 1, 110–28, 2003.
28. Ribeiro Dos Santos, I. *et al.*, Microencapsulation of protein particles within lipids using a novel supercritical fluid process. *Int. J. Pharm.*, 242, 1–2, 69–78, 2002.
29. Reeves, L.A. *et al.*, Electrostatic application of powder material to solid dosage forms, in: *PCT Int Appl*, Wo: Phoqus Limited, UK, UK, 2001.
30. Wang, J.Z.Y. and Bogner, R.H., Solvent-free film coating using a novel photocurable polymer. *Int. J. Pharm.*, 119, 1, 81–89, 1995.
31. Troster, S.D. and Kreuter, J., Influence of the surface properties of low contact angle surfactants on the body distribution of 14C-poly(methyl methacrylate) nanoparticles. *J. Microencapsulation*, 9, 1, 19–28, 1992.
32. Tröster, S.D., Müller, U., Kreuter, J., Modification of the body distribution of poly(methyl methacrylate) nanoparticles in rats by coating with surfactants. *Int. J. Pharm.*, 61, 1, 85–100, 1990.
33. Kreuter, J., Nanoparticles—A historical perspective. *Int. J. Pharm.*, 331, 1, 1–10, 2007.
34. Illum, L. and Davis, S.S., Effect of the Nonionic Surfactant Poloxamer 338 on the Fate and Deposition of Polystyrene Microspheres Following Intravenous Administration. *J. Pharm. Sci.*, 72, 9, 1086–1089, 1983.
35. Illum, L. *et al.*, The organ distribution and circulation time of intravenously injected colloidal carriers sterically stabilized with a block copolymer poloxamine 908. *Life Sci.*, 40, 4, 367–74, 1987.
36. Bose, S. and Bogner, R.H., Solventless pharmaceutical coating processes: A review. *Pharm. Dev. Technol.*, 12, 2, 115–131, 2007.
37. Aulton, M., Cole, G., Hogan, J., *Pharmaceutical Coating Technology*, p. 504, CRC Press, London, 1995.
38. Shokri, J. and Adibkia, K., Application of cellulose and cellulose derivatives in pharmaceutical industries, in: *Cellulose-medical, pharmaceutical and electronic applications*, IntechOpen, Florida, USA, 2013.

39. Embuscado, M.E. and Huber, K.C., *Edible films and coatings for food applications*, vol. 9, Springer, New York, USA, 2009.
40. Rowe, R.C., Sheskey, P., Quinn, M., *Handbook of pharmaceutical excipients*, Libros Digitales-Pharmaceutical Press, Pharmaceutical Press, UK, 2009.
41. Kaur, G. *et al.*, Oral controlled and sustained drug delivery systems: Concepts, advances, preclinical, and clinical status, in: *Drug Targeting and Stimuli Sensitive Drug Delivery Systems*, pp. 567–626, Elsevier, William Andrew Publishing, New York, USA, 2018.
42. Venkatesan, N. and Ichikawa, H., *Pharmaceutical Polymers, in Synthesis and Applications of Copolymers*, A. Parthiban (Ed.), pp. 307–333, John Wiley and Sons, New Jersey, 2014.
43. Jones, D., *Pharmaceutical applications of polymers for drug delivery*, vol. 15, iSmithers Rapra Publishing, Shawbury, UK, 2004.
44. Ma, H., Hsiao, B.S., Chu, B., Hydroxypropyl Cellulose (HPC) Membrane, in: *Encyclopedia of Membranes*, E. Drioli and L. Giorno (Eds.), pp. 1–2, Springer Berlin Heidelberg, Berlin, Heidelberg, 2015.
45. Jadhav, N.R. *et al.*, Glass transition temperature: Basics and application in pharmaceutical sector. *Asian J. Pharm. (AJP): Free full text articles from Asian J. Pharm.*, 3, 2, 82–89, 2014.
46. Nakamura, K., Hatakeyama, T., Hatakeyama, H., Heat capacities of carboxymethylcellulose-nonfreezing water systems at around glass transition temperature. *Kobunshi Ronbunshu/Jpn. J. Polym. Sci. Technol.*, 53, 12, 860–865, 1996.
47. Pillai, O. and Panchagnula, R., Polymers in drug delivery. *Curr. Opin. Chem. Biol.*, 5, 4, 447–451, 2001.
48. Tarcha, P.J., *Polymers for controlled drug delivery*, CRC Press; Taylor & Francis Group, Florida, USA, 1990.
49. Wen, H. and Park, K., *Oral controlled release formulation design and drug delivery: Theory to practice*, John Wiley & Sons, New Jersey, USA, 2011.
50. *Applications of Polymers in Drug Delivery*, Smithers Rapra, United Kingdom, 2014.
51. Aulton, M., Cole, G., Hogan, J., *Pharmaceutical coating technology*, Taylor & Francis, Florida, USA, 1995.
52. Alger, M., *Polymer science dictionary*, Springer Science & Business Media, Berlin, Germany, 1996.
53. Narang, A.S. and Boddu, S.H., Excipient applications in formulation design and drug delivery, in: *Excipient Applications in Formulation Design and Drug Delivery*, pp. 1–10, Springer, New York, USA, 2015.
54. Shin-Etsu Chemical Co., L., *Cellulose & Pharmaceutical Excipients Department USP Hypromellose Phthalate*, Tokyo, Japan, March 31, 2012, Available from: http://www.metolose.ru/files/hpmcp.pdf.
55. https://www.chemicalbook.com/ProductChemicalPropertiesCB9257146_EN.htm.

56. Holmes, P.F., Bohrer, M., Kohn, J., Exploration of polymethacrylate structure–property correlations: Advances towards combinatorial and high-throughput methods for biomaterials discovery. *Prog. Polym. Sci.*, 33, 8, 787–796, 2008.
57. Parthiban, A., *Synthesis and applications of copolymers*, John Wiley & Sons, New Jersey, USA, 2014.
58. Liu, F., *Update on polymers for oral drug Delivery*, Smithers Rapra, Shawbury, UK, 2011.
59. Eerikäinen, H., *Preparation of nanoparticles consisting of methacrylic polymers and drugs by an aerosol flow reactor method*, VTT Technical Research Centre Ltd, Finland, 2005.
60. Hashmat, D. *et al.*, Development of enteric coated flurbiprofen tablets using Opadry/Acryl-Eze System—A technical note. *AAPS Pharmscitech*, 9, 1, 116, 2008.
61. Lundqvist, R., Molecular weight studies on hydroxypropyl methylcellulose II. Intrinsic viscosity. *Int. J. Polym. Anal. Charact.*, 5, 1, 61–84, 1999.
62. Sakellariou, P., Rowe, R., White, E., The thermomechanical properties and glass transition temperatures of some cellulose derivatives used in film coating. *Int. J. Pharm.*, 27, 2–3, 267–277, 1985.
63. Rose, S. and Nelson, J., A continuous long-term injector. *Aust. J. Exp. Biol. Med. Sci.*, 33, 4, 415–419, 1955.
64. Gupta, B.P. *et al.*, Osmotically controlled drug delivery system with associated drugs. *J. Pharm. Pharm. Sci.*, 13, 4, 571–588, 2010.
65. Stuti, G. *et al.*, Osmotic pumps: A review. *Int. J. Compr. Pharm.*, 2, 1–8, 2011.
66. Verma, R.K., Krishna, D.M., Garg, S., Formulation aspects in the development of osmotically controlled oral drug delivery systems. *J. Control. Release*, 79, 1, 7–27, 2002.
67. Keraliya, R.A. *et al.*, Osmotic drug delivery system as a part of modified release dosage form. *ISRN Pharm.*, 2012, 528079, 2012.
68. Shanbhag, A. *et al.*, Application of cellulose acetate butyrate-based membrane for osmotic drug delivery. *Cellulose*, 14, 1, 65–71, 2007.
69. Russo, E. *et al.*, A focus on mucoadhesive polymers and their application in buccal dosage forms. *J. Drug Delivery Sci. Technol.*, 32, 113–125, 2016.
70. Möschwitzer, J. and Müller, R.H., Spray coated pellets as carrier system for mucoadhesive drug nanocrystals. *Eur. J. Pharm. Biopharm.*, 62, 3, 282 287, 2006.
71. Bose, S. and Bogner, R.H., Solventless pharmaceutical coating processes: A review. *Pharm. Dev. Technol.*, 12, 2, 115–31, 2007.
72. Matsuo, M. *et al.*, Delayed-release tablets using hydroxyethylcellulose as a gel-forming matrix. *Int. J. Pharm.*, 138, 2, 225–235, 1996.
73. Matsuo, M. *et al.*, Evaluation of hydroxyethylcellulose as a hydrophilic swellable material for delayed-release tablets. *Chem. Pharm. Bull.*, 43, 2, 311–314, 1995.

74. Takeuchi, H. *et al.*, Spray-dried composite particles of lactose and sodium alginate for direct tabletting and controlled releasing. *Int. J. Pharm.*, 174, 1, 91–100, 1998.
75. Takeuchi, H. *et al.*, Spray-Dried Lactose Composite Particles Containing an Ion Complex of Alginate-Chitosan for Designing a Dry-Coated Tablet Having a Time-Controlled Releasing Function. *Pharm. Res.*, 17, 1, 94–99, 2000.
76. Hadfield, P.J., Holmes, D.S., Yarwood, R.J., The Potential use of Compression Coating in the Blinding of Clinical Trial Supplies. *Drug Dev. Ind. Pharm.*, 13, 9–11, 1877–1890, 1987.
77. Joachim, J. and Cauture, E., A hot-melt coating agent for controlled release. 8, 3, 24–27, 1996.
78. Ghebre-Sellassie, I., *Multiparticulate Oral Drug Delivery*, CRC Press; Taylor & Francis Group, Florida, USA, 1994.
79. Obara, S. *et al.*, Dry coating: An innovative enteric coating method using a cellulose derivative. *Eur. J. Pharm. Biopharm.*, 47, 1, 51–9, 1999.
80. Pearnchob, N. and Bodmeier, R., Dry polymer powder coating and comparison with conventional liquid-based coatings for Eudragit) RS, ethylcellulose and shellac. *Eur. J. Pharm. Biopharm.*, 56, 3, 363–9, 2003.
81. Zheng, W. *et al.*, Properties of theophylline tablets powder-coated with methacrylate ester copolymers. *J. Drug Delivery Sci. Technol.*, 14, 4, 319–325, 2004.
82. Luo, Y. *et al.*, Dry coating, a novel coating technology for solid pharmaceutical dosage forms. *Int. J. Pharm.*, 358, 1–2, 16–22, 2008.
83. Yang, Q. *et al.*, An update on electrostatic powder coating for pharmaceuticals. *Particuology*, 31, 1–7, 2017.
84. Thies, C. *et al.*, A supercritical fluid-based coating technology 1: Process considerations. *J. Microencapsulation*, 20, 1, 87–96, 2003.
85. Torchilin, V.P., Multifunctional, stimuli-sensitive nanoparticulate systems for drug delivery. *Nat. Rev. Drug Discovery*, 13, 11, 813, 2014.
86. Torchilin, V.P., Micellar nanocarriers: Pharmaceutical perspectives. *Pharm. Res.*, 24, 1, 1, 2007.
87. Jones, M.-C. and Leroux, J.-C., Polymeric micelles–a new generation of colloidal drug carriers. *Eur. J. Pharm. Biopharm.*, 48, 2, 101–111, 1999.
88. Schärtl, W., Current directions in core–shell nanoparticle design. *Nanoscale*, 2, 6, 829–843, 2010.
89. Mora-Huertas, C.E., Fessi, H., Elaissari, A., Polymer-based nanocapsules for drug delivery. *Int. J. Pharm.*, 385, 1–2, 113–142, 2010.
90. Ai, H., Jones, S.A., Lvov, Y.M., Biomedical applications of electrostatic layer-by-layer nano-assembly of polymers, enzymes, and nanoparticles. *Cell Biochem. Biophys.*, 39, 1, 23, 2003.
91. Bruschi, M.L., *Strategies to Modify the Drug Release from Pharmaceutical Systems*, Elsevier Science, New York, USA, 2015.

92. Kumar, N. *et al.*, Controlled release technology, in: *Encyclopedia of polymer science and technology*, 2002.
93. Jantzen, G.M. and Robinson, J.R., Sustained and controlled-release drug-delivery systems, in: *Modern Pharmaceutics*, G.S. Banker and C.T. Rhodes (Eds.), Marcel Dekker, USA, 2005.
94. Swarbrick, J., Drug Delivery: Controlled Release, in: *Encyclopedia of Pharmaceutical Technology*, Informa health care, pp. 1082–1104, London, UK, 2007.
95. Rathbone, M.J., Hadgraft, J., Roberts, M.S., Osmotically Controlled Tablets, in: *Modified-release drug delivery technology*, pp. 101–14, CRC Press, Taylor & Francis, Florida, USA, 2003.
96. Theeuwes, F., Elementary osmotic pump. *J. Pharm. Sci.*, 64, 12, 1987–1991, 1975.
97. Martin, A.N., Sinko, P.J., Singh, Y., *Martin's Physical Pharmacy and Pharmaceutical Sciences: Physical Chemical and Biopharmaceutical Principles in the Pharmaceutical Sciences*, Lippincott Williams & Wilkins, Philadelphia, USA, 2011.
98. Tovey, G.D., *Pharmaceutical Formulation: The Science and Technology of Dosage Forms*, Royal Society of Chemistry, London, UK, 2018.
99. Aulton, M.E. and Taylor, K.M.G., *Aulton's Pharmaceutics: The Design and Manufacture of Medicines*, Churchill Livingstone/Elsevier, London, UK, 2013.

13

Self-Healing Polymer Coatings

Sathish Kumar Palaniappan[1]*, Moganapriya Chinnasamy[2], Rajasekar Rathanasamy[2] and Samir Kumar Pal[1]

[1]*Department of Mining Engineering, Indian Institute of Technology Kharagpur, Kharagpur, India*

[2]*Department of Mechanical Engineering, Kongu Engineering College, Erode, India*

Abstract

The critical characteristics of self-healing polymeric coatings are covered in this chapter. An outline of progress in existing self-healing coating methods and realistic frameworks of polymeric coatings are provided after elaborating the introduction and benefits on self-heal concepts in polymer coatings. Self-regeneration of surfaces and anti-corrosive protective layer fractures were discussed. Eventually, a current industrial applications and possible potential activity for the future prospect will also be discussed. This review aims to explore the fast expanding area of self-healing polymeric coatings by analyzing the main mechanisms that were effective in automatic repair over the past decade. Furthermore, some issues relating to the transition from laboratories to valid industrial application of these self-healing technologies were addressed.

Keywords: Coatings, self-healing, polymers, anti-corrosion, fractures, reliability, degradation, thin films

13.1 Introduction

Over decades, coatings have been considered as a desired material of importance. A layer or scheme of strata is used over a particular substratum mainly for three primary reasons: shielding, functionalization, and of cultural importance. There are a variety of plastics that need to be protected

**Corresponding author*: sathishiitkgp@gmail.com

Inamuddin, Rajender Boddula, Mohd Imran Ahamed and Abdullah M. Asiri (eds.) Polymer Coatings: Technology and Applications, (319–332) © 2020 Scrivener Publishing LLC

with coatings during their corresponding service life. Metal substrates, rocks, concrete, wood, leather, fabrics, polymers, composites, plastics, textiles, etc., are the most common coated materials. There are numerous coating formulations that can be used depending on the material to be protected and the essential coating properties. Inorganic, polymer, and artificial formulations are included in this case. It is not very obviously described about the extreme origin of coating materials. Many formulations, for example, lacquers, varnishes, enamels, bee wax, mineral clays, animal oil, and many more natural products, are renowned for millennia. The most common instant characteristics of coatings are decoration, aesthetic feel, luminosity, repellent, safety from corrosion, heat barrier, anti-fouling, wear resistance, and better lubrication.

Over the past few years, coating engineering has experienced enormous developments that will certainly remain in the close future. Today, there are many formulations, and they can simply not count the provided features. However, there are still inconveniences and fresh difficulties arises everyday as traditional and innovative high-tech functions require more functionality and shorter lives of coated products. Coatings for the corrosion safety of concrete plastic products have a considerable financial effect, because the coated layer determines the importance of the product in operation and also in its efficiency. For certain phases of moment, a safety layer must be sustainable, by maintaining its physical strength and operational characteristics. However, coatings are prone to fading during certain practical applications and the primary characteristics are progressively degraded. The degradation speed relies on several parameters, such as the physico-chemical design of the covering and substrate, the fabric size, and the internal environmental aggression. As a result, regular servicing is required necessarily, usually recoating or removing the exposed harmed components and replacing them. Currently, coating products with minimum internal interference that can last forever are ahead of the latest technology. There has been a number of study attempts to develop intelligent fabric products, particularly those with clever functions and self-healing characteristics. Intelligent self-curing coatings are an extremely interesting category of material. These categories of products with disruptive accomplishments have been the subject of numerous attempts in the last centuries. Among them, intelligent nanocontainer products are one of the most successful techniques for the development of self-healing coatings.

13.2 Self-Healing: Introduction and Benefits

The capacity to automatically and autonomously cure (recover/repair) injury can be characterized as the capability of a self-healing product, which is without any external interference. To identify such an asset in products, numerous popular words such as self-reparation, autonomy healing, and auto-reparation are used. Most often, it is impossible to incorporate the self-healing characteristics into human products without an exterior cause. So, the two kinds of self-healing are as follows:

- autonomous (no interference)
- non-autonomous (naturalintervention/external triggering necessary)

Self-healing can be described in coating techniques as the complete or temporary restoration of matrices harmed by aging or unforeseen violent occurrences [1]. This idea has been expanded, including the activation of oxidation in metal surfaces. Most of the suggested alternatives recognized as highly efficient therapeutic products are considered to be extremely important. In order to bring self-healing functionality into organic or hybrid coatings, two key approaches were developed: (i) intelligent matrix response, enabling mitigation of failures created in the matrix because of the existence of artificial species, and (ii) portable and synthetic organisms encapsulation, giving special functionality. Among other species, they include monomers, polymers, inhibitors of corrosion, antifouling molecules, antibacterial drugs, etc. Outdoor or interior activities can trigger self-healing functions. Quick coating is thus a category of products that can detect harm and react to particular causes which lead to the discharge of active organisms that mitigate harm. The renowned self-healing techniques have been anticipated for smart responsive polymer matrix or for restricted discharge of efficient species from micro- or nano-containers susceptible to explicit triggers.

The fundamental concepts of a self-healing process in a safety layer are illustrated in Figure 13.1. A fracture is instigated in Figure 13.1(a), which expands so that the triggering procedure can be activated automatically. As a result, the curing agent is liberated which heals the harm caused by the layer, as seen in Figure 13.1(b). In order to reaffirm the barrier properties of the protective coating, this process is highly effective and provides

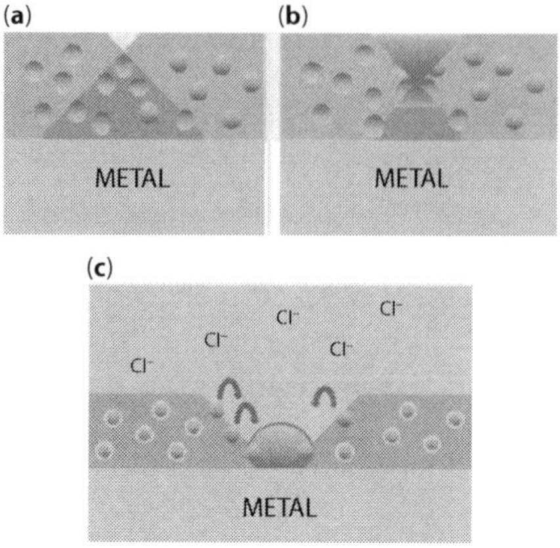

Figure 13.1 Schematic view of crack-induced self-healing processes. (a) A crack appears and develops affecting an internal area of the coating. The stress induced by the cracks breaks the reservoirs that release a mobile phase. (b) The mobile phase reacts and fills the crack, restoring the coating barrier properties. (c) The coating is damaged, exposing the bare metal, and a specific trigger (e.g., chloride ions activity) induces the release of the active healing agents stored inside the containers that repair the damaged areas. [With kind permission obtained from RightsLink, Elsevier License]

repair of the cover matrix. Figure 13.1(c) expresses the fully damaged coating layer and exposes the bare metal to natural environmental aggression. Containers packed with functional/repair organisms are caused by the existence of chloride ions because of continuing chemical and electrochemical interfacial modifications. The curative is discharged in the region affected by harmful occurrences, which inhibits the continuing decay procedures and cures the harmful region very effectively. Preferably, a demand/release-based series of occurrences should only operate when the harmful occurrences exists. The encapsulated object must be placed until the next occurrence after the harm has been cured.

Self-healing resources are not a myth anymore and the fact that man-made materials can recover structural reliability if they collapse. The fractures in walls can cover themselves or the marks on car bodies can restore their glossy unique exterior. Yes, this can be seen by everyone as wounds and cuts in living organisms are naturally healed. Almost all materials are susceptible and degradable in time to natural or artificial degradation. Roughly every product is prone and time-consuming to natural or artificial degradation. The long-term process of deterioration leads to microcracks that cause failure in case of construction materials. Repairing is therefore necessary to improve quality and material life. While researchers are motivated by usual blood clotting procedure or the repair of fractured bones, it does not really take account of the difficult environment of cure procedure in human being or further animal systems [2–7]. Nevertheless, Nissan's latest declaration that scratch-healing paints should be released commercially on automobile cover has drawn communal attention in a marvelous estate of such products [8].

13.3 Summary of Progress in Self-Healing Coating Technology

13.3.1 Coatings for Self-Regeneration

The reduction in specific utility during a certain instance deteriorates adhesion between coat and substrate, UV-protection failure, unexpected weight alteration, anticorrosion function loss for metal substrate, and so on, can only be illustrated by the damage of a coating, as stated earlier. Therefore, the rebound of damaged connections may not be essential, since any layer does not benefit from failure or decrease of mechanical force. Rather, the harm to the coating layer must be cured by restoring the coating feature in a smart manner. The fact that a species may be a small molecule or else a loop linked en route for cross-linked arrangement gives some covering features. For example, small ground layer wettability arises from fluorinated or silicone-based strings; an effective corrosion inhibitor can be introduced for the benefit of anti-corrosion purposes.

To regain the role of coating, the particular species must be transferred or they may transfer to the location where the layer feature decreases in a "self-regenerating" way. In particular, surface energy, climate modifications, and others can be the guiding power of these particular species and should also be able to benefit from a certain degree of flexibility for comparatively fast transport within a coating system. One instance of the self-regenerating coating is a corrosion-resistant coating in which the scratch agent replenishes at the layer/metal interaction [9]. Some other researchers [10] have followed layer-by-layer installation procedure to pinch the wind controller on silica nanoparticles and have added nano-servoirs to the ZrO_2 and SiO_2 solar gel placed on an aluminum substrate in epoxy-functionalized coatings. During corrosion of the aluminum alloy, benzotriazole is released through pH modifications.

Another instance is the self-regeneration coating with low adherence. The coated layer is very crucial and regulates the weight resistance and refriction. The ground structure of a layer is well recognized to be considerably different from its core structure. Low adhesion coatings have been used extensively today as their water/oil repellency allows them to be cleaned readily. Fluorine or silicone containing species at the film layer provide the low-surface stress. The ground segregation of fluorinated species has already produced low adherence coatings [11–15]. However, the fluorinated layer was shown to be very thin [14] and may not have poor adhesion to mechanical abrasion. Lately, independent regenerating strategy was created to maintain adherence properties as minimum as possible for fluorinated polyurethane coating [16]. However, the two main characteristics of this strategy are as follows: (1) fluorinated tails miscibility with the same polyester structure, and (2) the distance is supported by poly (pa-caprolactone) (PCL), which allows sufficient flexibility of the fluorinated body when integrated into the interconnected networks. In the event of surface fracture leading to the failure of upper parts of the coating, fluorinated tails from sub-layers can be redirected to reduce the interface energy, because of the PCLs flexibility. Controlled functional procurators of polyester model have been produced through perfluoroalkyl liquor which regulates ring-opening polymerization of β-caprolactone [17]. In order to achieve films of minimum surface energy, suitable precursors be healed by means of cross-linking polyisocyanate. Fluorinated species with self-regeneration performance was proved in conjunction with microtoming through XPS analysis [16].

Figure 13.2 shows that the F/C molar proportion throughout the film seems consistent, which means that the allocation of fluorinated species

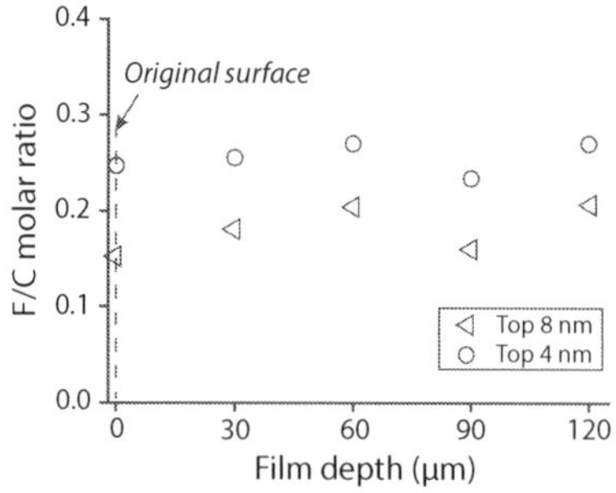

Figure 13.2 F/C ratios at two different probe depths (4 and 8 nm) for thin slices cut from a polyurethane coating containing perfluorooctyl-PCL (Rf8-PCL). The coating examined was prepared from a reaction mixture (Rf8-PCL, a PCL based polyol, and a polyisocyanate cross-linker). The coating was about 200 μm thick and contained 1wt% of fluorine. Thin slices of about 30 μm thick were cut by microtoming and the top surfaces of the slices were then examined by XPS.

in most of the film is comparatively homogenous. Another important thing is that 4 nm on upper side has 60% higher ratio on F/C for each slice compared to 8 mm. This obviously indicates that fluorinated tails are replenished following by microtoming prior to XPS experiment. Since the coating temperature of a glass transition region is approximately −20°C, self-regeneration of materials requires microtoming on chamber heat. Fluorinated tissue regeneration relies on the flexibility and heat of the polymeric networks. If the polymeric network has a temperature at glass transition region near the room temperature, fluorinated species may be regenerated simultaneously with the formation of a new surface. The regenerating behavior can be considered to be independent in this regard. This auto-recovery approach can be expanded to other active coatings like antimicrobial, biocidal, etc.

13.3.2 Anti-Corrosion Protective Layer Fractures

Arrangements and representations of the newly developed shape memory by Xiao fan Luo *et al.* [18] helped to self-heal the SMASH coatings (Figure 13.3). The coatings integrated a phase-installed morphology with the subjective transmission of electro-spun thermoplastic fiber (PCL) threads into a storage epoxy lattice. Mechanical damage to coatings may usually be caused by warming, resulting in two occurrences: first the matrix's restoration of form in space, and secondly, the dissolution and streams of PCL fibers to rebound from the break. In supervised trials, the damaged coatings basically and for all reasons repair it carefully. This correspondence shows the diagram and course of action for self-recovery

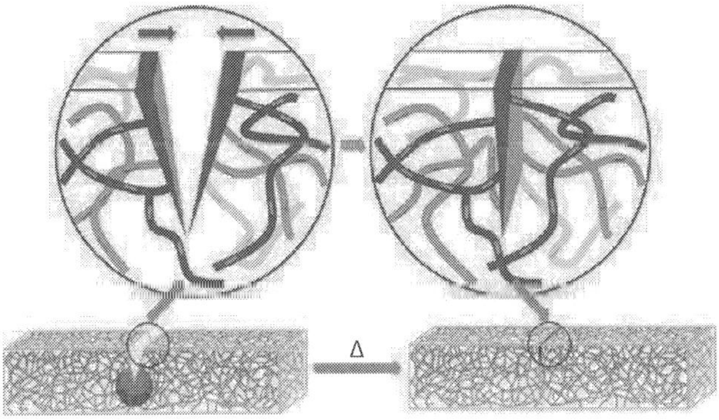

Figure 13.3 Schematic illustration of the coating morphology and the shape memory assisted self-healing (SMASH) concept.

systems with a self-regenerating component. The self-recovery was demonstrated by a fully reinstated block protection. In addition to this material development and composition analysis, there are different possible applications.

Mikhail L. Zheludkevich *et al.* [19] propose a different methodology of disintegration on request, which provides insightful self-recovery of the mutt sol-gel protected films. Controllable transportation in quarter and quarter of pretreatments is simplified, mixing the nano-levels of the benzotriazole inhibitor. The nano-compartments are restricted by silica nanoparticles securely LbL with nano-inhibitor polyelectrolyte ions. However, silica zirconia hybrid film containing 2024 aluminum compound provides an anti-corrosion effect. Silica nanoparticles protected layer-by-layer with polyelectrolyte strands and inhibitor materials which were placed in half-breed films. The impregnation of half-rate sol-gel films produces shield for the entire execution of metallic materials disintegration.

Dmitry O. Grigoriev *et al.* [20] controlled the discharges of polyelectrolyte compounds which can be used to manufacture vibrant coatings with fast cover. In specific, the capability to discharge anticorrosive actions of main coating is demonstrated by self-regenerating layers with complicated coating materials (Figure 13.4). The advantages of polyethylenimine (PEI) in antagonistic loaded SiO_2 nanoparticles were obtained by incorporating 20 ml of 15 wt.% SiO_2 solution. Nanocomposite multi-layer levels, colloidally templated lithers and nanosticks with solidified inhibitors are used to prevent new protected products from damage. As an assistant, inorganic aid or in-porous inorganic particle is interlinked with the disintegration inhibitor.

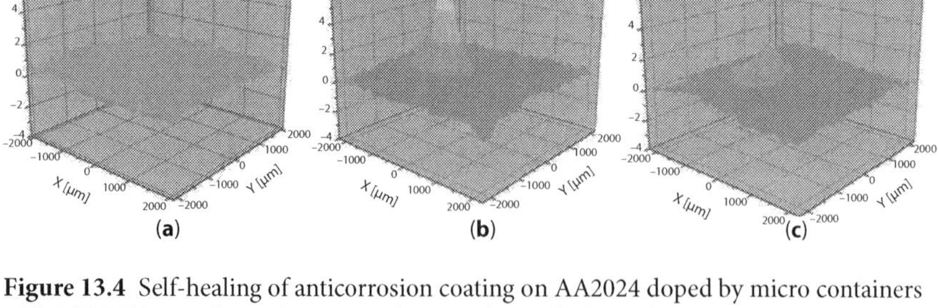

Figure 13.4 Self-healing of anticorrosion coating on AA2024 doped by micro containers with PSS/BTA complexes. Scanning vibrating electrode (SVET) measurements data collected: (a) immediately after immersion of AA2024 specimen into 0.1 M NaCl solution; (b) 42 h after immersion; (c) 60 h after immersion. Z-coordinates in all plots represent the local current density reflecting the development of the corrosion process.

13.4 Realistic Frameworks of Self-Healing Polymeric Coatings

Numerous self-healing method outlined within this analysis be confined to thin films, large polymers, coatings, and composites. So, thin layers can be applied to polymers that cure by UV-initiated bond. Healing incentive of UV irradiation has been entailed in these polymers. In addition, artificial UV resources cannot have penetrating lengths reaching the center of bulk polymers. Also, additional polymers with self-healing nature have also been recognized to comprise comparatively poor heating performances or fragile technical characteristics, which may be restricted to film and coating applications [21].

In contrast, a number of self-cure processes can able to employ only on polymeric composites. Most of this is due to the mass and quantity demands of certain therapeutic agents which have a higher thickness than typical thin film thicknesses [22, 23]. For comparable purposes, the healing product packed with hollow fibers is also confined to mass polymers. However, the self-healing polymers based on the microcapsules are not restricted as like the systems of microvascular and hollow fibres. The self-healing polymer microcapsules are most efficient at a scale of 100–200 μm, generally considered to be excessively huge on behalf of thin films [24]. However, current stories regarding nano-sized microcapsule-incorporated healing agents [25] possibly will lead to thin coatings in the area of microcapsule-related self-curing purposes. Even though it is likely adapted for mass implementations, the production in thin films of self-healing structures comprising phase-separated thermoplastic polymers may not be as impeded as several of the above-mentioned processes. Even if not yet published on in the literature, conventional methods of handling freeze could probably reduce EMAA and epoxy substances [26–28]. The PCL reactor stimulated phase-separation epoxy-polymers be particularly adaptable into applications of thin film, since the inherent PCL phase distribution divides on a micron level [29], eradicating the necessitate on behalf of further handing out measures on the road to diminish the amount of therapeutic agent.

13.5 Potential Historic Activity

Design and growth of intelligent self-healing nanocontainer polymer lacquers for multi-substrate functionality and safety is a difficult area of studies that offers many possibilities. One of these is certainly protective against corrosion of metals. Current industrial sector needs more efficient coatings to ensure that metal components are protected for longer periods of time. The goal is to decrease the cost of servicing and production activity. Other characteristics like superhydrophobicity and antimicrobial are necessary in relation to corrosive prevention. The capacity to automatically cure plastic damages, corrosion startups, improved mechanical characteristics, superhydrophobic, or anti-icing characteristics mixing long-term obstacle characteristics are beyond the focus, for instance, to safeguard costly asset in offshore wind power generation.

On another side, the tendency is to reduce the weight of parts in many applications by merging lightweight resources with increased effectiveness. The solution to this requirement is to use multi-metallic or non-metallic components combined with metallic materials. These mixes involve entirely a bright ways of protection against corrosion. The use in metallic or non-metallic products was commonly researched, but the latest research to advance the approach in multi-material mixes is not found to be feasible. Nowadays, standard and recent hi-tech applications often utilize multi-material assemblies, which require entirely clean layer architectures accommodating substrates with very distinct characteristics. Numerous fields like biomedicine, sport, car industry, and aircraft division; this is certainly an area of extreme importance. Thinner coatings are screened to reduce number of layers and its density. To maintain the necessary efficiency, coatings must contain distinct functions which enable individual surfaces to show a variety of features and to meet several needs. Thinner coatings with improved protection against corrosion and a strong esthetic quality are very much required in the automotive industry. Additional features like superhydrophobicity, anti-dust and anti-fingerprint treatment, and heal and soft touch also apply to mobile devices, home appliance, and furniture. Smart self-healing coatings for unusual appliances like biomedical, power stove, cabling, electronics, and digital panel panels are also in a part of recent new developments. There is no doubt that the area of smart multi-substrates protective coatings will develop quickly in the coming years, with encapsulation technology taking a main part in resolving many critical problems.

13.6 Conclusions

The various limitations of different materials in engineering disciplines include high maintenance and repair costs. It is very difficult to repair at remote locations. Therefore, self-healing coatings have enormous potential to improve the durability of structural materials. Hence, a large number of educational and industrial research organizations were included in this ambitious profile to explore new concepts. All government and industrial financing agencies are also showing a growing trend in investment. This paper explores a comprehensive literature assessment on research and development in self-healing polymeric coating materials. Specific solutions to self-healing substances like creeping cavity and resistance to fatigue cracks are explained in detail. In order to understand their uses, the self-healing mechanism is also studied. Damaged problems related to materials and corrosion protections are also examined for surface cracks. The unique properties of self-healing materials provide the system a new choice in different applications. These coatings can be used in many applications including spaceships, nuclear reactors, turbine injectors, and special cutting tools. Thus, automatic crack detection and eventual healing are the main challenge for design of self-healing polymeric coating materials. This wonderful area of self-healing polymer coatings will undoubtedly continue to grow beyond the technologies discussed in the above sections and will be accessible for daily use.

References

1. Montemor, M.F., Functional and smart coatings for corrosion protection: A review of recent advances. *Surf. Coat. Technol.*, 258, 17–37, 2014,
2. Weiner, S. and Wagner, H.D., The material bone: Structure mechanical function relations. *Ann. Rev. Mater. Sci.*, 28, 271–98, 1998.
3. Zhou, B.L., Some progress in the biomimetic study of composite materials. *Mater. Chem. Phys.*, 45, 2, 114–19, 1996.
4. Fratzl, P. and Weinkamer, R., Hierarchical Structure and Repair of Bone: Deformation, Remodelling, Healing in: *Self Healing Materials: An Alternative Approach to 20 Centuries of Materials Science*, S. vanderZwaag (Ed.), pp. 323–35, Springer, Netherlands, 2007.
5. Vermolen, F.J., van Rossum, W.G., Javierre, E., Adam, J.A., Modeling of Self Healing of Skin Tissue in: *Self Healing Materials: An Alternative Approach to 20 Centuries of Materials Science*, S. vanderZwaag (Ed.), pp. 337–63, Springer, Netherlands, 2007.
6. Kessler, M.R., Self-healing: A new paradigm in materials design, Proceedings of the Institution of Mechanical Engineers Part G-Journal of Aerospace Engineering. 221, 4, 479–95, 2007.
7. Wool, R.P., Self-healing materials: A review. *Soft Matter*, 4, 400–18, 2008.
8. http://www.nissanglobal.com/EN/TECHNOLOGY/INTRODUCTION/DETAILS/SGC/index.html.

9. Palaivel, V., Huang, Y., Van Ooij, W.J., Effects of addition of corrosion inhibitors to silane films on the performance of AA2024-T3 in a 0.5 M NaCl solution. *Prog. Org. Coat.*, 53, 2, 153–68, 2005.
10. Shchukin, D.G., Zheludkevich, M., Yasakau, K., Lamaka, S., Ferreira, M.G.S., Möhwald, H., Layer-by-layer assembled nanocontainers for self-healing corrosion protection. *Adv. Mater.*, 18, 13, 1672–8, 2006.
11. Ming, W., Laven, J., Van der Linde, R., Synthesis and surface properties of films based on solventless liquid fluorinated oligoester. *Macromolecules*, 33, 18, 6886–91, 2000.
12. Ming, W., Tian, M., van de Grampel, R.D., Melis, F., Jia, X., Loos, J., Van der Linde, R., Low surface energy polymeric films from solventless liquid oligoesters and partially fluorinated isocyanates. *Macromolecules*, 35, 18, 6920–9, 2002.
13. Ming, W., Melis, F., Van de Grampel, R.D., Van Ravenstein, L., Tian, M., Van der Linde, R., Low surface energy films based on partially fluorinated isocyanates: The effects of curing temperature. *Prog. Org. Coat.*, 48, 2–4, 316–21, 2003.
14. Van Ravenstein, L., Ming, W., Van de Grampel, R.D., Van der Linde, R., DeWith, G., Loontjens, T., Thuene, P.C., Niemantsverdriet, J.W., Low surface energy polymeric films from novel fluorinated blocked isocyanates. *Macromolecules*, 37, 2, 408–13, 2004.
15. Dou, Q., Wang, C., Cheng, C., Han, W., Thüne, P.C., Ming, W., PDMS-modified polyurethane films with low water contact angle hysteresis. *Macromol. Chem. Phys.*, 207, 2170–9, 2006.
16. Dikic, T., Ming, W., Van Benthem, R.A.T.M., De With, G., Low adherence coating. *PCT Int. Appl. WO 2007046687*, 207, 2170, 2006.
17. Dikic, T., Ming, W., Van Benthem, R.A.T.M., De With, G., Synthesis of well-defined perfluoroalkyl-end-capped polymers by ring-opening polymerization. *International conference on fluorine and silicone coatings*, Manchester, UK, 2005, Paper 3.
18. Luo, X., Mather, P.T., Shape memory assisted self-healing coating. *ACS Macro Lett.*, 2, 2, 152–6, 2013.
19. Zheludkevich, M.L., Anticorrosion Coatings with Self-Healing Effect Based on Nanocontainers Impregnated with Corrosion Inhibitor. *Chem. Mater.*, 19, 402–411, 2006.
20. Grigoriev D, O., Polyelectrolyte complexes as a "smart" depot for self-healing anticorrosion coatings. *Soft Matter*, 5, 1426–1432, 2009.
21. Mauldin, T.C. and Kessler, M.R., Self-healing polymers and composites. *Int. Mater. Rev.*, 55, 6, 317–346, 2010.
22. Toohey, K.S., Sottos, N.R., White, S.R., Characterization of microvascular-based self-healing coatings. *Exp. Mech.*, 49, 707–717, 2009.
23. Williams, H.R., Task, R.S., Bond, I.P., Self-healing composite sandwich structures. *Smart Mater. Struct.*, 16, 1198–1207, 2007.

24. Rule, J.D., Sottos, N.R., White, S.R., Effect of microcapsule size on the performance of self-healing polymers. *Polymer*, 48, 3520–3529, 2007.
25. Blaiszik, B.J., Sottos, N.R., White, S.R., Nanocapsules for self-healing materials. *Compos. Sci. Technol.*, 68, 978–986, 2008.
26. Meure, S., Wu, D.Y., Furman, S., Polyethylene-co-methacrylic acid healing agents for mendable epoxy resins. *Acta Mater.*, 57, 4312–4320, 2009.
27. Cho, S.H., Andersson, H.M., White, S.R., Sottos, N.R., Braun, P.V., Polydimethylsiloxane-based self-healing materials. *Adv. Mater.*, 18, 997–1000, 2006.
28. Cho, S.H., White, S.R., Braun, P.V., Self-healing polymer coatings. *Adv. Mater.*, 21, 645–649, 2009.
29. Luo, X., Ou, R., Eberly, D.E., Singhal, A., Viratyaporn, W., Mather, P.T., A thermoplastic/thermoset blend exhibiting thermal mending and reversible adhesion. *ACS Appl. Mater. Interfaces*, 1, 612–620, 2009.

14

Polymer Coatings for Biomedical Applications

Tahir Farooq[1], Arruje Hameed[2], Muhammad Sajid Hamid Akash[3] and Kanwal Rehman[4]*

[1]Department of Applied Chemistry, Government College University Faisalabad, Pakistan
[2]Department of Biochemistry, Government College University Faisalabad, Pakistan
[3]Department of Pharmaceutical Chemistry, Government College University Faisalabad, Pakistan
[4]Department of Pharamacy, University of Agriculture, Faisalabad, Pakistan

Abstract
Over the years, polymer coating has evolved as a fascinating tool for surface functionalization of materials usable in a wide range of fields. The polymer coated surfaces have been used for infectious therapy, drug delivery system for therapeutic agent, and protective layer for implants and biomedical devices. The chapter highlights some key developments in the field of polymer coating which were trend-setting for its applications especially in biomedical science.

Keywords: Polymer coating, biomedical applications

14.1 Introduction

Over the last few decades, the increasing consciousness about disease-treatment related risks have surged the development of new risk-free treatment strategies and medical devices. The traditional style medical devices pose serious threats of infection when come in contact with human body. As a result of increasing concerns, the synthesis of novel materials for biomedical applications has become an attractive area of research. In this context,

**Corresponding author*: kanwalakash@gmail.com

Inamuddin, Rajender Boddula, Mohd Imran Ahamed and Abdullah M. Asiri (eds.) Polymer Coatings: Technology and Applications, (333–346) © 2020 Scrivener Publishing LLC

a lot of efforts have also been made to modify the surfaces of medical devices to induce health-safety features like biocompatibility, non-adhesion of microbes, and prevention of non-specific binding of proteins which otherwise cause fouling. Over the last few years, the aforesaid health-safety features are induced in biomedical devices and implants through polymer coatings on in-contact surfaces [1].

The polymer coating provides tailored surface chemistry which enhances corrosion protection and improves mechanical strength, electrical conductivity, and water resistance characteristics to materials. Over the last decade, the tremendous developments in the synthesis of stimuli-responsive, smart-, shape memory, and conductive polymers of natural and synthetic origin (Figures 14.1 and 14.2) have revolutionized the fabrication of implants and medical devices. Owing to these excellent developments, the polymer coatings have widely been used to acquire characteristic features in biomedical applications [2, 3].

In current scenario, the adhesion, proliferation, and differentiation of cells have become controllable phenomena due to the surface-mediated tunable properties of polymer coatings. The polymer coating-mediated

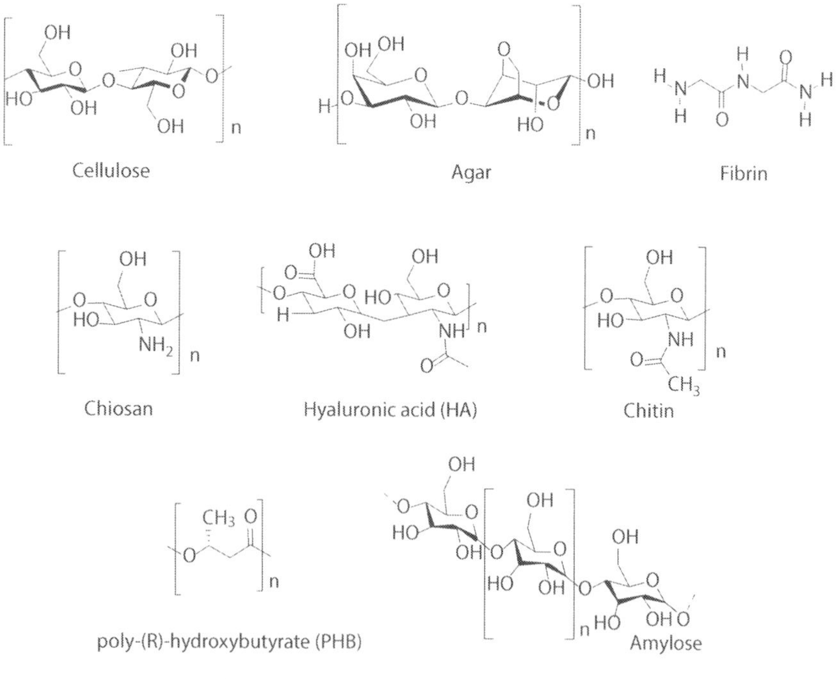

Figure 14.1 Natural polymers frequently used for coating.

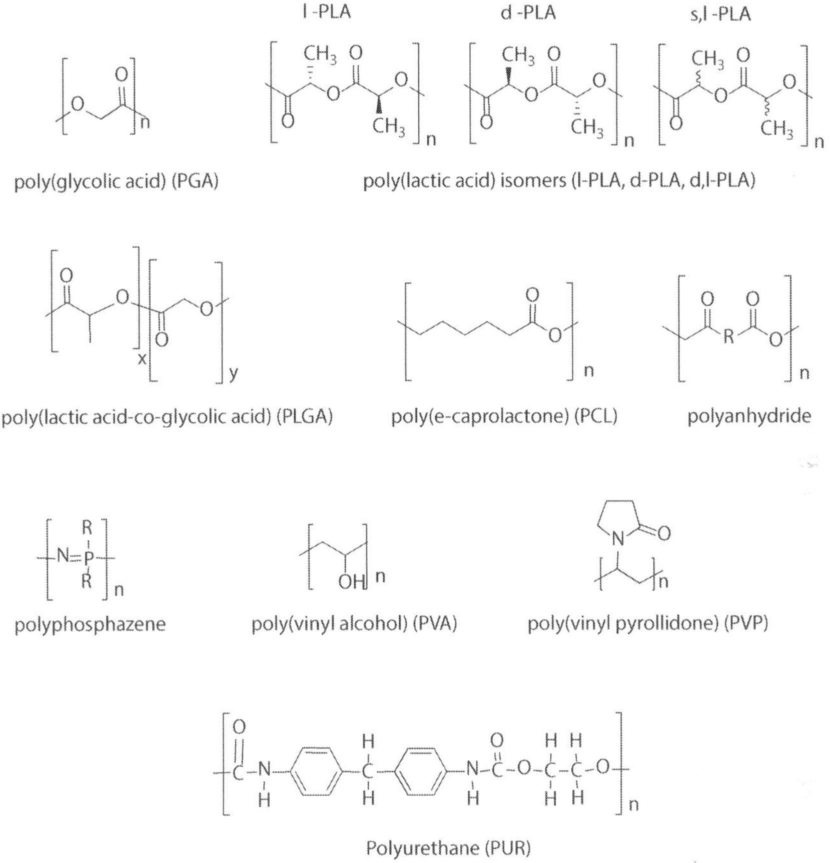

Figure 14.2 Natural polymers frequently used for coating.

guided cell/surface interactions have also opened up new horizons for guided tissue engineering. Using polymer coatings, the surface-mediated controlling of non-specific interactions are also used to produce anti-fouling properties on medical devices which often come in contacts with biological fluids. The approach has also gained considerable attention in context of sustained and controlled release of drugs. The polymeric coatings have also demonstrated their worth as potent drug delivery systems especially for those drugs which show promising therapeutic potential but low water solubility and bioavailability [4]. The recent developments in polymer coating have made it a new strategic tool with wide applications across many scientific domains. It has become a promising tool with ever increasing applications in orthopedics, bio-sensing, diagnostics, and targeted delivery of therapeutic agents.

The subsequent sections highlight some key developments and applications of polymer coating in biomedical field under the headings of tissue engineering, drug delivery, and antimicrobial surfaces.

14.2 Applications in Tissue Engineering

The elasticity and biocompatibility are the characteristic features of polymers which make them suitable candidates for applications in tissue engineering as coating materials. Over the few decades, the naturally occurring polymers including chitin, cellulose, alginate, gelatin, and cellulose has widely been employed as coating materials in tissue engineering due to above-said much needed features. Similarly, a number of synthetic polymers have proved their potential as superlative coating materials in tissue engineering.

The field of regenerative medicines has really been under high expectations to provide treatment strategies for diseased and damaged tissues through an easy generation of functional tissues and organs. There has always been a challenge to setup an ideal integration between material science approaches and principles of cell biology for three-dimensional growth of cells and tissues. Over the last few years, the *in vitro* generation of tissues has become a reality due to the recent advancements in both disciplines [5]. The engineered tissues show fully differentiated cells expressing protein profiles identical to the original ones. However, modulation of cell growth, tissue generation and morphology pose real-time challenges and require specialized hands to employ complex design and approaches. Very recently, chitosan-based coating has been used to modulate cell and tissue growth in *ex vivo* culture substrate. This facile, simple, and reproducible technique helped to control cell behavior, growth, and functions just by concentration of chitosan coating [6].

During the growth and tissue development stages, the cellular microenvironment experiences drastic changes in mechanical properties. The cells do sense the physical changes and mechanical stresses and show a responsive modulation in growth, organization, and tissue morphogenesis [7]. However, such mechanosensitive reactions of cells and their consequences on tissue morphology remain largely unknown. The understanding of such mechanosensitive responses of growing cells may help to control the regeneration of damaged tissues. Such mechanistic insights may also be adapted to regulate the size and morphology of artificially growing tissues. In one such study, the alginate-based polymeric hydrogel coating was used to create a biomimetic microenvironment with aim to evaluate the effects of physical cues on tissue morphology during *in vitro* tissue synthesis. Miyajima *et al.* found hydrogel-based microenvironment as guide, modulating the size and morphology of developmental submandibular gland (SMG). The softer gel enhanced the cleft formation and bud expansion of SMG while stiffer gel formulation attenuated them. This study presented an excellent example of polymeric coating applied as biomimetic cellular microenvironment for controlled tissue morphogenesis [8].

Sato *et al.* employed poly (L-lactic acid-co-e-caprolactone) coating as protective layer on luminal surface during artificial reconstruction of tracheobronchial in canine model. In fact, polymer coating protected the collagen layer and promoted the luminal epithelization. In the absence of such biodegradable protective coatings, the collagen layer undergoes early disintegration causing epithelization process incomplete [9].

In guided tissues regeneration technique, the barrier membranes are used as physical barriers for the deflection of growing tissues in defected regions. These physical barriers reduce the growth of competing connective tissue and only allow the fast regeneration of periodonatal ligament fibroblasts. Such membranes are expected to control the migration of epithelial cells, stabilize wound site and prevent infections. Further, they are require to show space maintaining ability, cell viability, biocompatibility, and bioactivity during guided tissue regeneration period [10, 11]. In this connection, collagen-based resorbable membranes and expanded polytetrafluoroethylene (ePTFE)-based membranes are commercially available. However, the applications of these available membranes have met limited success as collagen-based membranes cause localize inflammation and show fast degradation while ePTFE-based membranes need surgical operation for their removal [12].

Chen *et al.* used chitosan to prepare GTR-membrane with one side coated with alginate to induce biocompatibility with minimum toxicity. They performed cell adhesion test, degradation behavior, and mechanical strength of the uncoated and coated sides of the membrane to evaluate the feasible application of such membrane in GTR-technique. In cell adhesion test, 3T3 fibroblasts exhibited about 50% low adhesion on alginate-coated surface compared to non-coated side of the chitosan membrane. The alginate coated side displayed different surface morphology and hydrophilicity as it showed decreased contact angle and higher water contents. The alginate coated chitosan membrane was suggested as an ideal candidate for GTR for controlling the growth of competing hard tissue (alveolar bone) while promoting the growth of soft tissue (gingival tissue) [13].

In conditions of blood vessel fistulas and aneurysms, the vascular grafting with high safety profile has become the leading treatment option over the last few decades. Mainly healthy blood veins are removed and implanted to replace diseased arteries in patients as an autografting strategy. However, in cases where patient had already undergone autografting or pre-existing diseased vessels, the polymer-based materials are employed for the replacement of diseased vessels [14].

Such vascular grafts of medium to large diameter are generally prepared from polyethylene terephthalate (PET). The biological performance and long-term potency of vascular grafts depends on the porosity of the polymeric material. The high porosity of grafts causes blood leakage during implantation and considered as a big disadvantage of vascular grafting. Before implantation, the permeability of grafts is reduced by using certain sealant or preclotting with patient's own blood. Over the past few years, polymeric materials have become attractive choices to control the porosity of vessel grafts due to their biocompatible and biodegradable nature [15].

Manju *et al.* used oxidized alginate and gelatin to develop a polymeric hydrogel which was coated on the PET graft to control the porosity. The efficacy coated grafts was evaluated by investigating adhesion and proliferation of endothelial cells (EC), hemocompatibility, *in vitro* cell culture toxicity, burst strength, and water permeability. The results suggested that coating of PET reduced its hemolytic potential, posed no change in burst strength, and enhanced the adhesions and proliferations of EC. The biocompatible and biodegradable coating sealed the pores efficiently but did not affect the mechanical properties of graft [16].

In recent years, the bone tissue engineering has demanded the development of polymer-ceramic composites due to the highly complex composite nature of bone tissue. An ideal scaffold is expected to have well-harmonized structure of inorganic and organic components with ability to deliver cells, excellent osteoconductivity, and biodegradability. The ability to promote bone formation has made bioactive glass ceramics and silicate glasses as attractive entities in TE.

Considering the above facts, Erol *et al.* used boron-containing bioactive glass to develop novel scaffolds which were further coated with alginate cross-linked with Cu ions. This novel scaffold with controlled Cu-releasing ability was designed to enhance angiogenesis. The alginate coated scaffold higher compression strength, improved bioactive behavior, and elevated angiogenic potential compared to non-coated scaffold in *in vitro* bone regeneration studies.

14.3 Polymer Coating for Drug Delivery

In recent times, the excellent surface mechanical properties and controllable porosity have highlighted the role of polymeric scaffold in TE. As discussed earlier, such scaffolds are prepared from highly biocompatible biomaterials so they do not provoke any reaction against foreign invaders and there have always been excessive chances of infections. Clinically, injections are practiced regularly to reduce infection before it worsens the situation. Most of the times, patients face potential side effects due to higher drug concentration in blood. The biocompatible scaffold with extra advantageous of a local drug delivery system is required to control such unavoidable on-site infections without disturbing the bone regeneration process. Such local drug delivery systems are expected to show predictable therapeutic response, greater efficacy, and sustained drug release with ideal safety profiles [17, 18].

Huang *et al.* developed a novel scaffold carrying drug-loaded chitosan coating to confront infections during bone tissue regeneration. The nano-hydroxyapatite/polyamide66 (n-HA/PA66) scaffold was coated with chitosan loaded with berberine (antibacterial drug). Different ratios of CS and berberine were used to study drug releasing behavior, mechanical properties, and morphology. They suggested that the compression strength and elastic modulus of coated scaffold was much higher than the uncoated scaffold. Depending on the coating concentration, the burst drug release appeared in first 3 hrs and a continuous release was detected for 150 hrs in phosphate-buffered saline solution. The controlling of drug release by the coating concentration is highly suited requirement for infection therapy in bone tissue engineering. The MG63 cells exhibited enhanced adhesion, growth, and proliferation on coated scaffold which suggested the non-toxic nature of drug-loaded coating. The results suggested the potential applications of coated scaffolds in bone defect repair and TE [19].

Over the last few decades, curcumin, a natural lipophilic polyphenol, has established its worth in chemotherapy against a variety of cancers including breast, bone, gastrointestinal, lung, and prostate cancers. It is also known as multitherapeutic agent due to a range of bioactivities including antioxidant, anti-HIV, anti-amyloid, anti-inflammatory, and anti-malarial, etc. However, the clinical applications of this phenomenal drug has always been restricted due to poor water solubility, fast metabolic degradation, thus low bioavailability [20]. In last few years, the recent advances in nanotechnology have accelerated the development of nano-carrier systems for poorly water soluble drugs to enhance their stability and bioavailability [21, 22].

Quite recently, solid lipid nanoparticles (SLNs) have become attractive nanodrug delivery systems because of their ability to encapsulate lipophilic drugs in larger quantities. Ramalingam *et al.* enhanced the bioavailability and stability of curcumin by developing SNLs-coated with chitosan. The chitosan coated mucoadhesive SNLs enhanced the oral bioavailability and physical stability of curcumin [23].

In another study, the chitosan coating found to produce a thick layer on SLNs under gastric conditions thus enhanced their stability. In fact, mechanistic electrostatic repulsions and steric stabilization of chitosan coating ensure the stability of SLNs. It was concluded that chitosan coating improved mucoadhesion, cellular uptake, stability, and controlled delivery of hydrophobic drug [24].

The low molecular weight heparin (LMWH) is well-known for its anti-angiogenic activity and employed to inhibit progression of cancer and as therapeutic agent for rheumatoid arthritis. It also shows anticoagulation activity in the treatment of pulmonary embolism, deep vein thrombosis, and venous thromboembolism. Advantageously, it does not induce any toxic effects like bleeding and thrombocytopenia like heparin. However, poor penetration through intestinal wall, high negative charge, and high water solubility has restricted its applications through parental route only. Over the last few years, chemical conjugates and penetration enhancer have been employed to develop effective oral therapy for improved patient compliance [25].

Bagre *et al.* used LMW enoxaparin to develop oral delivery system as alginate-coated chitosan core shell nanoparticles (Alg-CS-NPs). Alginate is known for its non-risky ability for controlled and sustained oral drug delivery. The *in vitro* particulate uptake and permeation across intestinal epithelium, drug release profile, and pharmacokinetic studies were performed to evaluate the efficacy of Alg-CS-NPs. This new oral delivery system enhanced the oral bioavailability and gastrointestinal uptake of LMW enoxaparin thus reduced the above-said limitations in oral therapy [26].

The conventional cancer therapeutic options like chemotherapy always produce sever side-effects with low efficacy. In current era of nanotechnology developments, a number of drug delivery systems have been introduced for controlled drug delivery. The existing delivery systems for cancer therapeutics utilize liposomes, silica-based porous materials, dendimer-based micelles, and polymeric systems. These drug delivery systems still show limitations as they require complex preparatory procedures and highly skilled human resource so they are expensive for a large population of world [27, 28].

Over the last few years, the attentions have been diverted to explore novel vectors for sustained delivery of anticancer agents with minimum limitations. In this connection, Zhu *et al.* used bacterial vector as delivery agents for sustained release of polypeptide exotoxin A which blocks protein synthesis in eukaryotic cells thus cause cell death. The delivery system employed bacterium as nano-carrier, toxin PE66as anticancer agents, and combined coating of CP and polymyxin B (PLB) for ROS triggered sustained and controlled release. This system proved as a brand new dimension in encapsulated anticancer therapy for delivery and release of drugs [29].

The DNA-based vaccine therapy has emerged as an attractive field due to the capability to promote humoral and cellular immune responses, and lack of anti-vector immunity. However, the initial trials in human and non-human primates exhibited poor plasmid DNA immunization. A number of disadvantages are exhibited by both traditional needle-based administration of vaccines and advanced DNA vaccination technologies. The vaccination distribution has really become a problematic issue especially in developing world.

An approach can substantially enhance DNA vaccine delivery which can simultaneously induce potent immune response, promote non-toxic and sustained transfection, and target DNA in immune-cell-rich tissues. DeMuth *et al.* developed a polymer film loaded with vaccine carrying biodegradable polycations, immune stimulatory RNA and DNA. The vaccine-loaded film was implanted into the immune-cell-rich epidermis using releasable polyelectrolyte coated microneedles. In non-human primates, this DNA vaccination enhanced gene expression 140 folds compared to intradermal injections also increased memory T-cell generation. This new DNA vaccine also induced immune response against a model HIV-antigen. So, the new DNA vaccination strategy exhibited its potential against traditional as well as newly available techniques.

14.4 Polymer Coating as Antimicrobial Surfaces

In recent times, the poly(dimethylsiloxane) (PDMS) has emerged as an ideal choice to prepare contact lenses, catheters, pacemaker encapsulants, and surgical implants. Recently, its characteristic features like elastomeric properties, low toxicity, and biocompatibility have driven its applications in biosensors, DNA sequencing, diagnostics, and drug delivery. However, microbial adhesions and protein/hydrophobic analyte adsorptions cause biofouling problems due to low wettability of PDMS. Such safety issues restrict the PDMS-based devices especially which come in contact with microfluids like blood. In order to encounter such performance-limiting conditions, a number of physical approaches and chemical treatments have been adopted to induce hydrophilicity and anti-fouling characteristics on PDMS surface. Its surface state is changed via electrostatic or hydrophobic interactions by adsorption of coating material in physical methods. In some other physical treatments, its surface is activated by UV light, ozone or oxygen plasma treatment. However, all these physical treatments serve temporarily and the hydrophobic surface is recovered with passage of time. While chemical modifications changed the surface permanently and induce long-lasting anti-fouling characteristics [30].

As a part of chemical treatments, a number of polymer coatings could be applied to enhance wettability of PDMS surface. In this regard, the coating with hydroxyl-enriched polymers has been preferred for high hydrophilicity due to the abundant of hydrogen bond formation [31]. The anti-fouling performance is strongly correlated to the hydration layer formed on the surface. The PDMS microfluid chips were coated with polyvinyl alcohol (PVA). The PVA coating induced anti-fouling properties by preventing the adsorption of proteins both in acidic and basic media. The results suggested the applications of PVA coated PDMS-based implants while dealing with microfluids like blood, etc [32].

The microbial wound contamination is a main side problem with orthopedic operations under trauma and disease conditions. Such contaminations lead to infections which delay the recovery process putting impact on patient's health. In case of bone fractures, the internal fixation with infections can be dangerous while it helps to regenerate bone with stable fractures. Under such conditions, the implanted biomaterials could serve to provide infection therapy [33, 34].

Generally, bone fracture is fixed using internal fixation devices made up of stainless steel. Greene *et al.* coated the bone screw fixation with chitosan loaded with gentamicin, a known antibacterial agent. The chitosan coated bone screws were found biocompatible with stem cells and fibroblasts. The sustain release of drug was detected for 72 to 96 hrs and exhibited its antibacterial activity against Staphylococcus aureus. The results suggested the suitability of chitosan as potent bonding material for stainless steel orthopedic devices with extra advantage to carry therapeutic agents [35].

The biocompatibility, corrosion resistance, strength to weight ratio, and mechanical properties have increased the use of titanium and its alloys for the fabrication of biological implants. Especially, it is used to manufacture dental and orthopedic implants for longer service life. However, the so-called stable oxide layer on its surface gets rough with passage of time thus limits the life of implants in biological fluids. Such conditions ultimately cause infections and result in diseased conditions [36].

Swanson *et al.* developed chitosan loaded with vancomycin coating for titanium implants. They studied the drug releasing behavior of polymer coating with respect to the standard drug. The results suggested the tailoring of coating surface with anti-infectious features. They introduced concept of titanium-based safer biological implants with efficiency to prevent infections through local drug delivery by protective polymeric coating [37].

14.5 Conclusion

Functional polymer coatings have found wide spread application across many specific disciplines. It has emerged as a powerful tool capable of functionalizing the material surfaces with desired characteristics. The recent advances in polymer science have revolutionized the role of this fascinating strategy in biological science especially in biomedical applications. The chapter highlighted some key applications which were trend-setting for surface functionalization of implants and biomedical material. The recent development in polymer functionalization and coating technique reflects a promising feature of this fascinating tool.

References

1. Smith, J.R. and Lamprou, D.A., Polymer coatings for biomedical applications: A review. *Trans. IMF*, 92, 1, 9–19, 2014.
2. Vasilev, K. and Macgregor, M., Nanoengineered plasma polymer films for biomedical applications. *Adv. Mater. Lett.*, 9, 42–52, 2018.
3. Zhang, J., Jiang, X., Wen, X., Xu, Q., Zeng, H., Zhao, Y., Liu, M., Wang, Z., Hu, X., Wang, Y., Bio-responsive smart polymers and biomedical applications. *J. Phys.: Mater.*, 2, 3, 032004, 2019.
4. Lynge, M.E., van der Westen, R., Postma, A., Städler, B., Polydopamine—A nature-inspired polymer coating for biomedical science. *Nanoscale*, 3, 12, 4916–4928, 2011.
5. Ogawa, M., Oshima, M., Imamura, A., Sekine, Y., Ishida, K., Yamashita, K., Nakajima, K., Hirayama, M., Tachikawa, T., Tsuji, T., Functional salivary gland regeneration by transplantation of a bioengineered organ germ. *Nat. Commun.*, 4, 2498, 2013.
6. Kazi, G.A.S., Yamanaka, T., Osamu, Y., Chitosan Coating an Efficient Approach to Improve the Substrate Surface for *In Vitro* Culture System. *J. Electrochem. Soc.*, 166, 9, B3025–B3030, 2019.

7. Matsumoto, T., Yung, Y.C., Fischbach, C., Kong, H.J., Nakaoka, R., Mooney, D.J., Mechanical strain regulates endothelial cell patterning *in vitro*. *Tissue Eng.*, 13, 1, 207–217, 2007.
8. Miyajima, H., Matsumoto, T., Sakai, T., Yamaguchi, S., An, S.H., Abe, M., Wakisaka, S., Lee, K.Y., Egusa, H., Imazato, S., Hydrogel-based biomimetic environment for *in vitro* modulation of branching morphogenesis. *Biomaterials*, 32, 28, 6754–6763, 2011.
9. Sato, T., Araki, M., Nakajima, N., Omori, K., Nakamura, T., Biodegradable polymer coating promotes the epithelization of tissue-engineered airway prostheses. *J. Thorac. Cardiovasc. Surg.*, 139, 1, 26–31, 2010.
10. Jafari Sanjari, A. and Asghari, M., A review on chitosan utilization in membrane synthesis. *ChemBioEng Rev.*, 3, 3, 134–158, 2016.
11. Elgali, I., Omar, O., Dahlin, C., Thomsen, P., Guided bone regeneration: Materials and biological mechanisms revisited. *Eur. J. Oral Sci.*, 125, 5, 315–337, 2017.
12. Baek, Y.-j., Kim, J.-H., Song, J.-M., Yoon, S.-Y., Kim, H.-S., Shin, S.-H., Chitin-fibroin-hydroxyapatite membrane for guided bone regeneration: Micro-computed tomography evaluation in a rat model. *Maxillofacial Plast. Reconstr. Surg.*, 38, 1, 14, 2016.
13. Chen, T.W., Chang, S.J., Niu, G.C.C., Hsu, Y.T., Kuo, S.M., Alginate-coated chitosan membrane for guided tissue regeneration. *J. Appl. Polym. Sci.*, 102, 5, 4528–4534, 2006.
14. Li, S., Sengupta, D., Chien, S., Vascular tissue engineering: From *in vitro* to *in situ*. *Wiley Interdiscip. Rev.: Syst. Biol. Med.*, 6, 1, 61–76, 2014.
15. Qi, P., Maitz, M.F., Huang, N., Surface modification of cardiovascular materials and implants. *Surf. Coat. Technol.*, 233, 80–90, 2013.
16. Manju, S., Muraleedharan, C.V., Rajeev, A., Jayakrishnan, A., Joseph, R., Evaluation of alginate dialdehyde cross-linked gelatin hydrogel as a biodegradable sealant for polyester vascular graft. *J. Biomed. Mater. Res. Part B: Appl. Biomater.*, 98, 1, 139–149, 2011.
17. Goenka, S., Sant, V., Sant, S., Graphene-based nanomaterials for drug delivery and tissue engineering. *J. Controlled Release*, 173, 75–88, 2014.
18. Roseti, L., Parisi, V., Petretta, M., Cavallo, C., Desando, G., Bartolotti, I., Grigolo, B., Scaffolds for bone tissue engineering: State of the art and new perspectives. *Mater. Sci. Eng.: C*, 78, 1246–1262, 2017.
19. Huang, D., Zuo, Y., Zou, Q., Zhang, L., Li, J., Cheng, L., Shen, J., Li, Y., Antibacterial chitosan coating on nano-hydroxyapatite/polyamide66 porous bone scaffold for drug delivery. *J. Biomater. Sci. Polymer Edition*, 22, 7, 931–944, 2011.
20. Akbar, M.U., Rehman, K., Zia, K.M., Qadir, M.I., Akash, M.S.H., Ibrahim, M., Critical review on curcumin as a therapeutic agent: From traditional herbal medicine to an ideal therapeutic agent. *Crit. Rev.™ Eukaryotic Gene Expression*, 28, 1, 17–24, 2018.

21. Yallapu, M.M., Nagesh, P.K.B., Jaggi, M., Chauhan, S.C., Therapeutic applications of curcumin nanoformulations. *AAPS J.*, 17, 6, 1341–1356, 2015.
22. Rafiee, Z., Nejatian, M., Daeihamed, M., Jafari, S.M., Application of different nanocarriers for encapsulation of curcumin. *Crit. Rev. Food Sci. Nutr.*, 59, 21, 3468–3497, 2019.
23. Ramalingam, P., Yoo, S.W., Ko, Y.T., Nanodelivery systems based on mucoadhesive polymer coated solid lipid nanoparticles to improve the oral intake of food curcumin. *Food Res. Int.*, 84, 113–119, 2016.
24. Luo, Y., Teng, Z., Li, Y., Wang, Q., Solid lipid nanoparticles for oral drug delivery: Chitosan coating improves stability, controlled delivery, mucoadhesion and cellular uptake. *Carbohydr. Polym.*, 122, 221–229, 2015.
25. Jain, A., Mehra, N.K., Nahar, M., Jain, N., Topical delivery of enoxaparin using nanostructured lipid carrier. *J. Microencapsulation*, 30, 7, 709–715, 2013.
26. Bagre, A.P., Jain, K., Jain, N.K., Alginate coated chitosan core shell nanoparticles for oral delivery of enoxaparin: *In vitro* and *in vivo* assessment. *Int. J. Pharm.*, 456, 1, 31–40, 2013.
27. Dawidczyk, C.M., Russell, L.M., Searson, P.C., Nanomedicines for cancer therapy: State-of-the-art and limitations to pre-clinical studies that hinder future developments. *Front. Chem.*, 2, 69, 2014.
28. Bjornmalm, M., Thurecht, K.J., Michael, M., Scott, A.M., Caruso, F., Bridging bio–nano science and cancer nanomedicine. *ACS Nano*, 11, 10, 9594–9613, 2017.
29. Zhu, C., Yang, Q., Lv, F., Liu, L., Wang, S., Conjugated Polymer-Coated Bacteria for Multimodal Intracellular and Extracellular Anticancer Activity. *Adv. Mater.*, 25, 8, 1203–1208, 2013.
30. Zhang, H. and Chiao, M., Anti-fouling coatings of poly (dimethylsiloxane) devices for biological and biomedical applications. *J. Med. Biol. Eng.*, 35, 2, 143–155, 2015.
31. Trantidou, T., Elani, Y., Parsons, E., Ces, O., Hydrophilic surface modification of PDMS for droplet microfluidics using a simple, quick, and robust method via PVA deposition. *Microsyst. Nanoeng.*, 3, 16091, 2017.
32. Wu, D., Luo, Y., Zhou, X., Dai, Z., Lin, B., Multilayer poly (vinyl alcohol)-adsorbed coating on poly (dimethylsiloxane) microfluidic chips for biopolymer separation. *Electrophoresis*, 26, 1, 211–218, 2005.
33. Schmidt, A.H. and Swiontkowski, M.F., Pathophysiology of infections after internal fixation of fractures. *JAAOS-J. Am. Acad. Orthopaedic Surgeons*, 8, 5, 285–291, 2000.
34. Metsemakers, W., Kuehl, R., Moriarty, T., Richards, R., Verhofstad, M., Borens, O., Kates, S., Morgenstern, M., Infection after fracture fixation: Current surgical and microbiological concepts. *Injury*, 49, 3, 511–522, 2018.
35. Greene, A.H., Bumgardner, J.D., Yang, Y., Moseley, J., Haggard, W.O., Chitosan-coated stainless steel screws for fixation in contaminated fractures. *Clin. Orthopaedics Relat. Res.*, 466, 7, 1699–1704, 2008.

36. Khorasani, A.M., Goldberg, M., Doeven, E.H., Littlefair, G., Titanium in biomedical applications—Properties and fabrication: A review. *J. Biomater. Tissue Eng.*, 5, 8, 593–619, 2015.
37. Swanson, T., Cheng, X., Friedrich, C., Development of chitosan–vancomycin antimicrobial coatings on titanium implants. *J. Biomed. Mater. Res. Part A*, 97, 2, 167–176, 2011.

15

Antimicrobial Polymer Coating

Kanwal Irshad[1], Kanwal Rehman[2], Hina Sharif[1] and Muhammad Sajid Hamid Akash[1]*

[1]Department of Pharmaceutical Chemistry, Government College University Faisalabad, Faisalabad, Pakistan
[2]Department of Pharmacy, University of Agriculture, Faisalabad, Pakistan

Abstract
Antimicrobial polymers are the type of polymers having the ability to inhibit or kill the growth of microorganisms such as bacteria, protozoans, and fungi. These polymers mimic the action of antimicrobial peptides which are utilized by the immune systems of such living organisms to kill the bacteria. The main purpose is to develop a self-decontaminating coating that has bactericidal effect to kill the microorganisms when they are come to contact with the antimicrobial surface. Antimicrobial polymer coating is widely used in controlled drug delivery system and implants in dental restorative materials. Antibacterial surfaces may exhibit antibiofouling effector inactivate the cells when they come into contact with surface for repulsion or the resistance of initial attachment of bacteria. Because they have biocidal activity, metals are used in antimicrobial polymer as antimicrobial agents. Metals, copper, and silver, are highly toxic to bacteria even at low concentration. These antimicrobial agents release over time and prolong the lifetime of antimicrobial properties of system. The use of antimicrobial polymer is to treat the microbial infections which become more difficult due to resistant of microbial strains, to enhance the activity of some pre-existing antimicrobial agents and to prolong the effect of antimicrobial agents. The major purpose of antimicrobial coating is to combat antimicrobial resistance and infections.

Keywords: Mechanism of action, hydrophilicity medical device applications, charge density, medical applications

**Corresponding author*: sajidakash@gcuf.edu.pk

Inamuddin, Rajender Boddula, Mohd Imran Ahamed and Abdullah M. Asiri (eds.) Polymer Coatings: Technology and Applications, (347–358) © 2020 Scrivener Publishing LLC

15.1 Introduction

Microbes such as fungi, bacteria, and parasites are the source of infections. Pathogenic microbes are responsible for infectious diseases. Antimicrobial agents are utilized to kill the micro-organisms or inhibit microbial growth. Many antimicrobial drugs have been introduced to treat with infectious diseases [1]. Infection is critically associated with implanted medical devices. These infections include premature device failure, tissue destruction, and spread of infections to other areas. The adherence of microbes with biomaterial is a major key factor for the initiation of an infectious disease [2]. There are many ways through which infectious diseases can spread; approximately 50% of all types of microbial infections are due to the contaminated surfaces [3]. The biofilm formation on the surface is major issue related to infection. There is a need to develop the coating that prevent the formation of biofilm or eliminated the already constructed biofilm. Antimicrobial coating can kill the microbes before and after contact with surface and prevent adhesion of bacteria. Recent research studies also develop new antimicrobial polymer coating having the dual action biocidal action. Biocidal material and surfaces have better biocidal activity and effect for longer period. Antimicrobial polymer coating perform dual action as biocidal materials and surfaces by a simultaneously contact killing and biocidal released action. Quaternary Ammonium group covalently incorporated into polymeric substrate or polymeric chain is responsible for contact killing actions. Many polymeric materials have the ability to leach out biocidal agents. The example of such polymeric materials includes inorganic metal nanoparticles or some antibiotics that are responsible for released based biocidal action [4].

Metals are widely used as antimicrobial agents due to its biocidal activity in agriculture, industry, and in health care system. Non-essential metal such as silver is toxic to bacteria having biocidal activity at very low concentrations. Essential metal such as copper is lethal to bacteria. Nowadays, metal oxides and their salts based on copper and silver are used as antimicrobial agents. Silver is highly expensive metal so their salts can be used. Copper is a cheaper than silver but it prevents the corrosion. To overcome this problem, metal oxide can be used as antimicrobialagents [5].

Antimicrobial polymers are such substances that have ability to kill the microbes and inhibit the growth of microorganisms on any surface or any surrounding environment. Some antimicrobial agents have an inherited ability to exhibit the antimicrobial activity. The examples of such antimicrobial agents are chitosan, many compounds having the quaternary nitrogen groups, halamines, and poly-ε-lysine (ε-PL). Some antimicrobial polymers act as a backbone for incorporation of small biocides and these antibiotics also display their antimicrobial activity [4]. An ideal antimicrobial polymers should have the following characteristics a) broad spectrum for pathogenic microorganisms, b) exhibit the long term activity, c) should be stable and have no toxicity, d) can be easily synthesized and inexpensive, and e) can be regenerated when it lost its activity [6].

The antimicrobial polymers coating is an effective method for wide range of applications such as covalent and non-covalent bonding. Recent research has developed antimicrobial peptides-based antimicrobial hydrogels, antimicrobial polymer, and surfaces. In our daily life, we have to face many surfaces which are may be responsible for infections. Because such type of surfaces may act as highly potent reservoirs of many deadly pathogens and these pathogens form the biofilms which are core reasons for most of infections. For the limitation of surface associated infections, some strategy should be adopted such as preventing the adherence of microbes with biomedical devices and inhibit the colonization of microbes on the biomedical devices and other associated surfaces. This is possible if coating of surfaces is done by antimicrobial polymers. To inhibit the presence and colonization of the microbes on many surfaces, polymeric coating is a very effective strategy. Depending upon the applications, various types of polymeric substances can be used. The major purpose is to combat antimicrobial resistance and infections [3].

Polymers are used in the recent years as a substituent of many metal, wood, and glass in many applications because polymers have many physicochemical properties as well as economical. Table 15.1 shows the polymer medical device application [7].

15.2 Mechanism of Action

The technique for designing of antimicrobial polymers is to determine the structural characters of the outer most envelope of the bacterial cell. The outer envelope of the cells has negative charge. Gram-positive bacteria cell wall (CW) has teichoic acid, the outer membrane (OM) of Gram-negative bacteria has phospholipids and lipopolysaccharides, and the cytoplasmic membrane has phospholipids bilayer in which essential proteins are embedded. Based upon the structural characteristics of cell wall and cell membrane, antimicrobial polymer agents are designed. Antimicrobial polymer agents must have cationic hydrophilic and hydrophobic macromolecular systems which will be major target site for which the cytoplasmic

Table 15.1 Polymer medical device applications.

Polymer type	Medical device applications
Polyethylene	Orthopedic implants, containers, catheters, non-woven textiles
Polypropylene	Disposable items (e.g., syringes), non-woven textiles, membranes, sutures
Polyurethanes	Films, tubing, catheters
Polyvinylchloride	Catheters, tubing
Polyamides	Sutures, packaging, dental implants
Polyethylene terephthalate	Sutures, artificial vascular grafts

membrane will behave as active agents. These polymers which have a polar functional group will block membrane that bear cationic charge and a hydrocarbon functional group that are non-polar hydrophobic in nature block (or hydrophobic structure of a whole link), or random copolymers formed by a hydrophobic monomer and a hydrophilic co-monomer interactions with a functional group. Such polymers and copolymers structures has high affinity to bind with bacterial cell due to lipophilicity to cause the damage of structure and integrity of bacterial cell membrane which may cause the leakage of cytoplasmic content, cell lysis, and cell membrane disruption [8, 9].

15.2.1 Passive Action

A passive polymer layer prevents the attachment of bacteria by decreasing the adsorption of protein on surfaces. They repel the bacteria and they do not kill or interact with bacteria. Microbial membrane has hydrophobic character and negatively charged molecules then passive polymers may be hydrophilic, negatively charged, and low surface free energy [10, 11]. The mechanism of passive polymer is shown in Figure 15.1. Passive polymers are made up of the following. a) Surfaces that are self-healing and slippery in nature are mostly liquid infused highly porous surface, for example, poly(dimethyl siloxane). b) The polymers that have no charge include poly ethylene glycol and c) charged zwitterionic and polyampholytes polymers such as phospholipid polymers and phosphobetaine [1, 12]. Table 15.2 shows the example of antimicrobial polymers performing passive action.

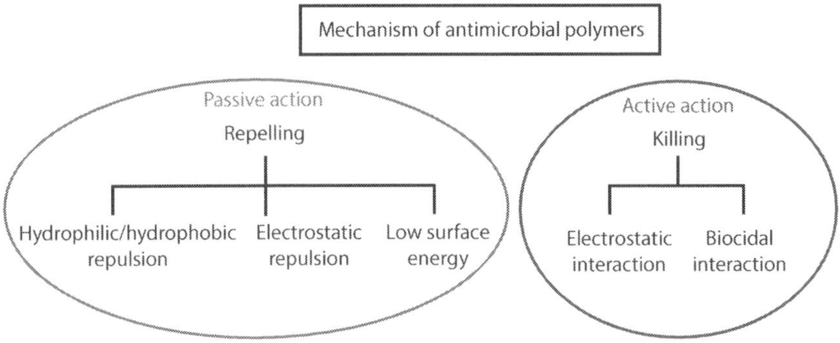

Figure 15.1 Schematic representation of mechanism of action of polymers.

Table 15.2 Polymers performing the passive action.

Polymer	Target	Remark	Reference
Poly[3-dimethyl (methacryloyloxyethyl) ammonium propane sulfonate-*b*-2-(diisopropylamino) ethyl methacrylate]	*Staphylococcus aureus*	Zwitterionic coronae and pH-responsive cores can impart bacterial anti-adhesive properties	[10]
Poly(2-methyl-2-oxazoline)	*Escherichia coli*	Dual-functional antimicrobial surface of poly(L-lysine)-graft-poly(2-methyl-2-oxazoline)-quarternery ammonium	[11]
Albumin, whey	*Bacillus subtilis, Escherichia coli*	No bacterial growth was observed on albumin-glycerol and whey-glycerol after 24 h inoculation	[12]

15.2.2 Active Action

Active polymers are capable to kill the microbes that stick to the surface of polymer. These polymers have the active agents which may be antimicrobial peptides, some cationic biocides, or other antibiotics that have ability to kill the bacteria when they are come into direct contact. The polymers kill the microbes by which mechanism depends upon the active agent present in the polymers. The mostly active antimicrobial polymers are usually activated with the positive charged specie such as quaternary ammonium, it may interact with the components of cell wall of bacteria and eventually cause the destruction of the cytoplasmic membrane of microbes, due to this destruction results, leakage of intracellular components takes place and consequently cell death occurs [1, 11]. There are polymers such as polyguanidine, polyethylenimine, and *N*-halamine that have capability to exhibit the antimicrobial activity. Polyethylenimine can cause the rupture of the cell membrane of bacteria by the development of electrostatic force of attraction between the polymer polyethylenimine and cell membrane of bacterial. Polyguanidine polymer has capability to inhibit the bacterial growth by the adhesion with cell membrane and subsequently destruction of Ca^{2+} salt bridges occur and eventually death of cell takes place. *N*-halamine can cause inhibition of cell growth and also responsible for inactivation of membrane due to presence of halogen group have oxidative effect which target such as thio and amino groups present in the receptors of cell [4]. Table 15.3 has the list of active polymers and applications of these antimicrobial polymers, showing that most of newly developed antimicrobials polymers are usually based upon quaternary ammonium group may be inform of salts.

15.3 Factor Affecting Activity of Antimicrobial

15.3.1 Polymers

There are many factors that have strong effect on the antimicrobial activities of the polymer such as molecular weight, charge density, hydrophilicity, counter ions, and pH.

15.3.2 Molecular Weight

Molecular weight has significant role in physicochemical characteristics of polymers. The molecular weight is an important factor for controlling

Table 15.3 Passive polymers perform antimicrobial activity.

Polymer	Target	Antimicrobial substance	Remark	Reference
Nisin-immobilized ozorganosilicon	*Bacillus subtilis*	Nisin	Superior antimicrobial activity, and resistant to several cleaning conditions	Yu, K., *et al.* [13]
Polyurethane containing quaternary ammonium	*Staphylococcus aureus*, *Escherichia coli*	Quaternary ammonium	Good antimicrobial activities against even at low concentrations (5 wt %)	[14]
Poly(*n*,*n*-diethylethylendiamine-co-yrosol-based acrylic)	*Staphylococcus epidermidis*, *Staphylococcus aureus*	Tertiary amine	Combination of two active compounds provide a synergistic action against biofilms and suppress reactive species oxygen	[15]

biocidal activity especially for the polymethyl acrylates and polyacrylates with biguanides group in side chain [4]. The molecular weight of polymer between 5×10^4 and 1.2×10^5 Da is optimal for the performing biocidal action. The antimicrobial properties increase with decrease the molecular weight from 5×10^4 Da while activity of antimicrobial polymer decrease with increase the molecular weight from 1.2×10^5 Da. The antimicrobial properties depend upon the permeability of polymer through the cell wall [8].

Gram-negative bacteria has greater diffusion rates for many antimicrobial polymers due to composition of cell wall but Gram-positive bacteria possess low diffusion rates [8]. In the case of ε-PL, alkyl chain length produces modifications. These have significant effects on the microbial activity and L-lysine chain length of residues that is present in the polymer having the optimum activity for the inhibition of growth of microbes [4].

15.3.3 Charge Density

Polymers have high positive charge clouds; they can develop the strong electrostatic force of interaction with cells of bacteria. In chitosan, charge density of polymer increase due to increasing of degree of deacetylation (DD), due to which electrostatic force of attraction increases as a result antimicrobial activity enhance [4]. At a higher rate of deacetylation, chitosan has higher antibacterial activity against *S. aureus* [13]. The polymers that have high charge density such as asparagine N-conjugated which is incorporated into chitosan structure for enhancing the antimicrobial action of polymer [14]. Chitosan structure is modified into guanidinylated chitosan which exhibit the higher antimicrobial action [15]. While chitosan having *N*-carboxyethyl can exhibit the antimicrobial action because of lack of free amino group [16].

15.3.4 Hydrophilicity

Hydrophilicity is a very essential factor for actions of antimicrobial agents [4]. Derivatives of water soluble chitosan that can be synthesized by the process of alkylation, acylation, quaternization, saccharization, and metallization show the higher antimicrobial activity [17].

15.3.5 Counter Ions

The quaternary ammonium/phosphonium compounds have counter ion effect. Counter ions which have high affinity towards quaternary compounds have lower antimicrobial action because of slow release of the free ions present in the medium. For quaternary ammonium compounds, chloride and bromide exhibit the high antimicrobial activity [4].

15.3.6 pH

The effect of pH on chitosan and antimicrobial polymers which are amphoteric in nature has been studied. Chitosan shows antimicrobial action depending upon its pH. Antimicrobial action of chitosan is limited to acidic environment because it dissolves in water only in the acidic conditions by such mechanism in which protonation of the amino group takes place. The antimicrobial activity and solubility of chitosan is increased over the entire range of pH by incorporating quaternary ammonium salts in chitosan backbone [18].

15.4 Medical Applications

The newly developed antimicrobial agents have the broad spectrum against the pathogens and also responsible for modulated immune responses [19–23]. The wound-dressings having antimicrobial agents are usually consist of cotton gauze and antimicrobial peptides for antimicrobial activity. These are combine with polyanion(sodium alginate) and polycation (chitosan) that show high antimicrobial activity towards *Klebsiella pneumonia* and *Staphylococcus aureus*. These types of wound dressings have also been proved as non-cytotoxic to the dermal fibroblasts of human [24]. The urinary tract infections associated with the use of catheter is the common reason for hospital-acquired infections. Urinary catheters impregnated with a mixture of triclosan, sparfloxacin, and rifampicin have been designed. The newly developed silicone catheters release the drugs about one month generally. The novel antimicrobial catheters can inhibit the colonization and adhesion of *Escherichia coli*, *Staphylococcus aureus*, and *Proteus mirabilis* approximately for a period of 7–12 weeks [25].

15.5 Conclusion

Recent advances show that antimicrobial polymers have gained the high attention in both industrial and academic research fields. This mini-review summarizes the mechanism of actions of polymers and factors that affecting the action of polymers. Recent advances made in antimicrobial polymers have been discussed. Passive antimicrobial polymers prevent the bacterial adhesion and growth of microorganisms that is alternative strategy rather than killing the microbes directly with active antimicrobial polymers agents. Despite of the recent advances in antimicrobial polymer substances, there is a need for the further classification of antimicrobial polymers depending upon their antimicrobial mechanism of action on the microbes. Biofilm-associated mechanisms especially require a very intensive effort for designing of new antimicrobial substances. The highly effective antimicrobial polymers can be formulated by the combination of different kinds of antimicrobial polymers having the different mechanism for exhibiting the antimicrobial activity. The more challenges in this field for the development polymers can exhibit their actions for the long period of time and have broad spectrum for antimicrobial activity. The major demand is that delivery of the drug on the desired site in a very controlled manner and for longer period.

References

1. Huang, K.S., Yang, C.H., Huang, S.L., Chen, C.Y., Lu, Y.Y., Lin, Y.S., Recent Advances in Antimicrobial Polymers: A Mini-Review. *Int. J. Mol. Sci.*, Sep 20, 17, 9, E1578, 2016.
2. Woo, G.L., Mittelman, M.W., Santerre, J.P., Synthesis and characterization of a novel biodegradable antimicrobial polymer. *Biomaterials*, Jun 21, 12, 1235–46, 2000.
3. Konai, M.M., Bhattacharjee, B., Ghosh, S., Haldar, J., Recent Progress in Polymer Research to Tackle Infections and Antimicrobial Resistance. *Biomacromolecules*, Jun 11, 19, 6, 1888–917, 2018.
4. Jain, A., Duvvuri, L.S., Farah, S., Beyth, N., Domb, A.J., Khan, W., Antimicrobial polymers. *Adv. Healthcare Mater.*, Dec 3, 12, 1969–85, 2014.
5. Palza, H., Antimicrobial polymers with metal nanoparticles. *Int. J. Mol. Sci.*, Jan 19, 16, 1, 2099–116, 2015.
6. Arora, S., Yadav, V., Kumar, P., Gupta, R., Kumar, D., Polymer based antimicrobial coatings as potential biomaterial: A review. *Int. J. Pharm. Sci. Rev. Res.*, 23, 2, 279–90, 2013.
7. Sedlarik, V., Antimicrobial modifications of polymers. *Biodegradation-Life of Science*, 187–204, 2013.
8. Kenawy el, R., Worley, S.D., Broughton, R., The chemistry and applications of antimicrobial polymers: A state-of-the-art review. *Biomacromolecules*, May 8, 5, 1359–84, 2007.

9. Timofeeva, L. and Kleshcheva, N., Antimicrobial polymers: Mechanism of action, factors of activity, and applications. *Appl. Microbiol. Biotechnol.*, Feb 89, 3, 475–92, 2011.
10. Francolini, I., Donelli, G., Crisante, F., Taresco, V., Piozzi, A., Antimicrobial polymers for anti-biofilm medical devices: State-of-art and perspectives. *Adv. Exp. Med. Biol.*, 831, 93–117, 2015.
11. Zhao, J., Ma, L., Millians, W., Wu, T., Ming, W., Dual-Functional Antifogging/Antimicrobial Polymer Coating. *ACS Appl. Mater. Interfaces*, Apr 6, 8, 13, 8737–42, 2016.
12. Liu, L., Li, W., Liu, Q., Recent development of antifouling polymers: Structure, evaluation, and biomedical applications in nano/micro-structures. *Wiley Interdiscip. Rev. Nanomed. Nanobiotechnol.*, Nov-Dec 6, 6, 599–614, 2014.
13. Takahashi, T., Imai, M., Suzuki, I., Sawai, J., Growth inhibitory effect on bacteria of chitosan membranes regulated with deacetylation degree. *Biochem. Eng. J.*, 40, 3, 485–91, 2008.
14. Jeon, Y.-J., Park, P.-J., Kim, S.-K., Antimicrobial effect of chitooligosaccharides produced by bioreactor. *Carbohydr. Polym.*, 44, 1, 71–6, 2001.
15. Hu, Y., Du, Y., Yang, J., Kennedy, J.F., Wang, X., Wang, L., Synthesis, characterization and antibacterial activity of guanidinylated chitosan. *Carbohydr. Polym.*, 67, 1, 66–72, 2007.
16. Yancheva, E., Paneva, D., Maximova, V., Mespouille, L., Dubois, P., Manolova, N. *et al.*, Polyelectrolyte complexes between (cross-linked) N-carboxyethylchitosan and (quaternized) poly [2-(dimethylamino) ethyl methacrylate]: Preparation, characterization, and antibacterial properties. *Biomacromolecules*, 8, 3, 976–84, 2007.
17. Ignatova, M., Manolova, N., Rashkov, I., Novel antibacterial fibers of quaternized chitosan and poly (vinyl pyrrolidone) prepared by electrospinning. *Eur. Polym. J.*, 43, 4, 1112–22, 2007.
18. Lim, S.H. and Hudson, S.M., Synthesis and antimicrobial activity of a water-soluble chitosan derivative with a fiber-reactive group. *Carbohydr. Res.*, Jan 22, 339, 2, 313–9, 2004.
19. Fu, T.H., Li, Y., Thaker, H.D., Scott, R.W., Tew, G.N., Expedient Synthesis of SMAMPs via Click Chemistry. *ACS Med. Chem. Lett.*, Sep 12, 4, 9, 841–5, 2013.
20. Kuroda, K. and Caputo, G.A., Antimicrobial polymers as synthetic mimics of host-defense peptides. *Wiley Interdiscip. Rev. Nanomed. Nanobiotechnol.*, Jan-Feb 5, 1, 49–66, 2013.
21. Thaker, H.D., Cankaya, A., Scott, R.W., Tew, G.N., Role of Amphiphilicity in the Design of Synthetic Mimics of Antimicrobial Peptides with Gram-negative Activity. *ACS Med. Chem. Lett.*, May 9, 4, 5, 481–5.
22. Liu, R., Chen, X., Chakraborty, S., Lemke, J.J., Hayouka, Z., Chow, C. *et al.*, Tuning the biological activity profile of antibacterial polymers via subunit substitution pattern. *J. Am. Chem. Soc.*, 136, 11, 4410–8, 2014.

23. Choi, H., Chakraborty, S., Liu, R., Gellman, S.H., Weisshaar, J.C., Single-Cell, Time-Resolved Antimicrobial Effects of a Highly Cationic, Random Nylon-3 Copolymer on Live Escherichia coli. *ACS Chem. Biol.*, Jan 15, 11, 1, 113–20, 2016.
24. Gomes, A.P., Mano, J.F., Queiroz, J.A., Gouveia, I.C., Incorporation of antimicrobial peptides on functionalized cotton gauzes for medical applications. *Carbohydr. Polym.*, 127, 451–61, 2015.
25. Fisher, L.E., Hook, A.L., Ashraf, W., Yousef, A., Barrett, D.A., Scurr, D.J. *et al.*, Biomaterial modification of urinary catheters with antimicrobials to give long-term broadspectrum antibiofilm activity. *J. Control. Release*, Mar 28, 202, 57–64, 2015.

16

Characterization Techniques for Polymer Coatings

Hina Sharif[1], Kanwal Rehman[2], Kanwal Irshad[1] and Muhammad Sajid Hamid Akash[1]*

[1]*Department of Pharmaceutical Chemistry, Government College University Faisalabad, Faisalabad, Pakistan*
[2]*Department of Pharmacy, University of Agriculture, Faisalabad, Pakistan*

Abstract
Polymer coating is a process of coating polymeric material to the external surface of dosage forms with certain intentions including protecting the active pharmaceutical ingredient (API) against the environmental hazards or against the body fluids, protecting body from adverse effects of API and modifying the release of API. In this respect, different coating types as film coating, enteric coating, extended release coating, and organic-inorganic nano-composites hybrid coatings are being used in pharmaceutical industry. Spray coating is a widely used technique for polymer coating. Irrespective to the reasons and types of polymer coating to the dosage forms, polymer films need evaluations for certain properties. This chapter has a collective knowledge about polymers, their coating technique, types of coating, and characterization techniques which include water vapor permeability, oxygen permeability, thermal characteristics, and mechanical and adhesive characteristics of the polymeric materials used for coating.

Keywords: Polymer coating, technique for coating, types of coating, film coating, extended release coating, water vapor permeability

16.1 Introduction

Polymer is derived from two Greek words, poly means many and mer means part. It is composed of monomers, the building blocks of polymer.

**Corresponding author*: sajidakash@gcuf.edu.pk

Inamuddin, Rajender Boddula, Mohd Imran Ahamed and Abdullah M. Asiri (eds.) Polymer Coatings: Technology and Applications, (359–370) © 2020 Scrivener Publishing LLC

According to the definition of IUPAC, a polymer is a substance composed of macromolecules. A macromolecule is a molecule of relatively higher molecular mass, the structure of which is comprised of many repeating units called monomers derived from molecules of relatively lower molecular mass [1]. Polymers have some unique characteristics including low density, good corrosion resistance, low coefficient of friction, good mould ability, low mechanical properties, poor tensile strength, and poor temperature resistance [2]. Many synthetic and natural polymers are extensively used as pharmaceutical excipients and for coating on the solid oral dosage forms as they mask some undesired physicochemical properties of drugs and hence improve the therapeutic effect. They allow increased solubility, swellowability, biodegradability, pH dependency, advanced coatings, and inhibition of crystallization [3].

Polymers can be classified in various aspects, as based on source, polymers may be natural, semisynthetic, or synthetic, and on the bases of structure, they may be linear, branched chain, or cross linked. Polymers have extensive applications in the field of pharmaceutical sciences as binders, disintegrating agents, and coating materials in the manufacturing of tablet, gelatin manufacturing in capsules, as suspending agents in suspensions, as emulsifying agents in emulsions, as film coating to mask bitter taste and modify drug release characteristics and in the drug delivery of various contraceptives and hormones [4].

16.2 Polymer Coating

Coating is a process of coating material (polymers) to the external surface of drug dosage form to achieve certain benefits and properties of the API as compared to uncoated variety, the benefits may include protection of API against the environmental hazards, protection against the harsh environment of the GIT or body fluids, protection of body itself against the adverse effects of API, or modification of the release characteristics of API [5]. Components for coating include polymers, pigments, plasticizers, opacifiers, glidents, binders, anti-tacking agents, surfactants, fillers, and extenders whereby polymer constitutes a major component. Important polymers used for coating are hydroxypropylmethylcellulose (HPMC), polyvinyl alcohol (PVC), sodium alignate, cellulose acetate phathalate, methacrylic acid co-polymers, etc [6].

Polymer coating has profound merits including (a) protection against environmental factors like temperature, sunlight (UV light), moisture, and toxic gases, (b) facilitation of swallowing, (c) mask unpleasant taste and odor, (d) improved appearance, (e) facilitate product identification during manufacture and packaging, (f) provide immediate release, sustained release, specific release, controlled release or targeted release of drug, and (g) improve mechanical strength of the dosage forms [6]. But coating may have some demerits like increase thickness and weight of tablets, increase cost of formulation, impact on surface smoothness due to uneven coating, altered pharmacodynamics of the API, sticking and picking, orange peel effect and bridging, i.e., inadequate spreading of coating solution [6, 7].

16.3 Technique for Coating

Spray coating is the most preferable technique for coating process which involves the formation of fine droplets of the coating solution, and then, this solution is deposited and coalesced on the surface of the substrate. Through the technique of atomization in spraying nozzle, the liquid which is used for coating is broken. Then, these droplets are deposited on the substrate from where they undergo coalescence to form a film [8].

Currently, ultrasonic atomization technique is preferably employed for the coating procedure which utilizes high frequency sound vibrations to form a fine mist of the coating solution. The benefits of spray coating include good transfer efficiency, simple and economical, good reproducibility and uniform coating on both planar and non-planar surfaces, and formation of tiny droplets which are in the micrometer range. Spray coating is dependent on some material and process parameters including (a) polymer solution's concentration, (b) solvent employed, (c) the drug and its dosage form, (d) flow rate of polymer solution, (e) temperature, (f) the distance between the spraying nozzle and the dosage from, (g) the speed of the moving nozzle, and (h) air pressure of the nozzle [9].

16.4 Types of Coating

16.4.1 Film Coating

It is a process of deposition of a thin polymeric film onto the solid dosage forms like tablets, capsules, or pellets [6]. Film formation of polymer coatings is very important with respect to the purpose of coatings of drugs like enteric resistance or modified release [10]. There may be free film or applied film. Casting method is generally used to obtain free films in which

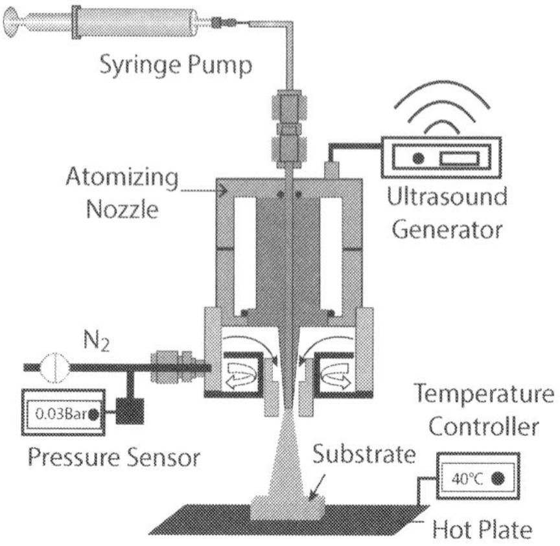

Figure 16.1 A schematic representation of apparatus used for spray coating. Adopted from Ref. [9].

the solution or dispersion of polymeric material is applied onto the substrate, and then, solvent is evaporated. A spray atomization technique is used for this purpose to avoid uneven film surface. But in case of applied film variable substrate, processing parameters, storage conditions, and coating formulations are required [11] Figure 16.1.

Film coating can be achieved using organic polymer solution or aqueous polymer dispersion [10]. While using organic polymer solution, after deposition, the dissolved polymer forms the film on the surface of the dosage form by undergoing sol to gel transition. When the organic solvent is evaporated, then the molecules of polymer come closer to each other and form a homogenous film with high penetration of polymer chain. Low polymer concentration and toxicity of the solvent which requires its recovery are the drawbacks of using organic polymer solution [10, 12].

To address these drawbacks, aqueous polymer dispersions are preferred in which the film is formed by the coalescence of the polymer molecules after the evaporation of water. This process is generally divided into three phases. In 1st phase, water evaporation occurs, and then, concentration of the dispersion is increased until dispersed particles form a closed-packed array. In 2nd phase, temperature is increased above the minimum film formation temperature (MFFT), and then, particles undergo deformation making polyhedral structures. In the last phase, the temperature is further increased above glass transition temperature (Tg) which leads to the inter-diffusion of polymer chains forming a continuous film without the distinguishable particles [10]. Here, the temperature plays an important role on the polymer diffusion rate and film formation [13].

16.4.2 Extended Release Coating

Extended release polymeric coating offers numerous advantages including controlled release of the therapeutics having short half-lives, inhibiting the non-specific distribution of drug and hence lowering the toxicities [14]. Extended release polymers are coated either by aqueous polymer dispersions or organic polymer solutions. Organic systems have some environmental and toxicological hazards whereas high temperature, high energy, and long processing time are the drawbacks of aqueous systems. In aqueous dispersion, high energy input is required to evaporate the dispersion medium. Moreover, less amount of coating material is dispersed in the large amount of dispersion medium which is then needed to be evaporated [5]. Films formed from aqueous dispersions undergo aging phenomena also. To address all these drawbacks, dry powder coating is used which does not use organic solvent and limited water is used. It is a time-saving process [10].

In dry coating, polymeric material is cast as a micronized powder directly onto the surface of dosage form with or without the spraying of liquid plasticizer [15]. Dry coating process occurs in two phases. The 1st phase is the coating phase in which the dry powder mixture of the polymeric material and plasticizer is applied to the substrates, whereas in 2nd phase known as curing phase, a film is formed onto the substrate [5].

16.4.3 Organic-Inorganic Nanocomposites Hybrid Coating

These coatings are nano-phase separated metal-oxo clusters which are connected to a continuous organic network of polymers through a phase coupling agent. Resistance to abrasion and scratching, heat stability, and improved mechanical properties are the merits of this coating technique. These hybrid materials are obtained by *in situ* sol-gel process in which silica domains are embedded into polymer network. Hydrolysable multifunctional alkoxysilane is used as precursor for the formation of inorganic domains which then undergoes a series of hydrolysis and condensation reactions. A suitable coupling agent is used to achieve an interconnected network without macroscopic phase separation; it provides bonding interactions between the organic phase and inorganic phases [16].

Currently, this coating technique is carried out by combining sol-gel process with a photo-polymerization process, which has advantages of both curing methods. In this technique, the formation of polymer is hastened through fast transformation of the liquid particles into a solid film [17]. As it is a solvent free process and carried out at the room temperature, so this technique is environment friendly and hence becoming increasingly important for coating [16].

16.4.4 Enteric Coating

Enteric coating allows release of API in the lower part of small intestine to protect it from the harsh acidic environment of stomach. It involves the use of biodegradable polymers such as azopolymer, dextrans, pectins, amylose, etc., as coatings in oral formulations [18].

16.5 Characterization of Coating System

16.5.1 Water Vapor Permeability

It is a parameter to assess the effectiveness of film coating against water. Various factors have an impact on water vapor permeability like film thickness, film composition, and technique employed for the film preparation. Fick's first and Henry's laws are utilized to access the water vapor permeability. Figure 16.2 is the schematic representation to assess water vapor permeability for free films. The containers shown are of glass or aluminum containing saturated solution of salt. The sample film is placed over the opening of container. Transmission cells are kept in desiccator whose humidity is known by creating a difference in water vapor pressure. By measuring weight change of the transmission cell over time, a graph is plotted between the weight change and time. From the slope of the line which is calculated by line of regression, the value of water vapor transmission rate (WVTR) is obtained [11].

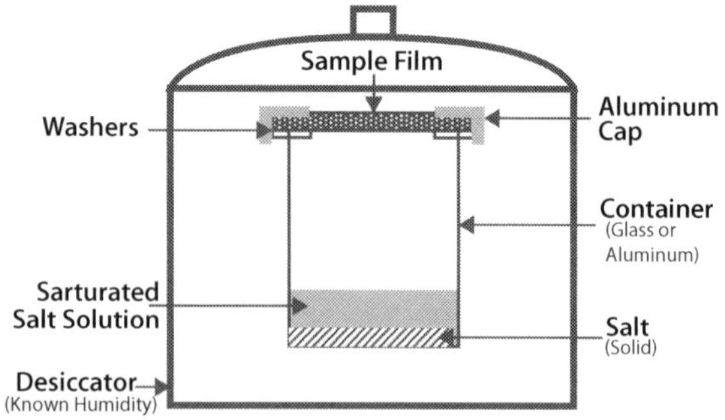

Figure 16.2 Schematic representation of assessment of water vapor permeability for free film. Adopted from Ref. [11].

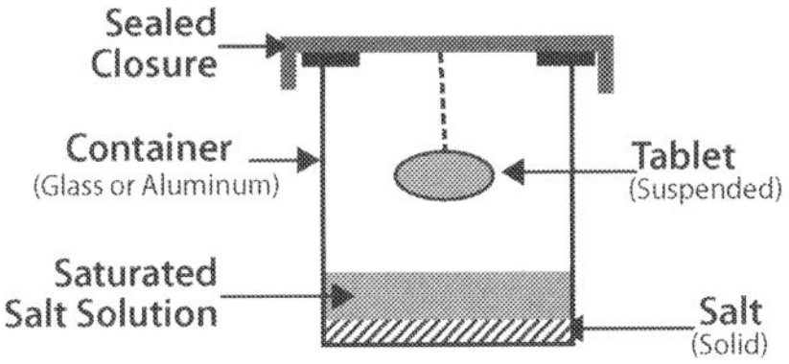

Figure 16.3 A schematic representation of assessment of water vapor permeability for applied films. Adopted from Ref. [11].

For applied films, the transmission cell apparatus for free film is modified. As shown is Figure 16.3, a wire loop is suspended over the saturated salt solution in a closed container made up glass or aluminum. The coated tablet under analysis is placed on that wire loop. The weight of tablet is calculated over time and a graph is plotted. From the slope of line, the value of WVTR is then determined [11].

16.5.2 Oxygen Permeability

After water vapor permeability, oxygen permeability is the most commonly studied parameter. This parameter is important for the drugs which are affected by oxidation. Oxygen permeability is actually the measurement of effectiveness of the coating polymeric material against oxygen and it is affected by film thickness and film composition. For free films as shown in Figure 16.4, test film is placed between the two chambers: inside chamber and outside chamber. Nitrogen gas is eliminated from the both chambers. Pure oxygen is passed through the one side of chamber and then using a coulometric sensor, diffused molecular oxygen is quantified [11]. For applied films, nondestructive paramagnetic

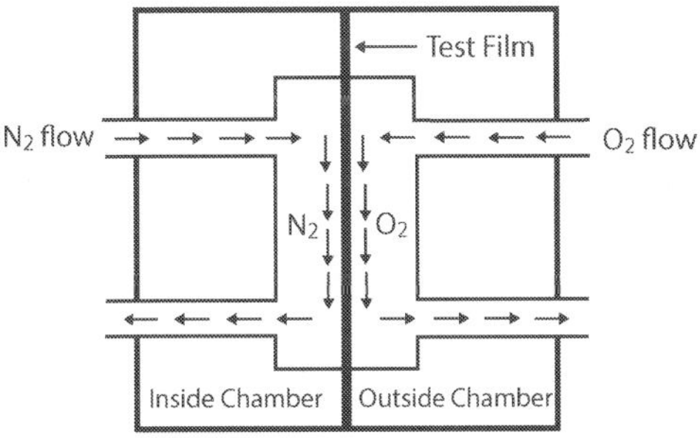

Figure 16.4 Schematic to quantify oxygen permeation for free films. Adopted from Ref. [11].

resonance spectroscopic technique is used in which lithium phthalocyamine crystal is inserted into the dosage form and then partial oxygen pressure (pO_2) is determined. A graph is plotted between pO_2 and time and then slope is calculated using line of regression which indicates oxygen permeation [11, 19].

16.5.3 Thermal Properties

16.5.3.1 Glass Transition Temperature (Tg)

Tg is the macro-manifestation of polymer chain's flexibility, an important parameter in case of amorphous polymer. It is defined as the range of temperature in which a polymer changes from rigid to soft material [20]. If temperature is less than Tg, the movement of polymer chain is restricted and a rigid film is formed and if temperature is increased above Tg, then the mobility of polymer chains is increased and crystallization of polymers is enhanced. Tg is measured through differential scanning calorimeter (DSC) [21]. DSC is a type of thermo-analytical technique that determines the heat and temperature flow either absorbed or released associated with the sample as a function of temperature and time [22]. In this technique, the sample pan containing polymeric material and reference pan which is kept empty are heated at the same rate simultaneously and the thermal transitions at which more energy is absorbed or released are determined. DSC is also used to detect polymorphism, determine melting temperature, evaluate polymer miscibility, and study oxygen degradation [11].

16.5.3.2 Minimum Film Forming Temperature (MFFT)

MFFT is the very important parameter for the film formation. It is defined as the minimum temperature at which a cast film becomes crackless or clear leading to coalescence of polymer which is applied to a substrate into a thin film [12, 23]. If temperature is less than MFFT, then a white opaque layer is formed or powdery material is appeared on the surface of the substrate but if temperature is equal to or higher than MFFT, a transparent film is formed. Hence, in the coating process, bed temperature should be above the MFFT. For the determination of MFFT, a temperature gradient plate is employed with drying air blown. The polymeric material is placed onto that plate and drying air is then passed through the sample. The lowest temperature at which a clear and transparent film is formed is inspected visually and this temperature is considered as the MFFT of that polymeric material [11, 21]. The value of MFFT is affected many factors such as presence of plasticizers or other excipients in the polymeric material [21].

16.5.4 Mechanical Testing

Polymer films must be mechanically strong enough so that they do not undergo breakage or fracture while processing, manufacturing, packaging, transportation, or storage. For this purpose, tensile strength, stress analysis, compression testing, puncture strength, and dynamic mechanical analysis are performed to evaluate the mechanical properties of polymeric material used for coating [11].

16.5.5 Polymer Adhesion

Poor adhesion between polymer and substrate may cause flaking or peeling effect and moisture can accumulate at the interface of film and substrate which can affect stability of drug and also the mechanical strength of the polymeric film. Polymer adhesion is related to both the internal stresses within the polymeric film and the interactions between the film and substrate. Strength of interfacial bond and the internal stresses within the polymeric film are the important factors which affect polymeric adhesion [24]. Several methods are employed to assess the polymer adhesion with the drug dosage forms. In peel test, a modified tensile tester is utilized. It peels the film from the surface of a tablet at an angle of 90°. Another methodology used is butt adhesion test (Figure 16.5) in which the whole film is removed from the surface of tablet. This method eliminates the variations

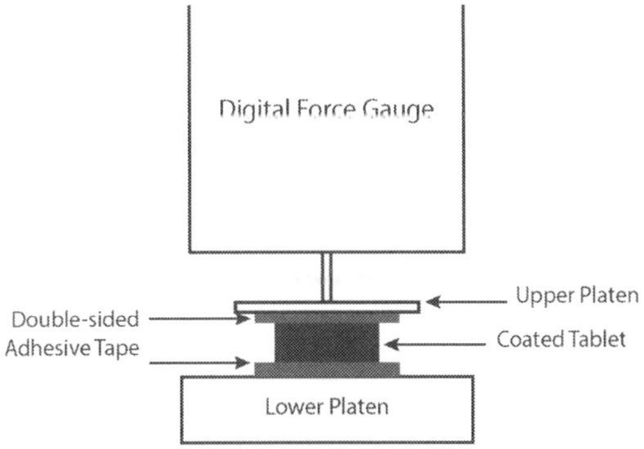

Figure 16.5 Schematic of butt adhesion test. Adopted from Ref. [11].

because of elasticity of film and is less affected by uniformity of tablets. This method utilizes flat faced tablets, but for concave tablets, a foam or rubber backing is added to ensure the uniform contact between the force gauge and film surface [11].

16.5.6 Surface Roughness

Rough surface of the coated tablet can influence drug release and it may affect the aesthetic appearance. Electron microscopy technique is employed to predict the qualitative analysis of surface morphology, whereas for quantitative analysis, profilometry is employed in which a sensing head is used which moves across the substrate. Atomic force microscopy is another quantitative method in which a sensor examines the small areas on substrate [11].

16.5.7 Film Thickness and Uniformity

Several approaches are applied to check film thickness and uniformity. Microscopic method can be used but it is tedious and time consuming. Another method is laser-induced breakdown spectroscopy, an elemental analysis technique, which is based on detection of atomic emission from a polymeric material. Computer-aided tomography (CAT) is a medical imaging technique which is now widely used. This involves the transmission of x-rays onto the sample, and then, the intensity of the transmitted x-rays is measured using a detector [11].

16.6 Conclusion

Polymer coating plays a key role in the drug design and modified drug delivery systems. This mini-review throws light on the type and properties of the polymeric materials and it also explains the characterization techniques that are needed to assess the mechanical and thermal properties of the polymers. All these techniques are required for the pharmacist or technician before coating processes and helps in the modification of drug release properties.

References

1. Wikipedia, Polymer, Available from: https://en.wikipedia.org/wiki/Polymer, 2015.
2. Slideshare, Polymers and their properties, Available from: https://www.slideshare.net/ripestone_ho/polymers-and-their-properties, 26-2-2012.

3. Debotton, N. and Dahan, A., Applications of Polymers as Pharmaceutical Excipients in Solid Oral Dosage Forms. *Med. Res. Rev.*, Jan, 37, 1, 52–97, 2017.
4. Slideshare, Polymer, Available from: https://www.slideshare.net/krk1995/polymer-ppt-33956909, 25-4-2014.
5. Kablitz, C.D., Kappl, M., Urbanetz, N.A., Parameters influencing polymer particle layering of the dry coating process. *Eur. J. Pharm. Biopharm.*, 69, 2, 760–8, 2008.
6. Sharma, D.P.H., Review on polymers used for film coating. *Asian J. Pham. Technol. Innov.*, 1, 2, 1–16, 2013.
7. Slideshare, tablet coating process, Available from: https://www.slideshare.net/raheemthayyil/tablet-coating-process, 2018.
8. Tobiska, S. and Kleinebudde, P., Coating uniformity: Influence of atomizing air pressure. *Pharm. Dev. Technol.*, 8, 1, 39–46, 2003.
9. Bose, S., Keller, S.S., Alstrom, T.S., Boisen, A., Almdal, K., Process optimization of ultrasonic spray coating of polymer films. *Langmuir : The ACS journal of surfaces and colloids*, Jun 11, 29, 23, 6911–9, 2013.
10. Kablitz, C.D. and Urbanetz, N.A., Characterization of the film formation of the dry coating process. *Eur. J. Pharm. Biopharm.*, 67, 2, 449–57, 2007.
11. Felton, L.A., Characterization of coating systems. *AAPS PharmSciTech*, Dec 21, 8, 4, E112, 2007.
12. Lecomte, F., Siepmann, J., Walther, M., MacRae, R.J., Bodmeier, R., Polymer blends used for the coating of multiparticulates: Comparison of aqueous and organic coating techniques. *Pharm. Res.*, May, 21, 5, 882–90, 2004.
13. Oh, J.K., Tomba, P., Ye, X., Eley, R., Rademacher, J., Farwaha, R. *et al.*, Film formation and polymer diffusion in poly (vinyl acetate-co-butyl acrylate) latex films. Temperature dependence. *Macromolecules*, 36, 15, 5804–14, 2003.
14. Kamaly, N., Yameen, B., Wu, J., Farokhzad, O.C., Degradable Controlled-Release Polymers and Polymeric Nanoparticles: Mechanisms of Controlling Drug Release. *Chem. Rev.*, Feb 24, 116, 4, 2602–63, 2016.
15. Terebesi, I. and Bodmeier, R., Optimised process and formulation conditions for extended release dry polymer powder-coated pellets. *Eur. J. Pharm. Biopharm.*, 75, 1, 63–70, 2010.
16. Sangermano, M., Amerio, E., Epicoco, P., Priola, A., Rizza, G., Malucelli, G., Preparation and Characterization of Hybrid Nanocomposite Coatings by Cationic UV-Curing and the Sol-Gel Process of a Vinyl Ether Based System. *Macromol. Mater. Eng.*, 292, 5, 634–40, 2007.
17. Weiss, P., Photopolymerization of surface coatings, C. G. Roffey, Wiley-Inter-science, New York, 1982, 353 pp. *J. Polym. Sci.: Polymer Letters Edition*, 21, 4, 310–, 1983.
18. Marvola, M., Nykanen, P., Rautio, S., Isonen, N., Autere, A., Enteric polymers as binders and coating materials in multiple-unit site-specific drug delivery systems. *Eur. J. Pharm. Sci.*, 7, 3, 259–67, 1999.

19. Felton, L.A. and Timmins, G.S., A nondestructive technique to determine the rate of oxygen permeation into solid dosage forms. *Pharm. Dev. Technol.*, Feb, 11, 1, 141–7, 2006.
20. Becker, H. and Locascio, L.E., Polymer microfluidic devices. *Talanta*, Feb 11, 56, 2, 267–87, 2002.
21. Porter, S.C. and Felton, L.A., Techniques to assess film coatings and evaluate film-coated products. *Drug Dev. Ind. Pharm.*, Feb, 36, 2, 128–42, 2010.
22. Gill, P., Moghadam, T.T., Ranjbar, B., Differential scanning calorimetry techniques: Applications in biology and nanoscience. *J. Biomol. Tech.: JBT*, Dec, 21, 4, 167–93, 2010.
23. Corrosionpedia, Minimum Film Forming Temperature (MFFT), Available from: https://www.corrosionpedia.com/definition/5696/minimum-film-forming-temperature-mfft, 2019.
24. Felton, L.A. and McGinity, J.W., Adhesion of polymeric films to pharmaceutical solids. *Eur. J. Pharm. Biopharm.*, 47, 1, 3–14, 1999.

17

Polymer Coatings for Corrosive Protection

Gobinath Velu Kaliyannan[1], Mahesh Kumar Karavalasu Velusamy[2], Sathish Kumar Palaniappan[3*], Mohan Kumar Anandraj[1] and Rajasekar Rathanasamy[1]

[1]*Department of Mechanical Engineering, Kongu Engineering College, Erode, India*
[2]*Department of Mechatronics Engineering, Kongu Engineering College, Erode, India*
[3]*Department of Mining Engineering, Indian Institute of Technology Kharagpur, India*

Abstract
Corrosion plays a predominant role in all the major industries that use corrosive materials mainly of chemical industries. Mainly, the corrosion may be controlled by adopting corrosion control methods. There are various types of corrosion control methods among those coating conductive polymers over the medium mainly generates corrosion resistance to the base material, which contains toxic and hazardous materials that damage the environment, to an extent. These major commitments have led the researchers for the development of innovative protective coating materials and methods which adopts carbon-based materials and conductive nanocomposites. Recent developments indicate that mainly the conductive polymers have increased the attention of the industrial and research community due to its economic viability and improved applications. Generally, polymers hold a long carbon linkage chain which therefore capable to block wider area of the metal surfaces over corrosion which is due to the adsorption property. The deposition of thin coating over the corrosive surface behaves as a barrier between both the metal surface and the environment for protecting such surfaces. This review article discusses various methods that are adopted for protecting metals from corrosion, mainly about nanomaterials, conducting polymers, nanocomposites, and also carbon-based materials for fighting against corrosion behavior.

Keywords: Metal substrate, corrosion, polymer coatings, conducting polymers

**Corresponding author*: sathishiitkgp@gmail.com

Inamuddin, Rajender Boddula, Mohd Imran Ahamed and Abdullah M. Asiri (eds.) Polymer Coatings: Technology and Applications, (371–396) © 2020 Scrivener Publishing LLC

17.1 Introduction

Corrosion control is a significant topic of growing concern in the electronic and metallurgical industries. Metal corrosion is one of the most significant issues in the global arena. The most common technique of corrosion prevention is to apply paint or organic layer to the substrate of metal. The organically coated materials can preferably provide corrosion resistance that lasts for longer time. Moreover, this advantage relies on the capacity of the coating to protect the covered metal layer when failures occur during its service life [1]. The most commonly used technique could be to carry out polymer coatings [2]. Conducting polymers as film developing corrosion substances or in a protective coating has attracted greater and greater attention owing to the excellent anti-corrosion capability and friendship with the environment [3]. The benefit of conducting polymers and their composite materials over other coatings, like paints, is that they do not produce environmental damaging. Furthermore, conducting polymers exist as both a physical barrier and an electronic barrier, improving the protection provided by other materials which act only as physical barriers [4].

The conducting polymers perform redox reactions and thus enable the bonding and removal of the counter ion (doping) in reaction to variations in the surface of metal potential caused through local electrochemical operations owing to corrosion. Doping, which could be incorporated or removed by conducting polymer based on local corrosive environment, is mostly a preventative inhibitor of local corrosion after discharge [5, 6]. This is one of the techniques described in the exploitation of the benefits given when a conducting polymer has been used as a primary ingredient of corrosion-resistant layer [7]. The use of conducting polymer for protecting metal from corrosion has drawn excellent concern significantly over the past 35 years. Despite the extensive work on anti-corrosive characteristics of conducting polymers, there are still many difficulties for conducting polymer-based coatings to meet the physical-electrochemical and mechanical demands of high-performance anti-corrosive coatings through a wide range of practical conditions. Disadvantages of conducting polymers as anticorrosive coatings are including (i) irretrievable usage of the load contained in the conducting polymer, which can oxidize the base metal as a result of creating passive oxide layer; (ii) anion-exchange characteristics; (iii) porous composition and bad barrier impact; and (iv) bad adherence to the parent metal.

In particular, all of the above-notified drawbacks are enlarged within extreme corrosion environment as it is in the existence of chlorides that may enter the steel surface either because of permeability of the conducting polymer layer or because of its anion-exchange characteristics when chlorides replace the conducting polymer layer anions. One of the effective methods for eliminating the above drawbacks is to consider conducting polymer-based composite structures generally containing a conducting polymer in which metal oxides which are distinct inorganic fillers were embedded. These conducting polymer-based composite materials mix the redox characteristics and therefore the self-healing function of conducting polymer with inorganic filler characteristics. As a result, conducting polymer composites have exposed stronger mechanical as well as physico-chemical characteristics enhancing barrier effect, adhesion property, and possibly hydrophobic characteristics. Nanocomposite conducting polymer-based coatings appear to blend the characteristics of conducting polymers and organic polymers more effectively with those of inorganic metals [8, 9]. This chapter examines the implementation of conducting polymer-based coatings to prevent the corrosion of metals and alloys described in latest developments.

17.2 Basics of Corrosion

Corrosion is commonly occurring defect that can be defined as the deterioration or degradation of a particular substance and also the properties of the same which is usually metal by using it for a longer period of time due to the environmental disclosure [10, 11]. This is an exergonic cycle because the metal falls to the lowest possible state of heat. So the materials like steel and aluminum have a general capacity to reform into lower state of energy when exposed with water and oxygen to form iron and aluminum oxides. These corroded materials are the final processed metals that degrade over a period of time when exposed to environment. Thus, the cycle of the industrial products from its processed ores and or its original state is indicated in Figure 17.1 with the consideration of both environment and materials.

Identifying these two elements can evaluate effective strategies to counteract these destructive processes. The metal corrosion is broadly classified into three groups.

- Wet oxidation in a corrosive atmosphere of dissolved bacteria, the electrolyte is usually a solvent and the process which is electrochemical.

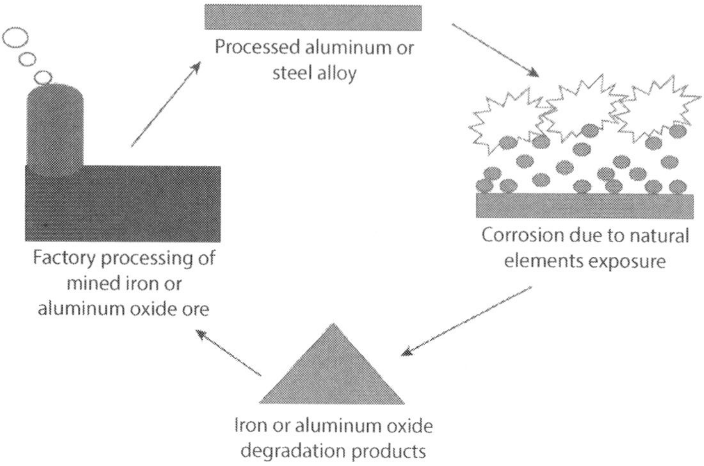

Figure 17.1 Schematic representation of the corrosion cycle of metal alloy [11].

- In other liquids such as fused salts and/or fused metals, oxidation.
- Dry corrosion where high-temperature gas in dry condition is the corrosive condition.

The wet corrosion behavior is extensively discussed in this article which mainly discusses the wet corrosion processes which affect the metals and its alloys and also the techniques that exist for controlling the corrosion behavior and also its effects.

Corrosion is mainly categorized into three groups, *viz.*,

- Group I - Nature of corrodent,
- Group II - Appearance of metal that is corroded, and
- Group III - Mechanism of corrosion.

Group I is generally intended as wet or dry condition, where wet corrosion occurs in liquid medium usually water which means freshwater or saline. The aqueous generally termed as wet corrosion process comprises three most essential elements that are required for the corrosion process to happen which are cathodic reaction, anodic reaction, and electrolyte solution. Group II implies that how the corrosion takes place through electrochemical or through direct chemical reaction. When referred to Group III, it describes which type of corrosion affects the metals which may be uniform corrosion, corrosion over the entire surface or in only small areas that affect the corrosion process generally called localized corrosion.

17.2.1 Essentials of Corrosion

The metal's anodic reaction and oxidation result in metal breakdown, which is transferred as Mn+ ions. Also, the cathodic reaction engrosses oxygen. Reducing the oxygen content is mostly predominant cathodic reaction in natural environments such as freshwater, seawater, atmosphere, and soil. This method forms electrical circuit devoid and charges accumulation. The anodic releases electrons that further got inducted through metal to the cathode which in turn is consumed by the cathodic reaction. This process requires an ionically conducting liquid that should be in contact with metal. The ion conduction through the electrolyte closes the electrochemical circuit, also all the three elements should be present for wet corrosion to take place. Normally Mn^+ ions are conducted towards OH– ions and when both these gathered produce a metal hydroxide that gets deposited on the metal surface. Copper and iron metals also follow similar trend of corrosion process where water is used as an electrolyte in the presence of dissolved oxygen (Figure 17.2).

As depicted in Figure 17.2, the two main features that is the source of corrosion are the oxygen and electrolyte. When hydroxide gets deposited onto the metal surface, there may be a reduction of oxygen available for continuation of corrosion process. This continuous metal hydroxide layer reduces oxygen reduction reaction which in-turn reduces corrosion rate. Because when metal dissolution or oxygen reduction rates are equal, the reduction in any of these will result in decrease in other. Here, the rate of corrosion is considered under cathodic control. This is one

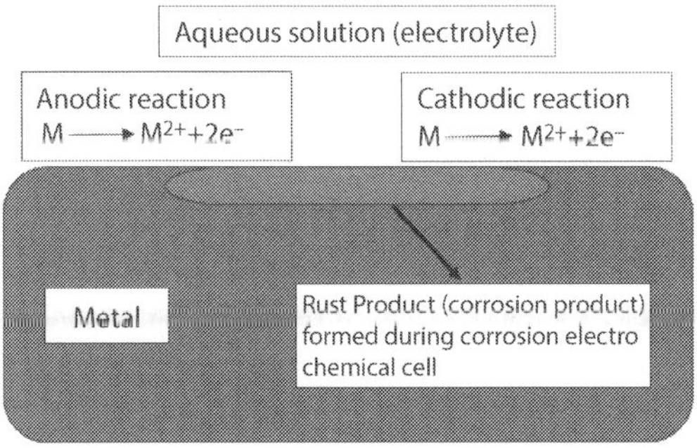

Figure 17.2 Anodic and cathodic reaction.

of the widely adopted methods for controlling corrosion by nature. This approach is commonly used in the design and management of corrosion naturally. In many cases, the corrosion products form a continuous layer of oxides that are similar to crystallographic composition. The generated films through this process prevent conduction of metal ions. The corrosion rate was very slow as the corrosion rate is under anodic control. This is called as passivation and is occurred for materials like aluminium and steel under natural conditions.

17.2.2 Methods of Coatings

Although the variation in pre-treatment and coating development is almost limitless, instances of popular methods used by paint coatings for corrosion control of steel substrates are explained below.

17.2.2.1 Zinc-Rich Coating

Generally, this zinc-rich coating comprises zinc ions and thus, by a sacrificial mechanism or cathodic protection, protects the steel. These layers must comprise 85%–95% w/w of zinc in order to have adequate electrical contact with the metal. Furthermore, the subsequent zinc corrosion components, which shape selectively when exposed to atmosphere, that intends to block pores in the coating which further enhances the coating's resistance characteristics. Furthermore, through this method, mainly two techniques of corrosion protection, sacrificial abided by enhanced resistance characteristics as the zinc is consumed.

17.2.2.2 Inhibitive Coating

Usually utilized as primers, inhibitive coatings are acquired by incorporating pigments which discharge materials that inhibit corrosion. The content of moisture consumed by the coated material helps the pigments to react and passive the steel. The overall efficient inhibitors like strontium chromates, indeed, have hazardous impacts over the environment. Because leaching is the basis for the discharge of inhibitors, coatings are to be extremely pigmented ensure adequate corrosion. Hence, the harmful substances were consistently discharged into atmosphere even if this is not required.

To summarize, paint coatings are important to prevent corrosion of steel materials and the requirement for environmentally friendly and non-toxic formulations to substitute standard paint content. Lately, the uses of polymers-based conducting material have demonstrated this promise

[7, 12–15]. The eco-friendly aspect of conducting polymers together the redox and strength characteristics leaves them excellent coating materials to protect metal alloys against corrosion.

17.3 Conducting Polymer-Based Coatings for Protection Against Corrosion

There are several conducting polymers accessible in commercial form. These generally include polyaniline (PANI), polythiophene (PTh), polypyrrole (PPy), and others like polythiophene, polyindole, polyrhodanine, poly(phenylenediamine), and poly(carbazole) [16–22]. The synthesis of conducting polymers is based on chemical or electrochemical oxidation techniques [23, 24]. Conducting polymers have discovered their applications in the sector of metal and metal alloy corrosion protection since the 1980s. Table 17.1 shows the chemical structure of conducting polymers.

Table 17.1 Chemical structure and conductivity of widely used for metal protection.

Conducting polymers	Chemical structure
Polyaniline (PANI)	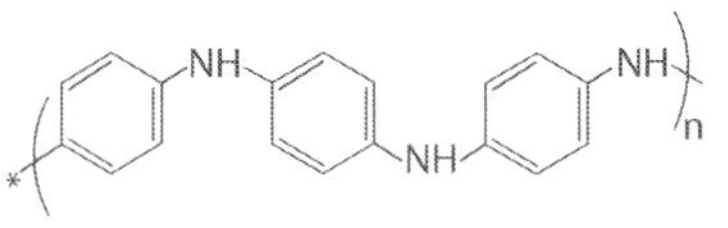
Polypyrrole (PAN)	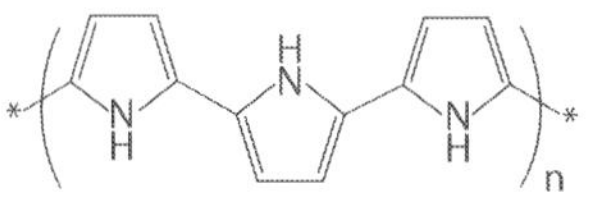
Polythiophene (PTh)	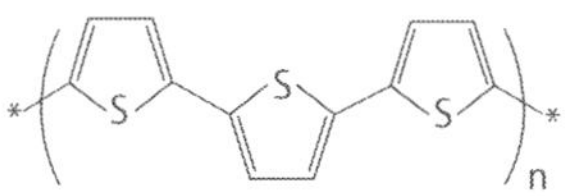
Poly(para-phenylene vinylene) (PPV)	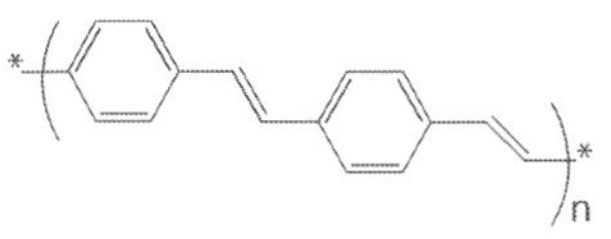
Polycarbazole (PCz)	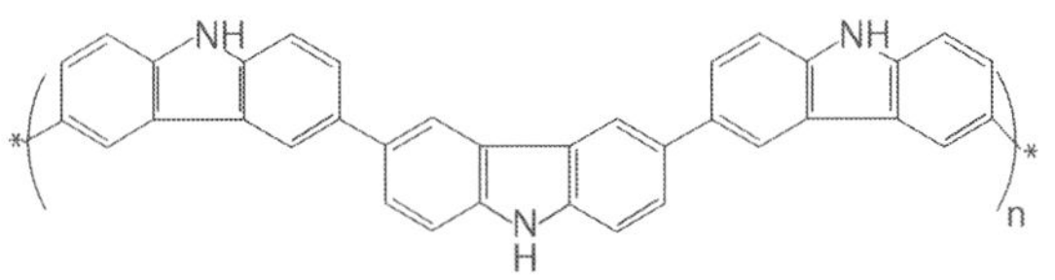

Their convenience of synthesis combined with electrical conductivity via both chemical and electrochemical techniques allows them for utilizing as anti-corrosive materials. Many different interaction processes can be used to perform conducting polymers on a variety of substrates. These processes depend on kind of substrate that is utilized, the use of particular conductive polymers, physical characteristics, its chemical composition, the quantity and type of dopant are using, the quantity of conductive polymer utilized, and its relative percentage to other active ingredients that are used for coating. The preparation of surface also has a significant impact on the conductive polymer functional characteristics. This becomes more essential when, under the layer, there are problems associated with the adhesion and development of potential corrosion products. The local area's exposed environment and pH also have a tremendous impact on these polymers performance as coatings. A conducting polymer can be used either as a primer coating on its own or used as a primary coating together with traditional base coat, or otherwise, the same could be mixed with polymer coatings or utilized as an anti-corrosive material for coating formulation. The above methods are discussed in the previous research under the category of different conducting polymers.

17.3.1 Chemical Oxidative Polymerization Technique

This method is widely used for the preparation of conducting polymer-based coatings as it is most simple way to produce large-scale conducting polymers. In the event of using conducting polymer as primers, drawbacks of chemical polymerization are as follows: the restricted solubility of conducting polymers, and particularly of PANI, in most prevalent organic-based solvents and the low adhesion between conducting polymers and steel substrates. For example, in almost all common solvents, PANI, the widely used conducting polymer in various applications, is generally insoluble, and numerous techniques are often utilized to enhance PANI process ability and solubility [25].

The existence of immobilized dopants in the polymer chain, such as sulfonic groups, improves the solubility of polar solvents [26]. Suitable binders are frequently used to enhance the adhesion of the PANI-based layer to metal substrates. A suitable oxidant initiator substitutes the enforced capacity needed in electrochemical polymerization in the event of chemical oxidative polymerization. Table 17.2 shows a range of such oxidants used with their standard oxidation potential for conducting polymer chemical synthesis, E0.

Table 17.2 Oxidants and their standard oxidation potential, E0, used by chemical polymerization for conducting polymer synthesis [25].

Oxidant	$E^0(V_{SCE})^a$
H_2O_2	1.538
$(NH_4)_2S_2O_8$	1.698
$Ce(SO_4)_2$	1.478
KIO_3	0.848
$FeCl_3$	0.528
$K_2Cr_2O_7$	0.988

Behind chemical oxidative polymerization, many different types of synthetic routes exist, such as catalytic enzyme polymerization, interfacial polymerization, and emulsion polymerization, widely used in the design of conducting polymer morphologies and characteristics need for particular applications. While the number of conducting polymer-based protecting coatings studies are dominated through the preparing of composition with primary components PANI, PTh, PPy, and its derivatives, some other conducting polymers have also been utilized against prevention of metal corrosion, like poly (para-phenylene-vinylene) (PPV) variants through a precursor technique [27], which varies from those of the earlier listed conducting polymers.

17.3.2 Electro-Chemical Oxidative Polymerization Technique

This polymerization process is regarded to be the significant technique of preparing homogeneous films of conducting polymer for desired coating film thickness and also surface morphology adhering to a metal electrode or semiconductive. Over the past few years, this electrochemical polymerization has drawn significant attention in the development of several conducting polymer nanostructures [28, 29].

Electrochemical polymerization offers different benefits relative to chemical polymerization: (1) increased bonding with metal substrate, (2) instantaneous doping with required doping anion, that is often selected as anion of supporting electrolyte in polymerization blend,

(3) fast polymerization at room temperature, (4) regulate of film thickness and possibly morphology and molecular structure, and (5) prevention of solubility complications encountered on metal substrates with several chemically synthesized conducting polymers. The disadvantages of electrochemical polymerization occur where polymerization on oxidizable metals are needed from acidic alternatives (e.g., polyaniline processing), as metal electro-dissolution can happen, and second when electrode position on huge substrates like buildings, pipelines, and boats are needed. Formerly, it is indeed unwieldy and practically inconceivable to electrochemically polymerize. Chemical deposition is preferable in these instances in combination with other methods such as casting and mixing.

Electrochemical polymerization could be deposited, potentiostatically or galvanostatically, using cyclic voltammetry (cyclic potential sweep). Figure 17.3 shows a typical 3-electrode cell used in electrochemical polymerization. The three electrodes, namely, reference electrode (RE), working electrode (WE), and counter electrodes (CE), are properly associated with a potentiostat/galvanostat processes reported by Bard & Faulkner *et al.* [30]. It is also possible to use an electrode cell of two numbers compatible with work and counter electrodes.

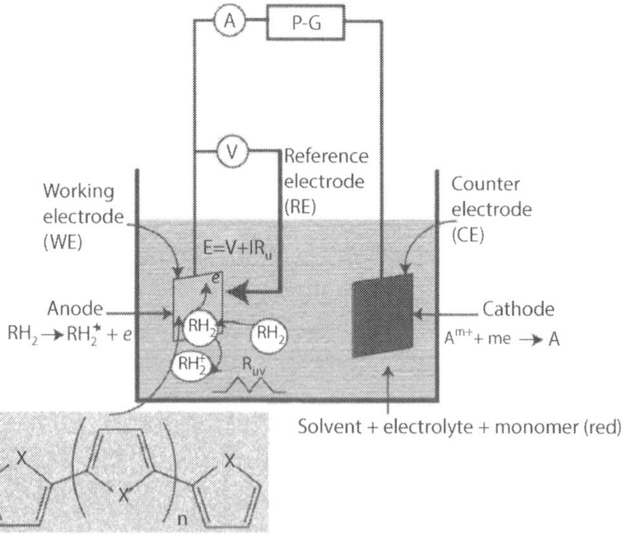

Figure 17.3 Typical 3-electrode cell used for electrochemical polymerization [32].

17.4 Synthesis of Conducting Polymer Commonly Used in Protection Against Corrosion

17.4.1 Synthesis of Conducting Polymer: PANI

PANI is indeed a conductive polymer belonging to the polymer class of semi-flexible rods. It has a p-type coupled spine group which allows electronic transmission while the polymer is partly corroded and protonated to the emerald state. The PANI most stable state is the emerald state. When PANI is completely lowered or oxidized, both pernigraniline and leucoemeraldine are similarly stable [31]. Manufacturing of printed circuit boards, electrostatic discharge coatings and also corrosion prevention are the major fields of application [32]. In PANI synthesis, two techniques are commonly used: chemical and electrochemical monomer polymerization in a suitable electrolyte. Chemical synthesis technique has to do with monomer oxidation by utilizing acid or alkaline intermediate oxidizing agent. Ammonium peroxydisulphate [33], ammonium persulphate [34], $(NH_4)_2S_2O_8$ in HCl [35], H_2SO_4 solution [36], etc., are frequently used oxidizing substances. The acidic medium's pH level is often held between 0 and 2 while the alkaline medium pH is maintained as 4 [31]. Moreover, the use of chemical techniques loses patronage, because the technique is dreary and the synthesized substance can be contaminated.

The electrochemical method includes monomer oxidative polymerization of aqueous solution like $LiClO_4$ containing acetonitrile [37], salicylate solution [38], tosylic acid solution [39], H_2SO_4 solution [40], and oxalic acid [41], etc. The technique of electrochemical synthesis has the benefit of being a one-step technique that effectively controls the polymer chemical and physical characteristics. Also, it is not necessary to extract the product produced from this method from the extraction scheme (solvent mixture, oxidant, monomer), thus avoiding contamination. Electropolymerization on distinct kinds of steel and other metals in which passivation easily precedes the polymerization method was carried out from distinct organic and inorganic acid environments arising in disciple PANI coatings with adequate protecting characteristics in corrosion environments [42, 43].

PANI is used in practice for reinforcement of concrete steel bars. Saravanan *et al.* [44] investigated the epoxy primary for steel bars comprising PANI. Based on the existence of PANI in the layer, it was discovered in this study that the base metal repassivates. Epoxy PANI coating shows a greater incidence of chloride, which in the steel bar is one of the main symptoms of concrete corrosion. This was assessed by augmented investigations of corrosion like anodic polarization, salt spray testing, and cathodic disbondment experiment. Studies of chemical resistance of all these coatings proved stronger strength to alkali, providing much greater protection against concrete alkaline environment. Time was conducted to crack the analysis of coated and uncoated concrete bars. In the case of uncoated bars, maximum current was achieved within 4 days, whereas for those coated with epoxy PANI primers, it took almost 50 days. Researchers conclude that repassivation triggered by PANI provided this extra protection.

Different dopants could be used for enhancing conducting polymers conductivity property and corrosion performance. Poly (methylmethacrylate-co-acrylic acid)-doped with PANI coating films have been utilized on AA3104-H19 aluminum compound through immersing the pre-treated surfaces into the polymers saturated ethyl acetate solution [45]. While these primers were non-uniform, porous and allowing electrolyte permeation, it was discovered in this research to provide stronger corrosion performance relative to a viable epoxy fabric (ICI-640C692). The sort of doping material and base coat (top coat) that is utilized has a powerful effect over corrosion protection [46].

Together with PANI, Kalendova *et al.* [47] investigated the effect of inorganic solvents such as zinc phosphate, calcium borate, and zinc powder to protect metals from corrosion. Figure 17.4 shows SEM images of PANI with different magnification.

They discovered a synergistic impact that was more effective than a simple defense mechanism in protecting metal. Loading of PANI at 2.5% w/w in water (H_2O) propagate alkydes in nano particulate shape led in enhanced corrosion productivity for metal substrates of these water-based alkyd materials [48]. PANI powder has been chemically produced and used as a pigment within a strength range of 10%–15% to

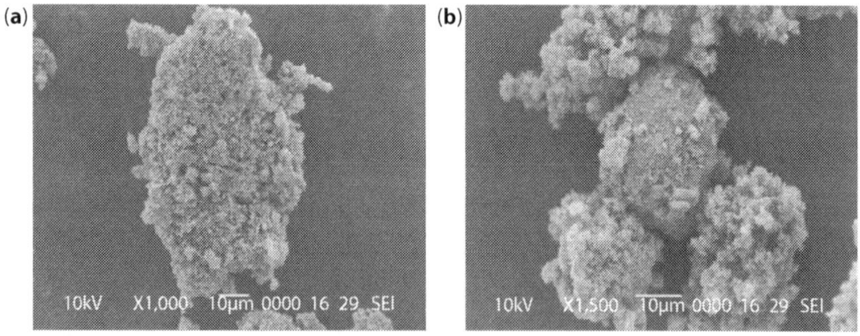

Figure 17.4 SEM images of PANI particles at different magnifications (a) 1,000 and (b) 1,500 [47].

improve the corrosion resistance of mild steel by 10–15 times in 3.5% NaCl [49]. It has been shown that duplex coating consisting of cerium alteration coating layer that act as compliance promoter and PANI utilized on aluminum component AA6063 exhibits efficient protection capability [31].

There are many difficulties with the direct application of conductive polymers coatings. Due to the insolubility and non-fusibility of polymers, their application on active metals becomes difficult. Several techniques were attempted to overcome problems of solubility, which includes chemical modifications and well-designed functional group attachment, resulting in solubility in a given medium. To increase overall conductivity, dopant anions were introduced. Several scientists have attempted to functionalize to render medium solubility in the last century. Other than solubility, surface functionalization also contributes to several characteristics being changed. Adding sulfonate groups to PANI polymer backbone changed the polymer's ion exchange characteristics, resulting in an impact of self-doping [50, 51]. Infrequently used solvents, an alkoxy and alkyl modification, were used commonly to perform polymer backbone, resulting in increased solubility. Consequently, several types of PANI and their changes can be used to protect metals and their alloys from corrosion [52].

17.4.2 Synthesis of Conducting Polymer: PPy

In the recent period, PPy has published several articles on the protection of metals from corrosion. PPy was incorporated in 1% w/w to epoxy polyamide coatings. Compared to the control sample, an increase in corrosion protection has been discovered, but greater proportions have not increased efficiency. It was noted that the effectiveness of conducting polymers relies on the application method and the circumstances under which the studies were conducted. Conducting polymers on their backbone can be synthesized with distinct anions. Dopant anion is added with conjugated polymer after oxidation of conducting polymers. When this polymer undergoes reduction, this dopant anion is released. These anions could be released as smart coatings in the coatings on demand. Dopant ions utilized in PPy synthesis have significant impact on the resulting PPy characteristics.

Kowalski *et al.* [53] investigated the bi-layered conducting polymer PPy coating on carbon steel. In this scheme, molybdophosphate and phosphate ions were doped into the inner layer, while naphthalenedisulfonate (NDS) particles were doped into the exterior coating. Molybdophosphate ions embedded into the inner layer hold passive oxide films over steel and the exterior layer doping by NDS big organic ions restrict the disintegration and discharge of molybdophosphate ions in interior layer. All these lead to an overall enhancement in the protection against corrosion. Conducting polymers electrochemical synthesis is a commonly used method for the deposition and synthesis of conducting polymers on the appropriate substrates. The corrosion behavior of electrochemically synthesized PPy and a poly(5-amino-1-naphthol) topcoat on PPy layer was investigated by Bereket *et al.* [54] using cyclic voltammetry. It was assessed that an increase in the performance of corrosion protection was accomplished as more layers of this blend were deposited.

Several scientists have made attempts for synthesizing electropolymerized PPy apart from aluminum and steel. Herrasti *et al.* [55] investigated about the PPy over copper surface and further studied. As a result, it was found that reduced porosity and consistent surface characteristics were present in the deposited layer. This was the result of conducting polymer film's constant growth rate. These deposits of PPy, formed on copper, proved great resistance to corrosion when in a mixture of 3% sodium chloride solution. It is also proved that for excellent barrier layer features and excellent redox properties, a greater monomer level of about 0.3 M should be used. Tuken *et al.* [56] attempted to electropolymerize PPy over brass and copper surfaces. They assessed superior adhesion of these PPy films to the oxidized copper, leading in stronger protection against corrosion relative to brass. They suggested a barrier effect protection mechanism exhibited by these coatings.

Hosseini *et al.* investigated the PPy, organophilic montmorillonite (MMT), along with epoxy nanocomposites and studied about the corrosion performance over aluminum surface. Epoxy/PPy/MMT nanocomposite was discovered to exhibit superior performance in corrosion referred to epoxy resin, epoxy resin along with PPy, and epoxy resin along with MMT coating. In attempt to develop increased performance, the characteristics of distinct materials can be mixed together [57]. In another researcher, in the existence of dihydroxybenzene, Tallman *et al.* [58] studied about the synthesis of nanocomposite through pyrrole polymerization over alumina nanoparticles. For coating, they used this material as filler in a ratio of less than 2%. Dihydroxybenzene was discovered to improve the adherence of coating to the surface of the aluminium. The research considerably shows the modifications caused by electrochemical impedance spectroscopy (EIS) in the shape of geometry and also volume proportion of voids occurred in the coating. The voids provide a direct effect on corrosive solution aging and penetration into the layer.

17.4.3 Synthesis of Conducting Polymer: PTh

There were few articles published on PTh to protect metal substrates from corrosion. A range of replaced PTh polymers can be synthesized. Compared to other conductive polymer derivatives, some of PTh derivatives perform very well. This eventually depends on the environment nature that is exposed to these conducting polymers. The combination of these two led to improved PTh and also its derivative could be produced by other conductive polymer substrates such as PPy with the proper application of voltage. The combination of these two led to improved performance in corrosion.

Kousik *et al.* [59] used acetonitrile as a medium to produce PTh coatings on mild steel (MS) layer. It was determined by AC impedance analyze and resulted that PTh film coated in MS was shielded by a process of triggered passivation by PTh redox operation. Studies of water absorption and the delamination region also verified electropolymerized PTh protective effect on mild steel layer. Ocampo *et al.* contrasted many marine anticorrosive paints with PTh derivative being added. It was discovered that adding a low poly (3-decylthiophene-2,5-diyl) content (0.2% w/w) improved the productivity of corrosion resistance of epoxy-based plastics.

The role of distinct layer treatment over the substrate was explored by Rammelt *et al.* [60]. On the surface of MS, homogenous and well adherence polymethylthiophene layer were developed subsequently treatment with 2(3-thienyl) ethylphosphono acids. The polymethylthiophene layer films were ultrathin and extremely organized (thickness was about 1 m). By using ultrathin films, they showed stronger protection with energy costs. These films have been discovered to efficiently distinguish oxygen decrease and iron dissolution electrochemical processes in the surface area.

17.5 Performance Improvement and Bulk Modifications of Conducting Polymers

The difficulty of insolubility about conducting polymers in sedimentary condition, the complexity allied with preparation, and low or medium anti-corrosive efficiency insisted some structural modifications and also synthesis characters. The several methods that improve the synergistic effect include doping, copolymerization, layering, and its combinations.

17.5.1 Doping

The process of doping which enhances the corrosion of conductive polymers arised from the intention of exchange between the anions like sulphates, chlorides, nitrates, etc., with polymers in the presence of electrolytic solution. Also, it is assumed that the failure of coatings arises due to the diffusion of corrosive species over the coating later.

17.5.2 Layering

Layering implies the formation of various films of polymer over the metal surface. The option of having a multilayer [54], trilayer [61], and bilayer [62] coatings of polymers over metals has been carried out by many researchers. The advantage of using this method has also been discussed by industrial ion interchange resins [63] and ions membranes [64]. In the layering method, the interior layer acts as anodic protecting where the exterior layer acts as barrier. Bilayers of conducting polymer polypyrrole with poly(5-amino-1-naphthol) [54], polyphenol [65], organosilanes [66], and cataphoretic [67]; polyaniline/polybithiophene [68], polyaniline/poly(N-methylaniline) [69], and polyrhodanine/isobutyltriethoxysilne [54] were discussed that they exhibit superior adhesion and effect of anti-corrosion over metal substrates when compared with the mono-layers.

17.5.3 Copolymerization

Based on the literature studies, it is evidenced that copolymerization elevates the properties of polymers [70]. Copolymerization is generally defined as a combination of two or more monomers when combined with each other for producing a new compound with different properties. It is generally an end product that connects the benefits of the monomers excluding the disadvantages of it. This copolymerization may overcome the drawbacks of poor solubility, porosity, and bonding property of conductive polymers and enhance the corrosion characteristics [71]. Srikanth *et al.* [72] identified that on copolymerizing aniline with N-methyl aniline, the resulted copolymer that protected the surface of carbon steel in hydrochloric acid medium with 81.2% as compared with 18.8% for polyaniline and similarly 56.3% for poly (N-methyl aniline). Another important parameter regarding copolymerization is about the modification of copolymerization characteristics as per the requirement of altering the ratio of both the monomers that is going to be adopted [73].

17.6 Conducting Copolymer Composites and Nanocomposites

Many investigations have indicated the capacity of utilizing copolymers as anticorrosive coatings. Generally, composites are the combinations of diverse materials with different functional properties for corrosion behavior of metals and its alloys. Various metals, metal oxide particles graphene, and carbon-based nanotubes particles can be integrated with PANI matrix for production of PANI based nanocomposite materials [74]. To evidence the improvement in protection capacity of PANI composites over a MS substrate, coating over PANI with zinc and its nanoparticles were made of iron by adopting solvent mixing technique [75]. The existence of Zn nanoparticles in PANI showed improved anticorrosion and conductive property when compared with coating of micro sized particles of Zn on PANI matrix. Also, the impact of filler materials has improved the properties of the base materials.

Another study indicates that on implementing chemical oxidative polymerization for production of PANI-Zn nanocomposites. The prepared nanocomposite was then adopted on iron substrate using solution casting method. It is also evidenced that 5% w/w loading maximum conductivity was achieved. Also, it was clear that the amount of zinc decides the conductivity property of PANI-Zn nanocomposite. The pores in the CPs are filled by corrosion products of Zn which results as barrier protecting medium along with electrochemical property of CPs [76].

PANI composites with carbon-based nanotubes were utilized as pigments on low carbon steel for superior anti-corrosion property [77]. It is also evidenced that this PANI-CNT-based paints drastically reduce corrosion property of steel in 3.5% sodium chloride solution evidenced through Tael polarization and Electrochemical Impedance Spectroscopy studies. Also, the PPy and ZnO nanocomposite coatings have depicted enhanced corrosion resistance property when compared with PPy coating. This ZnO utilized was in the form of nanorods which was clearly evidenced in studying morology which generally contributed to the corrosion properties of composites [78].

In some other investigation, PANI clay nanocomposite were fabricated using a chemical oxidative polymerization method [79]. The same was then added with zinc-rich ethyl silicate resin. Then, Electrochemical Impedance Spectroscopy measurements were done in 3.5% sodium chloride solution was noted for evaluation of corrosion. The protection barrier and also passivation of metal substrate owing to the existence of nanocomposite indicated improved corrosion resistance by coating PANI-clay nanocomposite when compared with pristine coat.

The conductive polymer that contains composite coating is newer materials that were developed for resolving the processability characteristics related to conducting polymer. PPy with aluminium flake composite have identified for protecting aluminium alloy from corrosion [80, 81]. This was identified that PPy wire forged in the surface of Al substrate resulting in formation of PPy/Al composite that provides corrosion protection to the substrate. Also, Phosphate and nitrate doped composite were applied on surface of the material which protected the substrate from corrosion. Also, redox property of PPy in combination with barrier properties of MIOX indicated superior corrosion protection [82].

17.7 Summary of Conducting Polymers-Based Protective Coatings

Different approaches that are adopted for development of conducting for protective coatings were summarized in Figure 17.5. This sorting helps for choosing conducting polymer formulation which depends on many factors which include period of protection requirement, structure, composition, and pH value of the corrosive environment. The latest studies reveal about the corrosive protection conducting polymer-based coatings mainly focuses on organic and inorganic composite coatings. It was also found that these protective coating offers long term superior protection to metal

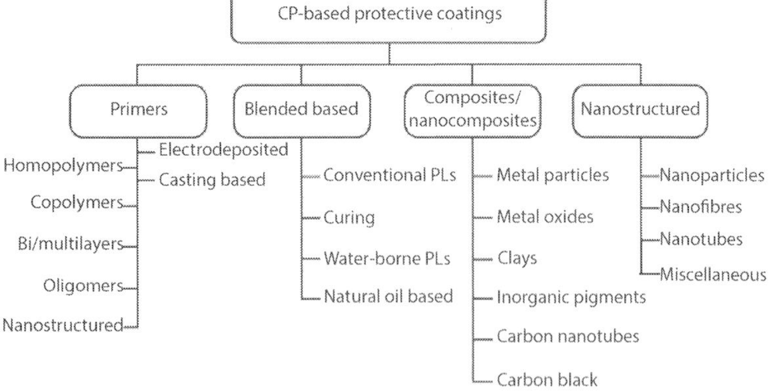

Figure 17.5 Different forms of conductive polymer-based protective coatings.

surface and also self-healing property by organic coating in concurrence with the reduced effect evidenced for inorganic pigments.

17.8 Conclusions

Various conductive polymer coatings have been utilized for protecting the corrosive medium. Different coating methods, *viz.*, zinc rich coatings, inhibitive coatings, conducting polymer-based coatings, chemical oxidative polymerization, and electrochemical polymerization methods that were used for corrosion resistance protection have been extensively discussed. Also, methods of synthesizing conducting polymers for corrosion protection were demonstrated. The performance improvement studies and also bulk modifications of the polymers which improve the effect that includes doping, copolymerization, and layering for corrosion improvement have been studied. Also, various forms of conducting polymers that were utilized for protection of metals and its alloys from corrosion have been studied. These conducting polymers are used for multilayer coating, pigments, and composites. Also, the corrosion protecting dopants can also be included over the backbone of conducting polymer for protection in particular conditions. The morphological properties influence the anticorrosion properties of such conducting polymers. Improved protective coatings have been developed for better performances of the substrate have been investigated. Enormous studies regarding the protective coating of conductive polymer base materials have been done for enhancement of anticorrosion behavioral properties.

References

1. Deshpande, P.P., Jadhav, N.G., Gelling, V.J., Sazou, D., Conducting polymers for corrosion protection: A review. *J. Coat. Technol. Res.*, 11, 4, 473–494, 2014.
2. Ates, M., A review on conducting polymer coatings for corrosion protection. *J. Adhes. Sci. Technol.*, 30, 14, 1510–1536, 2016.
3. Yao, B., Wang, G., Ye, J., Li, X., Corrosion inhibition of carbon steel by polyaniline nanofibers. *Mater. Lett.*, 62, 12-13, 1775–1778, 2008.
4. Herrasti, P., Recio, F., Ocon, P., Fatás, E., Effect of the polymer layers and bilayers on the corrosion behaviour of mild steel: Comparison with polymers containing Zn microparticles. *Prog. Org. Coat.*, 54, 4, 285–291, 2005.
5. Kinlen, P.J., Menon, V., Ding, Y., A mechanistic investigation of polyaniline corrosion protection using the scanning reference electrode technique. *J. Electrochem. Soc.*, 146, 10, 3690–3695, 1999.

6. Kinlen, P., Ding, Y., Silverman, D., Corrosion protection of mild steel using sulfonic and phosphonic acid-doped polyanilines. *Corrosion*, 58, 6, 490–497, 2002.
7. Rohwerder, M., Conducting polymers for corrosion protection: A review. *Int. J. Mater. Res.*, 100, 10, 1331–1342, 2009.
8. Eftekhari, A., *Nanostructured conductive polymers*, John Wiley & Sons: United Kingdom, 2011.
9. Bhandari, H., Kumar, S.A., Dhawan, S., Conducting polymer nanocomposites for anticorrosive and antistatic applications, in: *Nanocomposites: New Trends and Developments*, pp. 73–96, 2012.
10. Davis, J., *Corrosion: Understanding the Basics*, ASM International, Materials Park, OH, 2000, There is no corresponding record for this reference. Google Scholar].
11. Makhlouf, A.S.H., *Handbook of smart coatings for materials protection*, Elsevier: United Kingdom, 2014.
12. Ahmad, N. and MacDiarmid, A.G., Inhibition of corrosion of steels with the exploitation of conducting polymers. *Synth. Met.*, 78, 2, 103–110, 1996.
13. Sitaram, S.P., Stoffer, J.O., O'Keefe, T.J., Application of conducting polymers in corrosion protection. *J. Coat. Technol.*, 69, 866, 65–69, 1997.
14. Spinks, G.M., Dominis, A.J., Wallace, G.G., Tallman, D.E., Electroactive conducting polymers for corrosion control. *J. Solid State Electrochem.*, 6, 2, 85–100, 2002.
15. Zarras, P., Stenger-Smith, J.D., Wei, Y., *Electroactive polymers for corrosion control*, American Chemical Society: Washington DC, 2003.
16. Düdükcü, M., Udum, Y.A., Ergün, Y., Köleli, F., Electrodeposition of poly (4-methyl carbazole-3-carboxylic acid) on steel surfaces and corrosion protection of steel. *J. Appl. Polym. Sci.*, 111, 3, 1496–1500, 2009.
17. Frau, A.F., Pernites, R.B., Advincula, R.C., A conjugated polymer network approach to anticorrosion coatings: Poly (vinylcarbazole) electrodeposition. *Ind. Eng. Chem. Res.*, 49, 20, 9789–9797, 2010.
18. Duran, B., Çakmakcı, İ., Bereket, G., Role of supporting electrolyte on the corrosion performance of poly (carbazole) films deposited on stainless steel. *Corros. Sci.*, 77, 194–201, 2013.
19. Tüken, T., Yazıcı, B., Erbil, M., Electrochemical synthesis of polythiophene on nickel coated mild steel and corrosion performance. *Appl. Surf. Sci.*, 239, 3–4, 398–409, 2005.
20. Düdükcü, M. and Avcı, G., Electrochemical synthesis and corrosion inhibition performance of poly-5-aminoindole on stainless steel. *Prog. Org. Coat.*, 97, 110–114, 2016.
21. Altunbaş, E., Solmaz, R., Kardaş, G., Corrosion behaviour of polyrhodanine coated copper electrode in 0.1 M H2SO4 solution. *Mater. Chem. Phys.*, 121, 1–2, 354–358, 2010.
22. Shanmugham, C. and Rajendran, N., Corrosion resistance of poly p-phenylenediamine conducting polymer coated 316L SS bipolar plates for proton exchange membrane fuel cells. *Prog. Org. Coat.*, 89, 42–49, 2015.

23. Chandrasekhar, P., *Conducting Polymers*, Fundamentals and Applications, Springer US, 1999.
24. Inzelt, G., *Conducting polymers: A new era in electrochemistry*, Springer Science & Business Media: New York, 2012.
25. Wallace, G.G., Teasdale, P.R., Spinks, G.M., Kane-Maguire, L.A., *Conductive electroactive polymers: Intelligent materials systems*, CRC Press: Boca Raton, 2002.
26. Freund, M.S. and Deore, B.A., *Self-doped conducting polymers*, John Wiley & Sons: United Kingdom, 2007.
27. Wessling, R., The polymerization of xylylene bisdialkyl sulfonium salts. *J. Polym. Sci.: Polym. Symp., Wiley Online Library*, 72, 55–66, 1985.
28. Cho, S.I. and Lee, S.B., Fast electrochemistry of conductive polymer nanotubes: Synthesis, mechanism, and application. *Acc. Chem. Res.*, 41, 6, 699–707, 2008.
29. Guo, Y. and Zhou, Y., Polyaniline nanofibers fabricated by electrochemical polymerization: A mechanistic study. *Eur. Polym. J.*, 43, 6, 2292–2297, 2007.
30. Bernard, M.C., Joiret, S., Hugot-Le Goff, A., Phong, P.V., Protection of iron against corrosion using a polyaniline layer: I. Polyaniline electrodeposit. *J. Electrochem. Soc.*, 148, 1, B12–B16, 2001.
31. Johansen, H.D., Brett, C.M., Motheo, A.J., Corrosion protection of aluminium alloy by cerium conversion and conducting polymer duplex coatings. *Corros. Sci.*, 63, 342–350, 2012.
32. Williams, G. and McMurray, H.N., Polyaniline inhibition of filiform corrosion on organic coated AA2024-T3. *Electrochim. Acta*, 54, 17, 4245–4252, 2009.
33. Abd El-Ghaffar, M., Abdelwahab, N., Youssef, E., Polyaniline and poly (m-toluidine) prepared by inverse emulsion pathway as corrosion inhibitors in surface coatings. *Pigm. Resin Technol.*, 39, 4, 228–236, 2010.
34. Pan, T. and Yu, Q., Comprehensive evaluation of anti-corrosion capacity of electroactive polyaniline for steels. *Anti-Corros. Methods Mater.*, 63, 5, 360–368, 2016.
35. Baldissera, A. and Ferreira, C.A., Coatings based on electronic conducting polymers for corrosion protection of metals. *Prog. Org. Coat.*, 75, 3, 241–247, 2012.
36. Nam, N., Kim, J., Lee, Y., Son, Y., Effect of thermal treatment on the corrosion resistance of polyaniline in H2SO4–HF acid mixture solution. *Corros. Sci.*, 51, 12, 3007–3013, 2009.
37. Tüken, T., Özyılmaz, A., Yazıcı, B., Erbil, M., Electrochemical synthesis of polyaniline on mild steel in acetonitrile–LiClO4 and corrosion performance. *Appl. Surf. Sci.*, 236, 1–4, 292–305, 2004.
38. Chaudhari, S. and Patil, P., Inhibition of nickel coated mild steel corrosion by electrosynthesized polyaniline coatings. *Electrochim. Acta*, 56, 8, 3049–3059, 2011.

39. Conroy, K.G. and Breslin, C.B., The electrochemical deposition of polyaniline at pure aluminium: Electrochemical activity and corrosion protection properties. *Electrochim. Acta*, 48, 6, 721–732, 2003.
40. Fang, J., Xu, K., Zhu, L., Zhou, Z., Tang, H., A study on mechanism of corrosion protection of polyaniline coating and its failure. *Corros. Sci.*, 49, 11, 4232–4242, 2007.
41. Özyilmaz, A., Erbil, M., Yazici, B., Investigation of corrosion behaviour of stainless steel coated with polyaniline via electrochemical impedance spectroscopy. *Prog. Org. Coat.*, 51, 1, 47–54, 2004.
42. Yağan, A., Pekmez, N.Ö., Yıldız, A., Poly (N-ethylaniline) coatings on 304 stainless steel for corrosion protection in aqueous HCl and NaCl solutions. *Electrochim. Acta*, 53, 5, 2474–2482, 2008.
43. Sazou, D., Kourouzidou, M., Pavlidou, E., Potentiodynamic and potentiostatic deposition of polyaniline on stainless steel: Electrochemical and structural studies for a potential application to corrosion control. *Electrochim. Acta*, 52, 13, 4385–4397, 2007.
44. Saravanan, K., Sathiyanarayanan, S., Muralidharan, S., Azim, S.S., Venkatachari, G., Performance evaluation of polyaniline pigmented epoxy coating for corrosion protection of steel in concrete environment. *Prog. Org. Coat.*, 59, 2, 160–167, 2007.
45. Oliveira, M., Moraes, J., Faez, R., Impedance studies of poly (methylmethacrylate-co-acrylic acid) doped polyaniline films on aluminum alloy. *Prog. Org. Coat.*, 65, 3, 348–356, 2009.
46. Dominis, A.J., Spinks, G.M., Wallace, G.G., Comparison of polyaniline primers prepared with different dopants for corrosion protection of steel. *Prog. Org. Coat.*, 48, 1, 43–49, 2003.
47. Kalendová, A., Veselý, D., Stejskal, J., Organic coatings containing polyaniline and inorganic pigments as corrosion inhibitors. *Prog. Org. Coat.*, 62, 1, 105–116, 2008.
48. Sakhri, A., Perrin, F., Aragon, E., Lamouric, S., Benaboura, A., Chlorinated rubber paints for corrosion prevention of mild steel: A comparison between zinc phosphate and polyaniline pigments. *Corros. Sci.*, 52, 3, 901–909, 2010.
49. Rout, T., Jha, G., Singh, A., Bandyopadhyay, N., Mohanty, O., Development of conducting polyaniline coating: A novel approach to superior corrosion resistance. *Surf. Coat. Technol.*, 167, 1, 16–24, 2003.
50. Shimizu, S., Saitoh, T., Uzawa, M., Yuasa, M., Yano, K., Maruyama, T., Watanabe, K., Synthesis and applications of sulfonated polyaniline. *Synth. Met.*, 85, 1–3, 1337–1338, 1997.
51. Şahin, Y., Pekmez, K., Yıldız, A., Electrochemical preparation of soluble sulfonated polymers and aniline copolymers of aniline sulfonic acids in dimethylsulfoxide. *J. Appl. Polym. Sci.*, 90, 8, 2163–2169, 2003.
52. Santos, J., Jr., Mattoso, L., Motheo, A., Investigation of corrosion protection of steel by polyaniline films. *Electrochim. Acta*, 43, 3–4, 309–313, 1998.

53. Kowalski, D., Ueda, M., Ohtsuka, T., Corrosion protection of steel by bi-layered polypyrrole doped with molybdophosphate and naphthalenedisulfonate anions. *Corros. Sci.*, 49, 3, 1635–1644, 2007.
54. Bereket, G. and Hür, E., The corrosion protection of mild steel by single layered polypyrrole and multilayered polypyrrole/poly (5-amino-1-naphthol) coatings. *Prog. Org. Coat.*, 65, 1, 116–124, 2009.
55. Herrasti, P., Del Rio, A., Recio, J., Electrodeposition of homogeneous and adherent polypyrrole on copper for corrosion protection. *Electrochim. Acta*, 52, 23, 6496–6501, 2007.
56. Tüken, T., Yazıcı, B., Erbil, M., The electrochemical synthesis and corrosion performance of polypyrrole on brass and copper. *Prog. Org. Coat.*, 51, 2, 152–160, 2004.
57. Hosseini, M., Raghibi-Boroujeni, M., Ahadzadeh, I., Najjar, R., Dorraji, M.S., Effect of polypyrrole–montmorillonite nanocomposites powder addition on corrosion performance of epoxy coatings on Al 5000. *Prog. Org. Coat.*, 66, 3, 321–327, 2009.
58. Tallman, D.E., Levine, K.L., Siripirom, C., Gelling, V.G., Bierwagen, G.P., Croll, S.G., Nanocomposite of polypyrrole and alumina nanoparticles as a coating filler for the corrosion protection of aluminium alloy 2024-T3. *Appl. Surf. Sci.*, 254, 17, 5452–5459, 2008.
59. Kousik, G., Pitchumani, S., Renganathan, N., Electrochemical characterization of polythiophene-coated steel. *Prog. Org. Coat.*, 43, 4, 286–291, 2001.
60. Rammelt, U., Nguyen, P., Plieth, W., Protection of mild steel by modification with thin films of polymethylthiophene. *Electrochim. Acta*, 46, 26–27, 4251–4257, 2001.
61. Aradilla, D., Ocampo, C., Armelin, E., Alemán, C., Oliver, R., Estrany, F., Application of multilayered particles formed by poly (3, 4-ethylenedioxythiophene) and poly (N-methylpyrrole) as anti-corrosive additives of conventional organic coatings. *Mater. Corros.*, 58, 11, 867–872, 2007.
62. Liang, C., Chen, W., Huang, N., Electrochemical characterization for polybithiophene film on porous oxide layer of 304 stainless steel formed by square wave passivation. *Mater. Corros.*, 63, 2, 155–160, 2012.
63. Orata, D. and Kariuki, B., Charge-state trapping at a conducting polymer-redox ion-exchanger interface—A bilayer electrode. *React. Funct. Polym.*, 28, 3, 287–295, 1996.
64. Zhang, H., Salinas, S., Sundaresan, V.B., Conducting polymer supported bilayer lipid membrane reconstituted with alamethicin. *Smart Mater. Struct.*, 20, 9, 094020, 2011.
65. Tüken, T., Arslan, G., Yazıcı, B., Erbil, M., The corrosion protection of mild steel by polypyrrole/polyphenol multilayer coating. *Corros. Sci.*, 46, 11, 2743–2754, 2004.
66. Correa-Borroel, A., Gutierrez, S., Arce, E., Cabrera-Sierra, R., Herrasti, P., Organosilanes and polypyrrole as anticorrosive treatment of aluminium 2024. *J. Appl. Electrochem.*, 39, 12, 2385–2395, 2009.

67. Ferreira, C., Aeiyach, S., Coulaud, A., Lacaze, P., Appraisal of the polypyrrole/cataphoretic paint bilayer system as a protective coating for metals. *J. Appl. Electrochem.*, 29, 2, 259–263, 1999.
68. Pekmez, N.Ö., Abacı, E., Cınkıllı, K., Yağan, A., Polybithiophene and its bilayers with polyaniline coatings on stainless steel by electropolymerization in aqueous medium. *Prog. Org. Coat.*, 65, 4, 462–468, 2009.
69. Narayanasamy, B. and Rajendran, S., Electropolymerized bilayer coatings of polyaniline and poly (N-methylaniline) on mild steel and their corrosion protection performance. *Prog. Org. Coat.*, 67, 3, 246–254, 2010.
70. Francois, E., Dorcemus, D., Nukavarapu, S., Biomaterials and, scaffolds for musculoskeletal tissue engineering, in: *Regenerative Engineering of Musculoskeletal Tissues and Interfaces*, pp. 3–23, Elsevier: United Kingdom, 2015.
71. Lebrini, M., Bentiss, F., Chihib, N.-E., Jama, C., Hornez, J.P., Lagrenée, M., Polyphosphate derivatives of guanidine and urea copolymer: Inhibiting corrosion effect of armco iron in acid solution and antibacterial activity. *Corros. Sci.*, 50, 10, 2914–2918, 2008.
72. Srikanth, A., Raman, V., Tamilselvi, S., Nanjundan, S., Rajendran, N., Electropolymerization and corrosion protection of polyaniline and its copolymer on carbon steel. *Anti-Corros. Methods Mater.*, 55, 1, 3–9, 2008.
73. Gopi, D., Karthikeyan, P., Kavitha, L., Surendiran, M., Development of poly (3, 4-ethylenedioxythiophene-co-indole-5-carboxylic acid) co-polymer coatings on passivated low-nickel stainless steel for enhanced corrosion resistance in the sulphuric acid medium. *Appl. Surf. Sci.*, 357, 122–130, 2015.
74. Ćirić-Marjanović, G., Recent advances in polyaniline composites with metals, metalloids and nonmetals. *Synth. Met.*, 170, 31–56, 2013.
75. Olad, A., Barati, M., Shirmohammadi, H., Conductivity and anticorrosion performance of polyaniline/zinc composites: Investigation of zinc particle size and distribution effect. *Prog. Org. Coat.*, 72, 4, 599–604, 2011.
76. Olad, A. and Rasouli, H., Enhanced corrosion protective coating based on conducting polyaniline/zinc nanocomposite. *J. Appl. Polym. Sci.*, 115, 4, 2221–2227, 2010.
77. Deshpande, P.P., Vathare, S.S., Vagge, S.T., Tomšík, E., Stejskal, J., Conducting polyaniline/multi-wall carbon nanotubes composite paints on low carbon steel for corrosion protection: Electrochemical investigations. *Chem. Pap.*, 67, 8, 1072–1078, 2013.
78. Hosseini, M., Bagheri, R., Najjar, R., Electropolymerization of polypyrrole and polypyrrole-ZnO nanocomposites on mild steel and its corrosion protection performance. *J. Appl. Polym. Sci.*, 121, 6, 3159–3166, 2011.
79. Akbarinezhad, E., Ebrahimi, M., Sharif, F., Attar, M., Faridi, H., Synthesis and evaluating corrosion protection effects of emeraldine base PAni/clay nanocomposite as a barrier pigment in zinc-rich ethyl silicate primer. *Prog. Org. Coat.*, 70, 1, 39–44, 2011.

80. Yan, M., Vetter, C.A., Gelling, V.J., Electrochemical investigations of polypyrrole aluminum flake coupling. *Electrochim. Acta*, 55, 20, 5576–5583, 2010.
81. Yan, M., Vetter, C.A., Gelling, V.J., Corrosion inhibition performance of polypyrrole Al flake composite coatings for Al alloys. *Corros. Sci.*, 70, 37–45, 2013.
82. Jadhav, N., Vetter, C.A., Gelling, V.J., Characterization and electrochemical investigations of polypyrrole/aluminum flake composite pigments on AA 2024-T3 substrate. *ECS Trans.*, 41, 15, 75–89, 2012.

18

Polymer Coating for Industrial Applications

Moganapriya Chinnasamy[1], Rajasekar Rathanasamy[1], Sathish Kumar Palaniappan[2*], Mahesh Kumar Karavalasu Velusamy[3] and Samir Kumar Pal[2]

[1]*Department of Mechanical Engineering, Kongu Engineering College, India*
[2]*Department of Mining Engineering, Indian Institute of Technology Kharagpur, Kharagpur, India*
[3]*Department of Mechatronics Engineering, Kongu Engineering College, Erode, India*

Abstract

This chapter presents extensive research studies reported by worldwide scientists and specialists in the area of polymer coatings for industrial applications. The authors discuss new and emerging industrial applications, including micro systems, oil and gas industries, electronics, biomedical, pipeline, automotive industries, micro bit storage systems, corrosion protective, and anti-bacterial coating. The authors address the potential to improve and increase the marketability of polymer coating applications in future research in these fields. To date, extensive research has been initiated in order to identify new biodegradable polymer materials which could fulfill all the features needed for the desired quality of the eco-friendly polymeric coating materials.

Keywords: Polymer coating, bio-medical, corrosion resistance, anti-bacterial coating, oil and gas

18.1 Introduction

The word "polymer" is often used to define the multiple repetitions of macro molecules of one or even more units united to offer a set of characteristics.

**Corresponding author*: sathishiitkgp@gmail.com

Inamuddin, Rajender Boddula, Mohd Imran Ahamed and Abdullah M. Asiri (eds.) Polymer Coatings: Technology and Applications, (397–414) © 2020 Scrivener Publishing LLC

They are categorized into natural and synthetic based on their origin. Cellulose, dextran, starch, Arabic latex, chitosan, etc., are classed as natural polymers [1]. The synthetic polymers include three distinct subdivisions, namely, thermoplastics, thermosets, and elastomers. Polymers are employed in broad range of applications, including automotive sector, furniture, toys, coatings, boat hulls, glues, etc. They have attractive characteristics that depict them as important applicants for interfacial modifications against corrosion of metals. This comprises accessibility, environmentally friendly, cost-effective, reliability, and numerous adsorption centers. Under the International Guideline, it is prohibited to use inorganic compounds such as hexavalent chromium and the defined three key criteria are (a) non-bio accumulative, (b) biodegradable, and (c) no or very low level of environmental toxicity for approval of inhibitors. Most of those criteria are met by polymers in general, which naturally occur and extensive research attention has been directed [2].

The polymer films have numerous applications, such as protective coatings, sculptures, liquid and gas isolation barriers, storage and processing materials, semi-conductors, micro fluidic systems, optical instruments, fluorescent sensors and drug delivery and encapsulation, acoustic devices, and biomedical applications. Owing to their phenomenal characteristics like electrical, chemical, and optical with strong chemical and mechanical strength, multilayered, and hybrid multi-layer have attracted a lot of research [1]. We thought the time is absolutely correct for a detailed analysis of different applications of polymers coating. This chapter will provide necessary information for the scientists working in this field, student's community, and industry.

18.2 Polymer Coating in Oil and Gas Industry

The petroleum industry has extensively employed several polymers to sustain, treat, and improve the process of drilling and extraction in oil and gas well. Polymers minimize heavy well accumulation, preserve fluid viscosity, dense air, reduce corrosion, and maintenance downtimes, but also allow for surface cleaning of machinery and wells. The coating of a proppant during hydraulic fracturing HF operations is one of the most common forms of polymer usage in oil and gas industry. Proppants are small areas which have to possess sufficient strength to effectively deal with high locking stresses. Polymers have been introduced in recent years as a coating for proppants and promise increased performance, taking into account both the significance of strength and flexibility. The coating of the proppant with a thin polymer layer leads to increased conductivity of fracturation, decreases fines and scaling considerably, and improves the quality of HF treatment as a result. Polymers (e.g., resin) can be used to enhance the performance of proppant and to keep together many of its coating grains. As shown in Figure 18.1, sand is compressed by increased closed pressure and fines are released when it crushes.

The result of this failure is fine development and restricted conductivity of fracture. Proposed backflow, fine generation, supportive scaling, and low conductivity to fracture constitute the most common problems caused by incorrect choice of proppant. Crush occurs when the stress of fracture closure reaches the maximum strength of proppant. This leads to fines that can transfer, relocate, and block a proppant pack, which diminishes the output of wells. Due to lower width, loss of fracture conductivity can be caused. It can lead to additional compounded flow back, fine formation, and scaling. By employing polymer coating, many of the fines may be encapsulated by the polymer, so it can lead to the treatment of HF successfully [3].

Polymeric products provide better resistance to temperature and corrosion; low friction, low wear resistance; low noise emissions; and low production cost [4–6]. Hence, it is appropriate material for the tribological applications. Severe corrosion could be effectively reduced in petroleum and gasoline industry [7–9], if polymer coatings were applied on the surface of steel. In broad aspect, their wear resistance is a problem because polymers, especially for the thin film covers, are soft materials. Aromatic polyester oligomers discovered in the middle of 1990s [10], which is produced as a result of curing cross-linked oligomers. The transition temperatures are diversified based on type of oligomers. In past tribological investigations on tilting pad layers and compressor settings [11, 12], ATSP polymer layers were shown to be extremely low wear levels, high temperature working, and excellent abrasive wear resistance [13, 14].

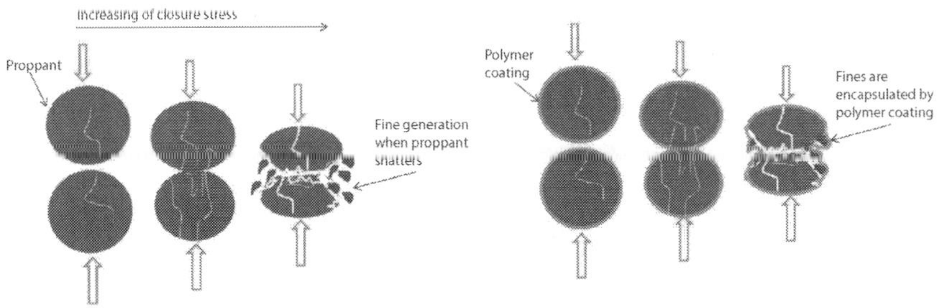

Figure 18.1 Polymer coating on proppant [3].

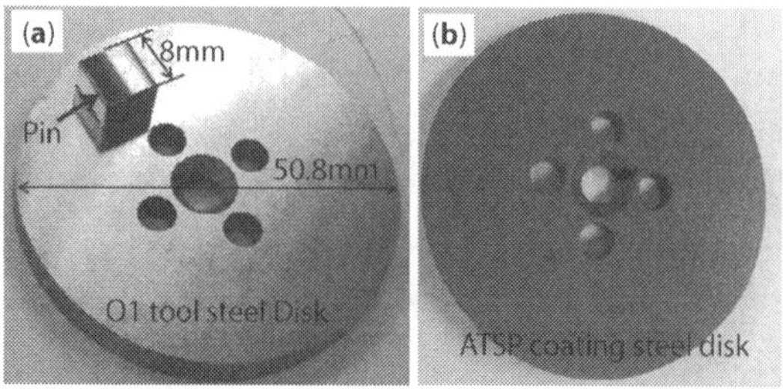

Figure 18.2 Photographs of samples, (a) O1 tool steel disk and pin, (b) ATSP coating on O1 tool steel substrate [15].

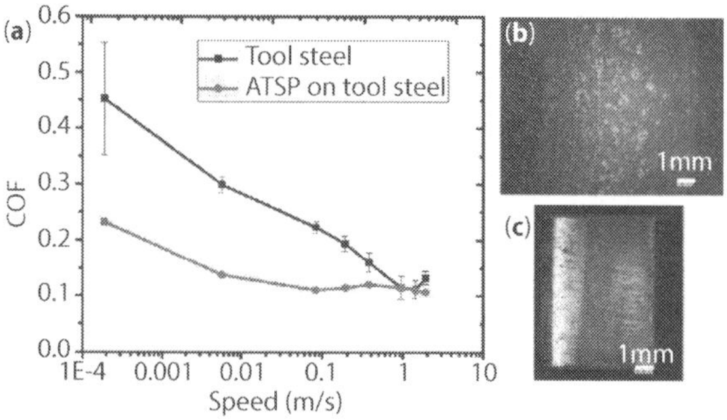

Figure 18.3 Experimental results of Stribeck curve analysis, (a) COF vs. sliding speed. Error bars designate one standard deviation, which in some cases are too small to show. Optical microscopy images of (b) ATSP coating, and (c) pin, after the experiments [15].

A research study on ATSP has been developed to mitigate COF as a cost-effective coating in oil and gasoline drilling operations. The test samples included in this research study were thin layer ATSP coated pure tool steel and curved pins as shown in Figure 18.2. In ASTM standard D2714 and standard for block-on-ring evaluation-D3704, the O1 tool steel is typically used. The powder sample of ATSP with the wt% 98.6 and PTFE with 1.4 was applied to 35 μm thick ATSP layer using an electrical spray deposition process, followed by a 30-minute cure at 270°C in warm air [15]. COF of ATSP coating predominantly depends on mechanical characteristics of coated layer and hence possesses a constant COF as depicted in Figure 18.3 [15].

18.3 Polymeric Coatings for Tribo-Technical Applications

Polymer coatings are now widely spread over the industrial process by refilling them with strengthening or usable inclusions. Forced, thermal, and other types of loading are applied to coatings deliberately designed for tribo-technical implementation. Coatings made from composite materials on a polymeric matrix are employed extensively because in combination with easy and quick sliding, they can offer better corrosion resistance and nominal friction. The phase of loading at the interface matrix inclusion reflects an individual deformation and fracture structural issues [16]. Coating wear is associated with material fractures in stress areas, but the surface is important because the load is applied locally on it [17]. Such concentration may be controlled by the following criteria: (1) the load is non-uniformly applied, (2) the substratum's surface is non-flat or specially shaped, (3) loaded polymer structure, and reinforcing inclusions serve as stress concentrators [18].

In experimental studies of a coating rendered from structurally and functionally modified UHMWPE filled with micro- and nano-fillers, the impact of elements at different level on the mechanical, physical, and tribological characteristics of UHMWPE was shown [19]. The polymeric matrix properties are influenced by the existence of amorphous and crystalline phases and should be taken into consideration when analyzing SSS variables of the UHMWPE-based polymer composite. This factor permits the computation on the fourth level of structure–Mesoscale III, as shown in Figure 18.4 [18].

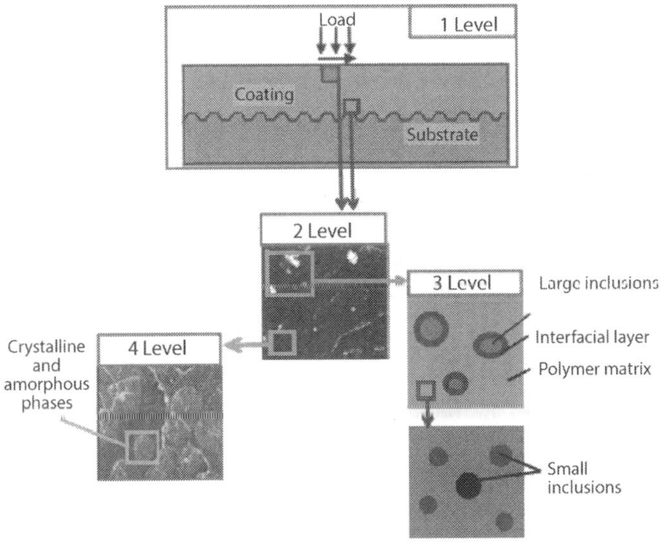

Figure 18.4 UHMWPE-based polymer composite [18].

18.4 Polymer Coating for Drug Delivery

Systematic and regulated drug delivery systems have many advances over traditional drug delivery methods, such as lower soluble drug delivery, very low dose toxicity, controlled drug release kinetics, which extend the reach of treatment [20, 21]. Because of these benefits, various medicines for the controlled release have been developed such as surface-deposited coatings for hydrogen [22–24], nanoporous, polymeric micelles, nanoparticles, water-soluble polymers [25, 26], liposomes, poly-(L-histidine) pH-reliable nano particles, etc. Over recent times, antimicrobial coatings have been developed from silver nano particle-based, amine-based materials. Plasma polymerized n-butyl methacrylate (PPBMA) coatings and n-hepylamine films are employed as a barrier layer for controlled and regulated drug delivery [27, 28].

PCL has been broadly explored as a core polymer membrane in 3D tissue engineering scaffolds and controlled with regulated drug delivery systems. Functional surfaces provide the possibility of increasing biological activity, but the lack of functional groups in the PCL-homo polymer interfaces bounds their application. Figure 18.5 shows the schematic of layer

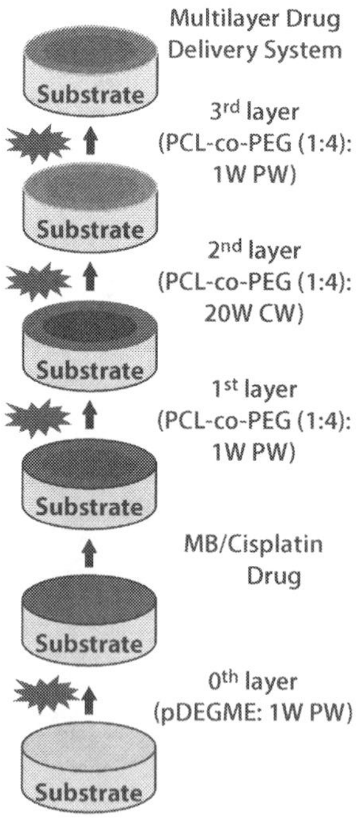

Figure 18.5 Multilayer coating for drug delivery [29].

deposition. Various pretreatments have been researched in order to introduce functional groups on PCLs like aliphatic polymer surfaces. Due to hydrophilicity, anti-toxicity, and lack of antigenicity and immunogenicity, selected PEG can be copolymerized with PCL [28, 29].

18.5 Polymer Coating for Corrosion Protection

Corrosion, with annual worldwide cost of USD $2.5 trillion, equivalent to 3.4% of the world's GDP, is a global challenge. Loss of corrosion is divided into direct and indirect. The costs of repairing corroded or broken equipment and machinery, painting, and refinishing rust prevention structures, the cost of other precautionary measures, the additional cost of utilizing corrosion-resistant materials and alloys, and the expenses of dehumidifying metal storage rooms are taken in consideration of direct losses [30]. Metallic surfaces are a very important concern in protecting them from corrosion. The development of surface treatment method, which would provide proper durability of corrosion protection of metals, is important depends entirely on the metal and surrounding in which it is exposed. In the past decades, a protective coating has been an important consideration for the corrosion protection of metals [31].

Conductive polymers (CPs) are effective in the area of corrosion protection because of their environmental reliability and can easily be developed by chemical or electrochemical approaches. The benefit of alternative methods is that the utilization of organic solvents minimizes the need for contaminants-free polymers [32]. The performance of the CP coating is influenced by a set of experimental factors, such as electro-synthesis, pH and composition of polymerization, that are implicated throughout electro-polymerization [33].

Research effort has been performed to electro-polymerize a pyroleum and aniline mixture known as composite, utilizing a cyclic voltammetry technique, on a pre-treated aluminum surface of Al2024-T3. The anti-corrosive properties of the developed coating were studied with respect to sweep frequency and concentration. Characterization of the coating carried out via cyclic (CV) voltametry and microscopy. A CV is an appropriate method for the electrochemical properties characterization of CPs, representing the systasis of polymer coating. The bibliography includes limited work on the deposition of specific polymers on aluminum substrates [33–36]. Some researchers have the deposition by potentiostatic or through CV of a combination of pyrrole and aniline, while others try to place sequential layers of the many polymers that create a sandwich-like structure [36].

18.6 Polymer Coating for Antibacterial Activity

The utilization of polymers containing the functionality of antimicrobials to improve the effectiveness of certain known antimicrobial agents, mitigate ecological problems with traditional agents, and increase lifespan of anti-microbial agents was extensively explored. The blend of antimicrobial polymers and nanosilver symbiotically improves the effect of bactericides, while polymers typically have a significantly weaker bactericidal efficacy than nanosilver. In order to increase antibacterial property, the polymers could alleviate and dissolve nanosilver [37, 38].

The probability of inhibiting accumulation of nanosilver and creating a uniform top surface coating on different materials is pronounced by polymeric materials with great adaptability and versatility. Such materials can also monitor silver iodine concentrations to mitigate cytotoxicity and persistent antibacterial impact. These can be designed more specifically to prevent adhesion to bacteria and to boost bactericidal properties. It is thus advantageous, for the antibacterial applications, to produce nanomaterial (nano silver) embedded in polymer matrix to develop multifunctional compounds. The composites of polymer/nanosilver with various forms, particularly the influence of shape of fiber had been thoroughly explored. Nanosilver can, on the other hand, play an important part in the medium-term antibacterial impact of developed polymer/nanosilver coatings with a stronger bactericidal property than polymers [39, 40], whereas bactericidal polymers can dominate in long or stable antibacterial coatings following depletion of polymers in silver ions. Figure 18.6 shows the procedure for deposition of amine groups embed with silver nano particles on substrates for anti bacterial activity [41].

The main categories of bactericidal polymers are the quaternary ammonium-based polymer compounds (QAC) [42, 43]. A polymer based on a QAC usually consists of three parts: the polar alkyl chain, cationic compounds, and spacers. The spacers shall ensure that antibacterial effects are transferred by cationic groups. The combination of the QAC and hydrophobic alkyl chain gives a potential antibacterial effects hole poking mechanism. The alkyl chain can have a significant impact on antibacterial properties through a high level of lipophilicity, which can cause effective structure damage and integrity of the cell membranes. A polymer-silver nanoparticle embedded coating on polyethylene terephthalate flat surfaces has been synthesized from N-hexyl-N-(4-vinylbenzyl)-4,4-bipyridinium di nitrate. The composite coatings are pretty highly antibacterial to E. coli

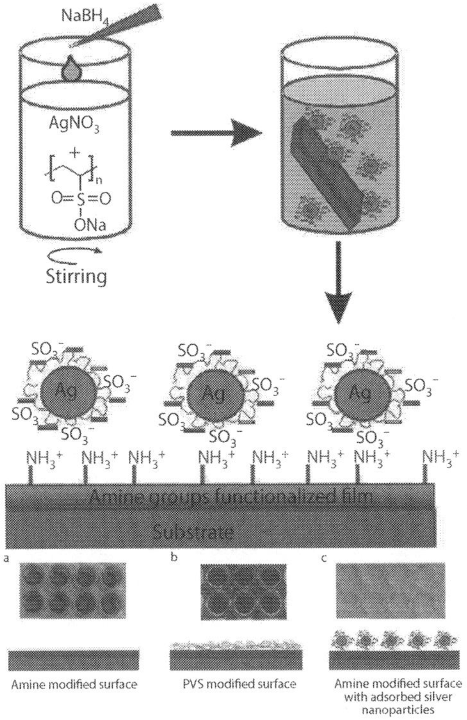

Figure 18.6 PVS modified amine coated substrates [45].

and stay stabilized after sustained immersion and ageing in the weathering channel in the phosphate buffer solution [44].

Chitosan is a functional and natural polymer extracted through deacetylation of chitin which exhibits enhanced anti-bacterial properties and biocompatibility. A number of potential mechanisms for interpreting its antibacterial activity have been suggested. Composite coatings for antibacterial applications of chitosan/heparin/nanosilver had prepared. E. coli has a good bactericidal effect in the composite coatings without substantial mammalian cell cytotoxicity. The effective antibacterial property may also last 1 month. There have been studies of other bactericidal polymers, including poly hexamethylene bi guanides, polymers incorporating furanones, polymers bound to antibiotics, and antibacterial peptides. While these polymers have significant potential for combination with nanosilver to produce composite polymer/nanosilver coatings, very little is still used. They should, therefore, bring enormous possibilities for future research [44, 45].

18.7 Polymer Coating for Micro Bit Storage

Currently, the versatility of associated processes and remarkable materials attracted attention from the work of vinylidene fluoride (VDF)-based ferroelectric polymers. As far as processes are concerned, ferroelectric film can be deposited through various means, such as physical, chemical vapor deposition, Langmuir-Blodgett, and solution methods. Their physical characteristics would often be modified by altering the molecular structures synthetically with distinct moieties. Mechanical adaptability and transparent optical visibility enable ferroelectric polymers to be used in transparent film applications [46].

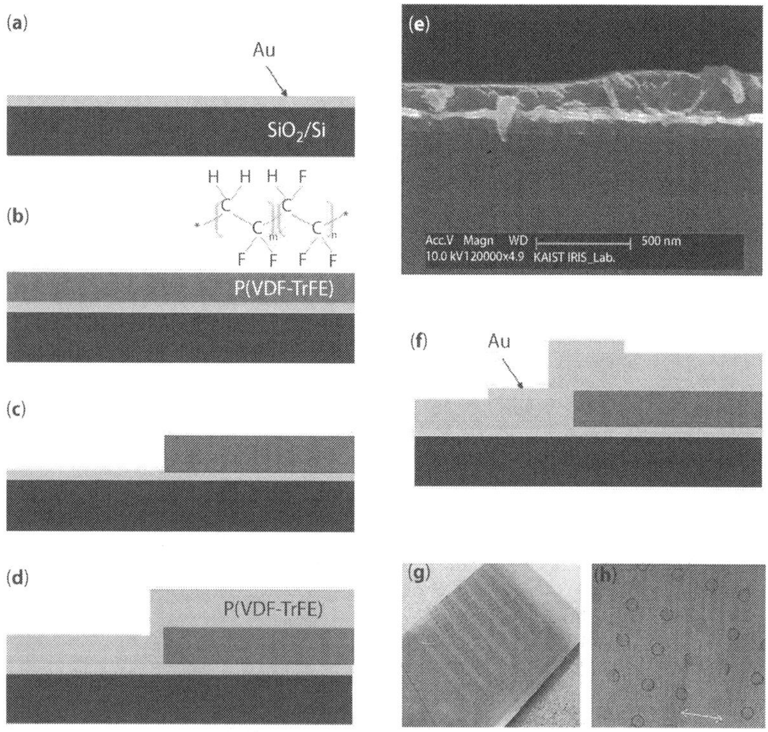

Figure 18.7 Process flow chart: (a) Au-deposited SiO_2/Si substrate; (b) ferroelectric polymer, P(VDF-TrFE), spin-coating; the chemical structure of P(VDF-TrFE) is depicted and the ratio of m and n is 0.75:0.25. (c) patterning of the ferroelectric polymer film; (d) conformal double-coating of the ferroelectric polymer film; (e) cross-sectional SEM image of sample after process (d); (f) deposition of the Au top electrode; (g) photographic image of the final sample; (h) microscopic image of the final sample. The white scale bar represents a distance of 500 μm and the area of the top electrode (circle) is 2.54×10^{-4} cm^2 [46].

The subsequent environmental inertness enables for a non-volatile storage medium. Applications such as multi-bit storage media are investigated using these characteristics of ferroelectric polymers. One research study found that a condenser consisting of a two varying thicknesses had been used to manipulate the morphology of ferroelectric polymer film by simple indentation with a rigid mould as depicted in Figure 18.7. This sort of condenser can be used as a multi-bit memory device if a surface electrode surrounds two areas simultaneously with varying thicknesses [46].

18.8 Polymer Coating for Micro Batteries

Three-dimensional micro batteries (3DMBs) explore new prospects for small energy storage solutions that can react to combined heavy power and energy density requirements. Three-dimensional micro batteries are a key element of this approach. In addition, it is necessary to identify appropriate electrolyte mediums that ion conduction and electron-isolation, conformal, or pinhole-free but nevertheless mechanically or electrochemically sustainable in contrast to the implementation of a latest 3D nano-architectural electrodes. Solid polymer electrolytes (SPEs) offer increasing protection and versatility in a range of architectures and electrode properties compared to standard liquid electrolytes.

An ionically conductive polymer surface subsequently deposited on these structures will represent an important step forward to finalize 3DMB. The oligomer poly(ether amine) (glyceryl poly(oxypropylene); PEA), that has surfactant characteristics, is to be mixed in a cross-linked polypropylene glycol (PPGDA) network to develop uniform and smooth electrolyte coatings on complex electrode surfaces, for practical utilization within 3DMBs. The work presented concentrates on the synthesis and evaluation of a unique solid polymer element that consists of statistically-replaced PEAs at one of ends of three oligomer chain and one methacrylic group at the average. This change incorporates a bi-functionality of a PEA monomer, which becomes polymerizable as well as a surfactant, as the substitution of a PEA/PPGDA electrolyte system [47].

18.9 Polymer Coating for Biomedical Applications

Anti-corrosion strength of biomaterials has been enhanced by employing coatings in many respects. There were anodized coatings but they have been found to decrease corrosion and prevent magnesium degradation. Other protecting materials have also been examined, like organic polymers and hydroxyapatite, which minimize corrosion while still allowing to some degradation. Moreover, some of these coatings are avoided from employing to implant materials due to lack of mechanical properties (especially in adherence and cracking).

No optimal coating that permits degradation at a suitable rate thus sustaining mechanical integrity has yet been established, regarding current research. Two-layer bio ceramic-polymer architecture is known to increase the adhesive capacity of the polymeric film, if the bio ceramic coating is well attached to the substratum and has a large surface area. The goal of a research study was to improve mechanical integrity and to control corrosion levels by applying polymer-bio ceramic composite films to magnesium. The bioceramic layer was chosen for bio active hydroxyapatite and could stimulate bone development. Polylactic acid (PAL) has been selected as the biodegradable polymer coating and is also provided with the additional benefit of FDA approval for application of implants [48].

Alternative layers of photozyme-graft copolymer, Chitosan-g-fluorescein and photocymestatic polymer, poly (sodium styrene-co-vinyl naphthalene -Co-3-dimethyl (methacryloyethyl) ammonium propane sodium sulfonate) -ZI, the nanoscale, multi-layered photo active films have been coated with genipin-crosslinked gelatin substrate by means of the LbL technique

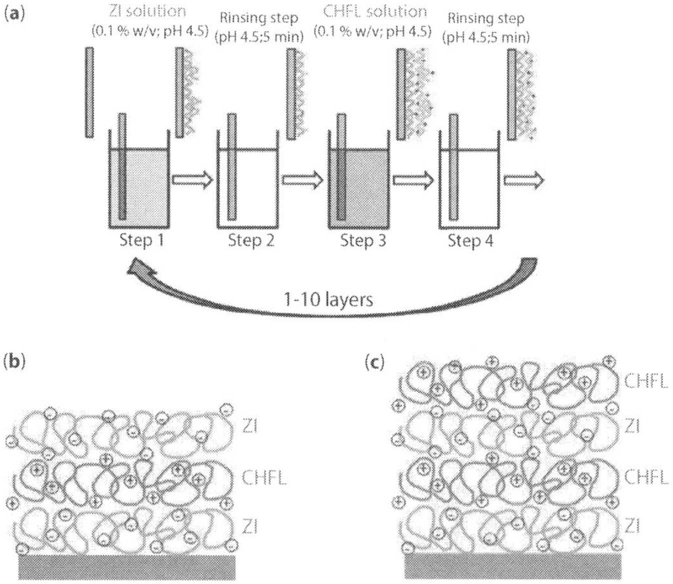

Figure 18.8 Scheme of (a) the method for the modification of G_GP substrate via LbL method, (b) an anti-microbial multilayer promoted by ZI photozyme and (c) anosteointegration improvement multilayer induced by CHFL photozyme [49].

as depicted in Figure 18.8. Nano coating was distinguished by morphological and physicochemical studies. Owing the photosensitizing characteristics of distinct sorts of photozymes, nanostructured materials produced can have significant applications as multifunctional coating for bone repair materials. In order to imitate the cycle of natural bio mineralization, the development of bone-like carbonates apatite layer on the engineered scaffolds includes bio-mimetic scaffolding for the one tissue engineering [49].

Certain biological liquids that surround hip joint do not damp the surface enough, resulting in increased friction and eventually release wear debris. It is probably due to HMWPE in aqueous biologically based body fluids being hydrophobic. Recent work has been investigated in order to develop a wettable layer, minimizing wear particles due to abrasion and the graft of hydrophilic, biocompatible polymers to the surface of HMWPE. The grafted polymers must therefore be deposited on the top surface of hip implant, not only for bio-compatibility, but robust enough for the anticipated friction movement during usage. For use in multiple applications of medical devices, polymer coatings can be developed to have unique chemical, mechanistic, and physical characteristics [50].

18.10 Polymer Coating for Pipe Line Applications

In recent times, polymeric materials were extensively used in numerous industrial applications, particularly in pipe coating. Such coatings are employed mostly for resistance to corrosion, but also as a mechanical damage protection. For these applications, several polymeric materials are used, including polypropylene, ethylene butyl acrylate, fusion bonded epoxy, and high density polyethylene. Currently, in the pipe industry, polyethylene coating has become progressively predominant, and Figure 18.9 portrays the pipe line polymer coating. Furthermore, it constitutes a reliable, elevated-performance, and economical layer of protection against metal [51].

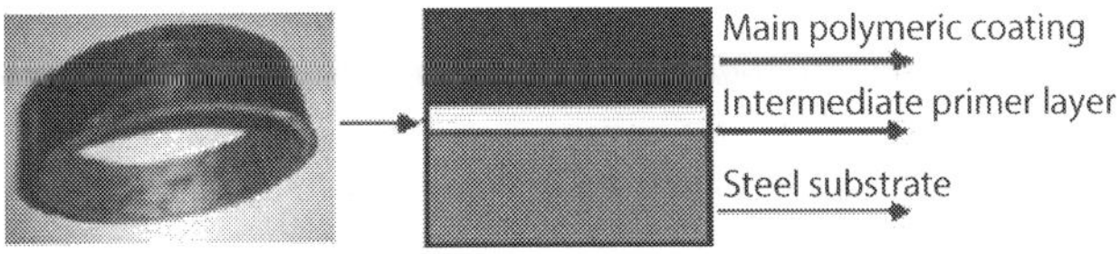

Figure 18.9 Polymer coating for pipe lines [51].

18.11 Conclusions

The chapter provides an overview of polymer coating and its industrial applications. Polymeric products provide better resistance to temperature and corrosion; low friction, low wear resistance; low noise emissions; and low production cost. Hence, it is appropriate material for the many industrial applications. The numerous applications of polymer coating such as protective coatings, sculptures, liquid and gas isolation barriers, storage and processing materials, semi-conductors, micro fluidic systems, optical instruments, fluorescent sensors and drug delivery and encapsulation, acoustic devices, and biomedical applications have been discussed elaborately in this chapter.

References

1. Umoren, S. and Solomon, M., Recent developments on the use of polymers as corrosion inhibitors-a review. *Open Mater. Sci. J.*, 8, 1, 39–54, 2014.
2. Umoren, S.A. and Solomon, M.M., Protective polymeric films for industrial substrates: A critical review on past and recent applications with conducting polymers and polymer composites/nanocomposites. *Prog. Mater. Sci.*, 104, 380–450, 2019.
3. Zoveidavianpoor, M. and Gharibi, A., Application of polymers for coating of proppant in hydraulic fracturing of subterraneous formations: A comprehensive review. *J. Nat. Gas Sci. Eng.*, 24, 197–209, 2015.
4. Holmberg, K. and Mathews, A., Coatings tribology: A concept, critical aspects and future directions. *Thin Solid Films*, 253, 1–2, 173–178, 1994.
5. Shipway, P. and Ngao, N., Microscale abrasive wear of polymeric materials. *Wear*, 255, 1–6, 742–750, 2003.
6. Repka, M. *et al.*, Lubricant-surface interactions of polymer-coated engine journal bearings. *Tribol. Int.*, 109, 519–528, 2017.
7. Guo, Y. *et al.*, Experimental research on the corrosion of X series pipeline steels under alternating current interference. *Eng. Fail. Anal.*, 78, 87–98, 2017.
8. Ning, J. *et al.*, The Role of Pyrite in Localized H2S Corrosion of Mild Steel, in: *CORROSION 2017*, NACE International - Texas, 2017.
9. Zheng, Y. *et al.*, Advancement in predictive modeling of mild steel corrosion in CO2-and H2S-containing environments. *Corrosion*, 72, 5, 679–691, 2016.
10. Frich, D. *et al.*, Novel high-temperature aromatic copolyester thermosets: Synthesis, characterization, and physical properties. *Macromolecules*, 29, 24, 7734–7739, 1996.

11. Lan, P. *et al.*, Advanced polymeric coatings for tilting pad bearings with application in the oil and gas industry. *Wear*, 354, 10–20, 2016.
12. Akram, M.W., Meyer, J.L., Polycarpou, A.A., Tribological interactions of advanced polymeric coatings with polyalkylene glycol lubricant and r1234yf refrigerant. *Tribol. Int.*, 97, 200–211, 2016.
13. Lan, P. *et al.*, A phenomenological elevated temperature friction model for viscoelastic polymer coatings based on nanoindentation. *Tribol. Int.*, 119, 299–307, 2018.
14. Lan, P. *et al.*, Three-body abrasive wear by (silica) sand of advanced polymeric coatings for tilting pad bearings. *Wear*, 382, 40–50, 2017.
15. Lan, P. and Polycarpou, A.A., High temperature and high pressure tribological experiments of advanced polymeric coatings in the presence of drilling mud for oil & gas applications. *Tribol. Int.*, 120, 218–225, 2018.
16. Kanerva, M., Sarlin, E., Saarela, O., Variation in mode II dominated interface fracture of stainless steel-epoxy bonds. Part 1: Mechanical testing. *Eng. Fract. Mech.*, 99, 147–158, 2013.
17. Dag, S. *et al.*, A surface crack in a graded coating subjected to sliding frictional contact. *Eng. Fract. Mech.*, 80, 72–91, 2012.
18. Lyukshin, B. *et al.*, A multilevel analysis of deformation and fracture of filled polymeric coatings for tribotechnical application. *Eng. Fract. Mech.*, 130, 75–82, 2014.
19. Panin, S.V. *et al.*, Antifriction nanocomposites based on chemically modified UHMWPE. Part 1. Mechanical and tribological properties of chemically modified UHMWPE. *J. Frict. Wear*, 32, 3, 199–204, 2011.
20. Colilla, M., Manzano, M., Vallet-Regí, M., Recent advances in ceramic implants as drug delivery systems for biomedical applications. *Int. J. Nanomed.*, 3, 4, 403–414, 2008.
21. Degim, I.T. and Çelebi, N., Controlled delivery of peptides and proteins. *Curr. Pharm. Des.*, 13, 1, 99–117, 2007.
22. Zilberman, M. and Elsner, J.J., Antibiotic-eluting medical devices for various applications. *J. Controlled Release*, 130, 3, 202–215, 2008.
23. Wu, P. and Grainger, D.W., Drug/device combinations for local drug therapies and infection prophylaxis. *Biomaterials*, 27, 11, 2450–2467, 2006.
24. Lin, C.-C. and Anseth, K.S., PEG Hydrogels for the Controlled Release of Biomolecules in Regenerative Medicine. *Pharm. Res.*, 26, 3, 631–643, 2009.
25. David, A. *et al.*, Design of a multivalent galactoside ligand for selective targeting of HPMA copolymer–doxorubicin conjugates to human colon cancer cells. *Eur. J. Cancer*, 40, 1, 148–157, 2004.
26. Dharap, S.S. *et al.*, Molecular targeting of drug delivery systems to ovarian cancer by BH3 and LHRH peptides. *J. Controlled Release*, 91, 1, 61–73, 2003.
27. Yuan, Y. *et al.*, Radio frequency plasma polymers of n-butyl methacrylate and their controlled drug release characteristics: I. Effect of the oxygen gas. *Surf. Coat. Technol.*, 201, 15, 6861–6864, 2007.

28. Moon, H.T. *et al.*, Improved blood compatibility by sustained release of heparin-deoxycholic acid conjugates in a PCL-PEG multiblock copolymer matrix. *J. Biomater. Sci., Polym. Ed.*, 13, 7, 817–828, 2002.
29. Bhatt, S. *et al.*, Plasma co-polymerized nano coatings–As a biodegradable solid carrier for tunable drug delivery applications. *Polymer*, 54, 18, 4820–4829, 2013.
30. Schreiber, H., Wertheimer, M., Wrobel, A., Corrosion protection by plasma-polymerized coatings. *Thin Solid Films*, 72, 3, 487–494, 1980.
31. Anagri, A. *et al.*, Nanocomposite coatings based on graphene and siloxane polymers deposited by atmospheric pressure plasma. Application to corrosion protection of steel. *Surf. Coat. Technol.*, 377, 124928, 2019.
32. Erdem, E., Saçak, M., Karakişla, M., Synthesis and properties of oxalic acid-doped polyaniline. *Polym. Int.*, 39, 2, 153–159, 1996.
33. Patil, D. *et al.*, An Mn doped polyaniline electrode for electrochemical supercapacitor. *J. Electrochem. Soc.*, 158, 6, A653–A657, 2011.
34. Rajagopalan, R. and Iroh, J.O., Development of polyaniline–polypyrrole composite coatings on steel by aqueous electrochemical process. *Electrochim. Acta*, 46, 16, 2443–2455, 2001.
35. Iroh, J.O. *et al.*, Electrochemical synthesis: A novel technique for processing multi-functional coatings. *Prog. Org. Coat.*, 47, 3–4, 365–375, 2003.
36. Herrasti, P. *et al.*, Effect of the polymer layers and bilayers on the corrosion behaviour of mild steel: Comparison with polymers containing Zn microparticles. *Prog. Org. Coat.*, 54, 4, 285–291, 2005.
37. Shi, Z., Neoh, K., Kang, E., Surface-grafted viologen for precipitation of silver nanoparticles and their combined bactericidal activities. *Langmuir*, 20, 16, 6847–6852, 2004.
38. Yuan, W. *et al.*, Self-assembled chitosan/heparin multilayer film as a novel template for in situ synthesis of silver nanoparticles. *Colloids Surf. B: Biointerfaces*, 76, 2, 549–555, 2010.
39. Lischer, S. *et al.*, Antibacterial burst-release from minimal Ag-containing plasma polymer coatings. *J. R. Soc. Interface*, 8, 60, 1019–1030, 2011.
40. Ho, C.H. *et al.*, Nanoseparated polymeric networks with multiple antimicrobial properties. *Adv. Mater.*, 16, 12, 957–961, 2004.
41. Zhang, W. *et al.*, Influence of dissolved oxygen on aggregation kinetics of citrate-coated silver nanoparticles. *Environ. Pollut.*, 159, 12, 3757–3762, 2011.
42. Lichter, J.A., Van Vliet, K.J., Rubner, M.F., Design of antibacterial surfaces and interfaces: Polyelectrolyte multilayers as a multifunctional platform. *Macromolecules*, 42, 22, 8573–8586, 2009.
43. Timofeeva, L. and Kleshcheva, N., Antimicrobial polymers: Mechanism of action, factors of activity, and applications. *Appl. Microbiol. Biotechnol.*, 89, 3, 475–492, 2011.
44. Dash, M. *et al.*, Chitosan—A versatile semi-synthetic polymer in biomedical applications. *Prog. Polym. Sci.*, 36, 8, 981–1014, 2011.

45. Guo, L. *et al.*, Polymer/nanosilver composite coatings for antibacterial applications. *Colloids Surf., A: Physicochemical and Engineering Aspects*, 439, 69–83, 2013.
46. Kim, W.Y. *et al.*, Solution-processed conformal coating of ferroelectric polymer film and its application to multi-bit memory device. *Microelectron. Eng.*, 160, 68–72, 2016.
47. Sun, B. *et al.*, Solid polymer electrolyte coating from a bifunctional monomer for three-dimensional microbattery applications. *J. Power Sources*, 238, 435–441, 2013.
48. Oosterbeek, R.N. *et al.*, Polymer–bioceramic composite coatings on magnesium for biomaterial applications. *Surf. Coat. Technol.*, 236, 420–428, 2013.
49. Chiono, V. *et al.*, Layer-by-layer coating of photoactive polymers for biomedical applications. *Surf. Coat. Technol.*, 206, 8–9, 2446–2453, 2012.
50. Farhat, S. *et al.*, Polymer coatings for biomedical applications using atmospheric pressure plasma. *Surf. Coat. Technol.*, 241, 123–129, 2014.
51. Guermazi, N. *et al.*, Aging effect on thermal, mechanical and tribological behaviour of polymeric coatings used for pipeline application. *J. Mater. Process. Technol.*, 203, 1–3, 404–410, 2008.

19

Formulations for Polymer Coatings

Mallesh Kurakula[1], N. Raghavendra Naveen[2] and Khushwant S. Yadav[3*]

[1]*Department of Biomedical Engineering, The University of Memphis, Memphis, Tennessee, United States*
[2]*Department of Pharmaceutics, Annamacharya College of Pharmacy, Rajampet, India*
[3]*Shobhaben Pratapbhai Patel School of Pharmacy & Technology Management, SVKM's NMIMS (Deemed to be University), Mumbai, India*

Abstract

Over a previous couple of decades, polymers are widely used as pharmaceutical excipients in many novel and conventional drug delivery systems from being merely used as gelatin shells to contribute massive formulation benefits such as controlled/sustaining the drug release and also targeting of drugs to the specific site, consequently hold a considerable clinical promise. Currently, numerous dosage forms are provided with diverse coating techniques, which may give out a variety of functions. Based on functionality, several polymer coatings are available commercially to offer incessant formulating choices. In this chapter, recent advancements in the usage of polymers for coating of different dosage forms such as tablets, capsules, implants, nanoparticles, and liposomes, etc., are discussed. Mechanisms of polymeric film formation and applications of polymer coatings in the different areas of biomedicine are clearly explained. Application of different polymers in various coating functions is extensively reported.

Keywords: Enteric coating, colon targeting, natural polymers, biomedicine, taste masking, moisture protection

**Corresponding author:* khush.yadav@gmail.com; khushwant.yadav@nmims.edu

Inamuddin, Rajender Boddula, Mohd Imran Ahamed and Abdullah M. Asiri (eds.) Polymer Coatings: Technology and Applications, (415–444) © 2020 Scrivener Publishing LLC

19.1 Introduction

The coating is a method by which an outer dry layer of coating material (containing polymers and other excipients) is applied to the dosage form surface (core) to confer specific benefits. Coating of pharmaceutical dosage forms is one of the regular unit operations in the industry. The basic concept of sugar-coating technology was borrowed from the candy industry in the mid-1800s to coat the tablets contain bitter drugs [1]. In recent days, film coating process predominantly replaced by the sugar-coating process by making use of polymers, which will produce a very thin film on applying to the substrates. By selecting the appropriate polymer-drug release profile can be tailored. Product differentiation, chemical protection, and separation of incompatible drugs are some of the other advantages. The coating process can also increase the packaging speed by reducing the dust generation and friability [2]. The applications of coating mechanism not only constrained to pharmaceutical filed but also to biomedicine and other engineering areas. This chapter summarizes application of different polymers for coating of different dosage forms like tablets, capsules, liposomes, pellets, etc. Application of coating process in different areas of biomedicine is clearly explained.

19.2 Film Coating

Film coating process involves deposition of very thin polymeric film (approximately 25 to 100 μm in thickness) on to the different solid surface substrates such as tablets, granules, or capsules to i) alter the drug release, ii) protect the drug from various environmental/physiological factors, iii) to enhance the physical and chemical properties of the substrate, iv) mask the unpleasant taste and odor, and also v) to avoid the interaction of various incompatible ingredients. Functional coating such as modified release, osmotic pressure-controlled drug delivery system, and enteric coatings can be done for advanced drug delivery systems. Film coating process ensures for batch to batch uniformity as well as good reproducibility [3]. Coating formulation majorly contains polymers, plasticizer, anti adherents, surfactants, and colorants.

19.2.1 Polymers for Film Coating

The film-forming polymer can be classified into two major classes depending on the solubility, aqueous soluble and insoluble polymers (pH-dependent soluble polymers). Majority of aqueous soluble polymers are cellulosic polymers such as MC (Methylcellulose), HPC (hydroxypropyl cellulose), HPMC (Hydroxypropyl methyl cellulose), CMC sodium (carboxymethyl cellulose sodium), polyethylene glycols, and polyvinyl pyrrolidone (povidone) [4]. The water-insoluble or pH-dependent soluble polymer is widely used for delay release (enteric coating) or some special controlled systems. Some of the common enteric polymers are methacrylic acid copolymers (Eudragit® E, Eudragit® S Eudragit® L), phthalates (cellulose acetate phthalate-CAP and hydroxypropyl methylcellulose phthalate-HPMCP), polyvinyl acetate phthalate (PVAP), in addition to EC (ethylcellulose), PVA (Polyvinyl acetate) Eudragit® RL and RS grades [4].

19.2.2 Plasticizer

During the coating process, on drying films prepared from pure polymers as often as possible are fragile. To deal with this insufficiency or to make the film more pliable, the polymer can be modified chemically by making use of some excipients such as plasticizers [4]. Plasticizer allows/helps the polymer molecule to move easily and also have the capability to weaken the intermolecular attractions [5]. This will result in enhancing the flexibility of the polymer, improving flowability, increasing the toughness, reducing brittleness, and improving the tear resistance of the film. Plasticizers can broadly classify into two major categories: internal and external plasticizers, based on inclusion or addition of plasticizer in the coating process. Internal plasticizers can modify the physical properties of basic polymer by altering the degree and type of substitution or polymer chain length of or both. The external plasticizer acts by enhancing the flexibility of the film by modifying the polymer-polymer interactions and by relieving molecular rigidity [4].

19.2.3 Polymer-Plasticizer Compatibility

The plasticizer must be compatible with the polymer, so as to form continuous film and to make the formulation successful. The foremost difficulties while formulating the coating solution are i) selection of proper plasticizer, ii) assessment of the interaction of polymer with plasticizer, and iii) perception of mechanism, how the plasticizer is improving the flexibility of film. Methods to predict the compatibility between plasticizers and polymers are on basis of i) solubility parameter calculation using molecular dynamics, ii) molecule charge density by using COSMO, iii) experimental DSC analysis, and iv) mesoscale simulation using endeavors.

The impact of plasticizer on polymer properties have been studied by many researchers, a few of them are given below. Results from DPD simulations confirmed the compatibility between polyethylene glycol (PEG) and PVP, as PEG mix well with the PVP. DSC results showed that there is a superior interaction between HPMC and PEG than between MCC and PEG, because of higher degree of miscibility and attraction of HPMC with PEG molecules in contrast to MCC [6]. The impact of the plasticizer (PEG) on mechanical strength and thermal behavior of coating films (HPMC) was studied by Laboulfie *et al.* [7] Jarray *et al.* [8] additionally examined the affinity of the base polymers with the coating in both dry and aqueous systems. Ould-Chikh [9] found that addition of the plasticizer (PVA) to the aqueous polymeric suspension noticeably minimizes the cracking or segmentation of coating films after drying.

19.2.4 Mechanism of Film Formation

Depending on the solubility of film coating polymers, film coating applications are mostly solutions or dispersions [10]. Film formation mechanism from dispersions is more complex in contrast to solutions [11]. On applying the polymer solution, a bond is formed between the coating polymer molecules because of the cohesion forces between them; whereas in case of dispersion, film will be formed as a result of coalescing between the polymeric particles.

Generally, film coating is applied by spray atomization process. On spraying, the formed droplets spread on the substrate, and with further drying, the solvent evaporates, and polymer chain interpenetrates and forms the film. The degree of spreadability of droplets relays on surface tension. Several surface tensions are involved in the coating process like a) liquid-vapor interfaces (LVI), b) solid-liquid interfaces (SLI), and c) solid-vapor interfaces (SVI). Spreading of the droplets can be described by replacement of SLI or SVI. The wetting mechanism or replacement of interfaces can be mainly of three types such as adhesion, spreading, and immersion wetting, Replacement or formation of interfaces is depends upon the type of wetting mechanism (Table 19.1). Figure 19.1 describes the surface tensions relevant and types of wetting mechanisms in coating process.

The coating process can be considered as a complex phenomenon, owing to the number of processing variables and their interactions, variables related to the coating formulation, and substrate properties.

Table 19.1 Replacement or formation of interfaces with the type of wetting mechanism.

S. no	Type of wetting mechanism	Formation/ disappearance of interfaces	Replacement of interfaces
1.	Adhesional wetting	LAI disappears	SAI replaced by SLI
2.	Spreading wetting	LAI forms	SAI replaced by SLI
3.	Immersional wetting	No change in LAI	SAI replaced by SLI

Liquid-vapor interfaces (LVI), solid-liquid interfaces (SLI), and solid-vapor interfaces (SVI).

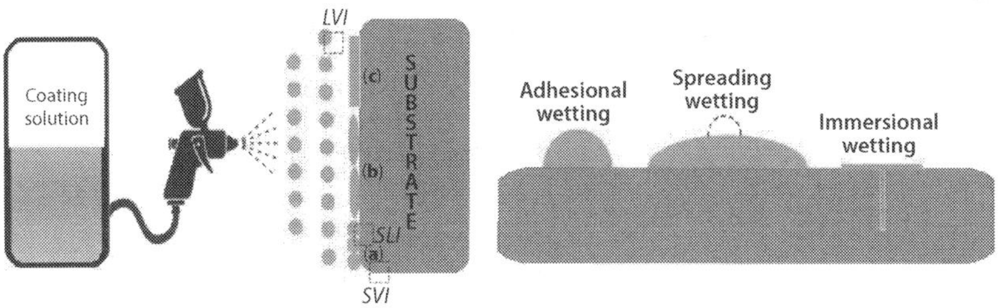

Figure 19.1 The surface tensions relevant, mechanism of film formation: (a) Contact, (b) Spreading and solvent evaporation, (c) Film formation] and types of wetting in the film coating process.

Process variables influencing wetting mechanism are summarized in Table 19.2 [12].

19.3 Functions of the Polymeric Coating

There are three types of coatings concerning the function of the coat, like nonfunctional, functional, and active coatings. Non-functional coatings are meant for improving compliance. Majority of the functional coatings can modify (delay release) or prolong the drug release apart from masking the bad taste/odor. Inactive coatings, drug layer (coating layer), will be deposited on to the inert substrate. These kinds of coatings are mainly applied to avoid the interaction of different drugs and also to merge different release characteristics in one dosage forms. Different polymers used

Table 19.2 Process variables influencing the wetting mechanism.

S. no	Type of variable	Factors
1.	Substrate related	a) Chemical properties of the drug b) Physicochemical properties of dosage form like strength; size/shape; surface properties, etc. c) Hydrophobicity d) Core polymer interactions e) Surface roughness f) Porosity
2.	Coating formulation related	a) Particle size b) Plasticizer c) Viscosity d) The solubility of coating polymer
3.	Coating process variables	a) Atomizing air pressure b) Type of spray c) Solvent evaporation rate d) Temperature and/or Humidity e) Distance between spray gun and substrate f) Post coating curing g) Droplet momentum

for protection and masking are given below in the following Tables 19.3 to 19.6 [13].

19.3.1 Application of Film Coating in Modified Release System (Enteric Release)

Enteric coating also called as gastro-resistant coating, have been used widely from mid of 19th century to resist the release of drug in stomach in order to protect the drug material from the acidic environment of stomach and also to target the release at specific intestinal segments [14]. This also helps from the irritating effect of few drugs. All these can be achieved by coating with the weak acid enteric polymers, which were in unionized form at the acidic pH and ionized form at higher alkaline pH of intestine. To target the different sections of intestinal segments, polymers were designed with different pH thresholds as shown in Table 19.7 [13, 15].

Table 19.3 Different aqueous soluble polymers for protection and taste masking.

Manufacturer	Trade name	Polymer
Colorocn	Opadry	PVA
Ashland Aqualon	Oxycellulose Natrosol	HEC
BASF	Kollidon Kollicot IR Kollicot IR protect	PVP PVA-PEG copolymer PVA-PEG: PVP
Wolff cellulosics	Walocel	Na CMC
Bio Grund	Aquapolish	PVA-PEG copolymer
Down chemical	Methocel	HPMC
Wolff cellulosics	Walocel	
Sensient	Spectracel	
Shin-Estu	Pharmacoat	

Table 19.4 Entero-soluble cationic polymers for protection and taste masking.

Manufacturer	Trade name	Polymer
Evonik	Eudragit E, E PO, E 12.5	Amino domethyl methacrylate copolymer
BASF	Kollicoat, Smart seal 30 D	

These polymers coated layer will get dissolve only after reaching to the lower GIT, where pH is more than their dissolution pH threshold [16]. Conversely, the in vivo performance of enteric-coated products largely depends upon several gastrointestinal physiology variables such as gender, feeding status, transit time, and gastrointestinal pH [17, 18]. Various polymers like CAP, HPMCP, PVAP, and hydroxypropyl methylcellulose acetate succinate (HPMCAS) and methacrylic acid copolymers, etc., can be used for delay release. Numerous solid dosage forms like pellets, tablets, microcapsules, etc., can be coated with these enteric polymers to achieve the required benefits. Extensive research has been done to coat the tablets by making use of several polymers like Eudragit L-30 D-55, Eudragit L-30 D-55 HPMCP, CAP, Acryl-EZE®HPMC-55, and Ethyl cellulose, etc.

Table 19.5 Enteric anionic polymers for protection and taste masking.

Manufacturer	Trade name	Polymer
Eastman	Cab Eastman	CAB (Cellulose acetate butyrate)
	Kollicoat MAE 30 DP/100 P Eastacryl 30 D NF Eudragit L 30 D-55/L 100-55	Type A Methacrylic acid copolymer
	Eastman C-A-P NF	CAP
	Eudragit L 100	Type B Methacrylic acid copolymer
Evonik	Eudragit L 100	Type B Methacrylic acid copolymer
	Eudragit FS 30 D	Methacrylic acid copolymer
	Eudragit S 100 Eudragit S 12, 5	Methacrylic acid copolymer, Type C
FMC biopolymers	Aquacoat CPD	CAP
	Keltone LV CR	Sodium alginate
SSB Pharma	SSB 55	Shellac
Aqualon	Akucell	CMC

Hard gelatin capsules can be coated with non-aqueous solutions of Eudragit E, L, and S 12.5 for lower GIT targeted delivery. Coated capsules are gained their importance because of the independency of coating composition on capsule filling [19].

19.3.2 Liposomal Coating

The retention time of liposomes can be increased by surface modification with fewer uptakes by RES. The polymer coating is an alternative method for this. The liposomal coating is a simple process of mixing liposome

Table 19.6 Insoluble polymers used for protection and taste masking.

Manufacturer	Trade name	Polymer
BASF	Kollicoat SR 30 D	PVA
Evonimk	Eudragit RS 30 D, RS 100/RS PO, RS 12.5	Ammonio methacrylate Type B: Poly(ethyl acrylate-co-methyl methacrylate-co-trimethylamminoethyl methacrylate chlorid) 1:2:0.2
	Eudragit RL 30 D, RL 100/RL PO, RL 12.5	Ammonio methacrylate Type A: Poly(ethyl acrylate-co-methyl methacrylate-co-trimethylamminoethyl methacrylate chlorid) 1:2:0.2
	Eudragit NM 30 D	Poly (ethyl acrylate-co-methyl methaacrylate) 2:1
	Eudragit NE 30 D	Poly (ethyl acrylate-co-methyl methaacrylate) 2:1
Eastman	Eastman CA	Poly (ethyl acrylate-co-methyl methaacrylate) 2:1
Colorcon	Surelease	Cellulose acetate
FMC	Aquacoat	Ethyl cellulose
Dow chemical	Ethocel	Ethyl cellulose

suspension with that of polymer solution. In this, there will be no chemical linking of the polymers to the lipid molecules of liposome. A reported by Hirofumi Takeuchi [20], there is a feasibility of coating the surface of liposomes with modified hydrophilic polymers such as PVA or PAA

Table 19.7 Threshold pH with target site of GIT.

S. no	pH thresholds	Target site in GIT
1.	~5–5.5	Duodenum
2.	~6	Jejunum
3.	~7	Ileum and Colon

(polyacrylic acid) having a hydrophobic anchor (s) (PVA-R or PAA-R). The residence time of these liposomes in the biological system can be increased by coating of liposomes with mucoadhesive polymers and thus enhancing the oral bioavailability of the drug [21]. Oral absorption of peptide drugs has been enhanced by chitosan-coated liposomes [22–24]. Other polymers that have been employed in the surface coating of liposomes include amylopectin, dextran, and pullulan [25].

19.3.3 Aerosol Coating

Aerosol performance of micronized drugs in carrier excipient drug blends was improved by excipient coating. Daniela Traini [26] investigated this use in the dry powder inhalers. Both EC (ethylcellulose) and PVP were used as coating agents. Carriers were prepared via sieve fractioning followed by spray drying, with and without polymer additive. After blending for 24 h, each coated and uncoated blended systems (salbutamol sulphate-SS) were evaluated for particle size, drug-carrier adhesion, morphology, and aerosolization performance. Particle adhesion between SS and aerosol performance (fine particle fraction; FPF) followed the rank: PVP coated > uncoated > EC coated lactose. Enteric coating of different dosage forms with enteric polymers are given in Table 19.8.

19.4 Polymeric-Coating Approaches to Targeted Colon Delivery

Colon-targeted drug delivery has been gained importance from many years owing to its prospective to progress the treatment of some of the local diseases like Ulcerative colitis (UC) and Crohn's disease (CD) and irritable bowel syndrome (IBS) and minimizing systemic side effects. A higher concentration of drug will reach to colon with negligible systemic absorption or without being absorbed in upper part of GIT. Enhancement of absorption of several drugs can be seen owing to longer retention of colonic content and colonic content nature and making this organ as an ideal site for drug delivery. A drug can be delivered to the colon via the oral, or the rectal route, among that oral route is most preferred than rectal due to their convenience [4]. Major approaches for targeting colon have been given in Table 19.9.

Table 19.8 List of polymers for the enteric purpose of several dosage forms.

S. no	Dosage form	Drug	Polymers used	Ref
1.	Tablets	Diclofenac and Prednisone	Blend of cellulose esters and enteric polymers (CAP, E L 100, HPMC-HP 55)	27
2.	Pellets	Duloxetine hydrochloride	HPMC P, HPMC ASMG, and HPMC ASMF	28
3.	Tablets	Esomeprazole magnesium trihydrate	HPMCP, Eudragit L-30 D-55, CAP	29
4.	Pellets	Duloxetine hydrochloride	Aqoat® AS-LF, Eudragit® L30D55 HPMCP-HP55.	30
5.	Tablets	Azithromycin Dihydrate	EC, Eudragit L-30 D-55, HPMC-55,	31
6.	Pellets	Duloxetine hydrochloride	Aqoat® AS-LF, Eudragit® L30D55 HPMCP-HP55.	32
7.	Tablets	Rabeprazole	Hydroxypropyal methyl cellulose-Phthalate CAP	33

(Continued)

Table 19.8 List of polymers for the enteric purpose of several dosage forms. (*Continued*)

S. no	Dosage form	Drug	Polymers used	Ref
8.	Capsules	Omeprazole	Methyl acrylic acid and ethyl acrylate	34
9.	Tablets	Prednisolone	Eudragit1 L 30 D-55. HP-DD	35
10.	Tablets	Diclofenac sodium	Hydroxypropyl methylcellulose (HPMC) E3 Eudragit L 100 Eudragit E 100	36
11.	Capsules	Paracetamol	HPMCAS-LF, Eudragit L100-55 and HP-55	37
12.	Tablets	Pantoprazole	CAP Eudragit L100	38
13.	Tablets	Ilaprazole	Eudragit L 100 HPMC P 50	39
14.	Tablets	Zileuton	Eudragit L-100 Ethylecellulose Eudragit S-100	40
15.	Pellers	Budesonide	Eudragit s 100	41
16.	Tablets	Lansoprazole	Eudragit® L100-55:	42

(*Continued*)

Table 19.8 List of polymers for the enteric purpose of several dosage forms. (*Continued*)

S. no	Dosage form	Drug	Polymers used	Ref
17.	Solid lipid nanoparticles	Enrofloxacin	Polyacrylic resin II	43
18.	Microsponges	5-FU	Eudragit S100/Eudragit L100 blend of EdS100	44
19.	compressed enteric coated pellets	–	Eudragit FS 30 D, Eudragit L30 D-55	45
20.	Microspheres	Duloxetine	Eudragit L-100	46
21.	Microporous Osmotic pump tablet	Quetiapine fumarate	HPMCP	47
22.	Microsphres	5 FC	Eudragit-S-100 Eudragit- L-100	48
23.	Tablets	Rifampicin and isoniazid	EudragitR L30D-55	49
24.	Microparticles	Cinnarizine	Eudragit® L100-55	50
25.	Nanoliposomes	Alendronate Sodium	Eudragit L100	51

(*Continued*)

Table 19.8 List of polymers for the enteric purpose of several dosage forms. (*Continued*)

S. no	Dosage form	Drug	Polymers used	Ref
26.	Microparticles	Biochanin A	Ethyl cellulose	52
27.	Microcapsules	Rabeprazole sodium	EC, HPMCK100M Eudragit L100	53
28.	Solid Lipid Nanoparticles	Alendronate sodium	Eudragit S100	54
29.	Microcapsules	Astaxanthin	Alginate gelatin	55
30.	Nanoparticles	5-fluorouracil	Eudragit S-100	56
31.	Tablets	Didanosine	HPMCP 5.5	57
32.	Nanoparticles	–	HPMC PVP, PVA–PEG, HPC	58
33.	SNEEDS	Insulin	Eudragit1 L100	59
34.	Liposomes	Sildenafil	Chitosan solution HPMC	60
35.	Osmotic pump capsule	Sinomenine hydrochloride	HPMCP HP 50, HPMCP HP 55. Eudragit L 100	61
36.	Tablets	Prednisolone	Zein and Kollicoat® MAE 100P	62

Table 19.9 Major approaches for targeting colon drug delivery [63].

Colon targeted drug delivery-Approaches			
Approach		**Basic feature**	
I. Chemical Approaches	i. Azo conjugates	The drug is conjugated	via an azo bond
	ii. Cyclodextrin conjugate		with cyclodextrin
	iii. Glycoside conjugate		with glycoside
	iv. Glucuronide conjugate		with glucuronate
	v. Dextran conjugate		with dextran
	vi. Polypeptide conjugate		with polypeptide
	vii. Polymeric prodrug		to the polymer
II. Pharmaceutical Approaches			
1. Polymeric coating			
	i. Film Coating by using pH-sensitive polymer	Coating with enteric polymers releases drug only at alkaline range	
	ii. Film Coating by using biodegradable polymer	After reaching the colon, degradation of polymer layer can be seen because of colonic bacteria action, followed by drug release.	
2. Embedding in matrices			
	i. Embedding in biodegradable polysaccharides	Swelling and biodegradable action can be seen with the polysaccharides and drug embedded in the matrices will be released.	
	ii. Embedding in pH-sensitive matrices	The pH-sensitive polymer will degrade in the target area of GIT and results in the release of drug content.	

(*Continued*)

Table 19.9 Major approaches for targeting colon drug delivery [63]. (*Continued*)

Colon targeted drug delivery-Approaches	
3. Timed released systems	
4. Redox-sensitive polymers	
5. Bioadhesive system	Substrates coated with the bioadhesive polymer which selectively gives the adhesion to the colonic mucosa.
6. Coating of microparticles	The Drug should get a release through the semipermeable membrane
7. Osmotic controlled delivery	Osmotic pressure

19.4.1 Enzymatically Degradable Film Coatings [64]

19.4.1.1 Film Coatings Based on Naturally Occurring Polysaccharides

Naturally occurring polysaccharides of animal, plant, algal or microbial origins such as chondroitin, pectin, chitosan, amylase, and galactomannan are widely used for colon delivery owing to their susceptibility to colonic degradation, safety, biocompatible nature, and also implications as food additive.

19.4.1.2 Film Coating on the Basis of Synthetic Azo Polymers

The probability of colon targeting by the utilization of azo reductive bacterial activities was first time used for the development of oral administration of peptide delivery. Dip coating was done by using copolymers of styrene and hydroxyethyl methacrylate (HEMA) (1:6), which is cross-linked with 1% or 2% of divinyl azobenzene. These were dissolved in methylene chloride or chloroform with methanol at concentration of 20 % v/v.

Hard gelatin capsule (size 5) containing bovine insulin was coated with film-forming material such as copolymer of HEMA and styrene cross-linked with substituted divinylazobenzene derivative. Copolymers of HEMA and methyl methacrylate (MMA), or elseterpolymers of the latter compounds and methacrylic acid, were prepared by using substituted divinylazobenzenes as azo cross-linkers and applied, as ethanolic solutions, to size 9 hard-gelatin mini capsules containing various tracer molecules, such as fluorescein, caffeine, ibuprofen, and theophylline.

19.4.2 pH-Sensitive Film Coatings

19.4.2.1 Film Coating on Basis of Enteric Solubility of Polymers

For targeting the colon, pH-sensitive film coatings were widely exploited, especially by using polymethacrylates. These materials having the enteric solubility nature, eg, Eudragit L (having the solubility at pH above 5.5 or 6.0) and Eudragit S (with higher dissolution pH threshold of 7.0), are widely used from many years.

19.4.2.2 Film Coatings on the Basis of Acid Solubility of Polymers

Acid soluble pH-responsive film coatings were also used apart from enteric for colon targeting. This will be coupled with gastro-resistant layer and will be normally expecting to dissolve as a result of direct or indirect microbial fermentation activity of flora present in the caecum and ascending colon areas.

19.5 Natural Polymers Applications in Modified Release Dosage Forms

Modified release oral solid dosage forms afford many specific advantages in drug release pattern in contrast to conventional drug delivery systems. Natural polymers are of special interest in formulating these delivery systems because of their safety, biodegradability, rich sources, and biocompatibility. Structure of the natural polymers could also be modified to achieve the desired effect [65]. Typically, cellulose-based carriers provided many choices for formulators in solid oral modified-release dosage forms such as matrix tablets and pellets [66, 67]. However, despite the fact that some natural polymers and modified natural polymers have been included in the pharmacopeia, the applications of most natural polymers in solid oral modified-release dosage forms need to be further explored and an updated review on the recent advances in this field is essential. Few examples of natural polymers for modified release are given in Table 19.10 [68].

Table 19.10 Modified-release dosage forms based on modified chitosan.

Polymer	Drugs	Dosage forms/description
Modified Chitosan S-protected thiolated chitosan	Antide	Coated matrix tablets Increase mucoadhesive properties, improve the bioavailability of peptides, using Eduragit L 30 D-55 as the coating material
Pectin based Physical mixtures of pectin-hypromellose	–	Compression coated tablets, scintigraphic evaluation of tablets in healthy volunteers
Physical mixtures of pectin-Comritol* 888 ATO	Mesalamine	Compression coated tablets, Box-Behnken experimental for optimization of formulations
Physical mixtures of pectin-hydroxyethylcellulose	Rhubarb	Matrix coated with Eudragit S 100 for further retarding the drug release
Guar gum-based polymer	Omdazole, Metronidazole	Dry coating (compression coating) *In vivo* evaluation in healthy volunteers
Guar gum-based polymer	Prednisolone	Dry coating (compression coating), *In vivo* gamma scintigraphic study in human volunteers
Guar gum-based polymer	Flurbiprofen	Dry coating (compression coating), *In vivo* study in healthy volunteers

*Compritol® 888 ATO is glyceryl behenate

19.6 Application of Polymer Coating in Biomedicine

Biomedicine (medical biology) is a branch of medical science that applies physiological and biological principles to clinical practice. It has been the dominant health system for more than a century. The use of polymer coatings in the field of biomedicine can be categorized on basis of its application as follows; i) orthopedic materials, ii) cardiovascular stunts, iii) anti-bacterial surfaces, iv) drug delivery, v) biosensors, and vi) tissue engineering. These polymer coatings majorly help to improve the mechanical strength, to enhance biocompatibility, for electrical conductivity, protection from deterioration, electrical conductivity, and to tailor the surface chemistry of compounds. Magnesium and its alloys are having vast fracture toughness, compressive yield strengths, low density, and similar elastic moduli relative to bone [69]. Despite the fact that magnesium and alloys are widely used materials for implants, but there is a need to reduce their corrosion rate. Luo *et al.* [69] used poly(3,4-ethylenedioxythiophene) (PEDOT) polymer by electropolymerization technique. Dextramethasone was incorporated into the liquid ionic electrolyte. The formed coating was lessened the rate of corrosion. Failures in orthopedic implants are because of poor osteointegration and bacterial infection. Indomethacin was incorporated along with biodegradable and anti-bacterial polymers (chitosan and PLGA) by modifying the Ti surfaces with titania nanotube (TNT). Delivering drugs to the bone for extended time has also been reported [70]. Growth factors were incorporated along with antibiotics to prevent infections and promote bone healing [71]. Application of polymer coating on other areas was given in Table 19.11.

19.7 Pellet Coating (Film Coating and Dry Coating)

Pellets serves as an excellent coating substrate owing to i) low surface area to volume ratio compared to powders and granule, ii) higher concentrations of active substances can be incorporated without forming large-sized particles, iii) high degree of flexibility in product development, and iv) therapeutic advantages such as division of dose strength, modification of drug release.

19.7.1 Pellets by Solution/Suspension Layering

Pellets can be prepared by solution or suspension layering using fluid bed granulators, coating pans, and Wurster coaters. Certainly, all the formulation components are dissolved or suspended during the solution or suspension layering.

Table 19.11 Biomedical applications of polymer coatings [72].

Area of application	Polymer used	Description
Orthopedic material		
	poly(L-lactide) (PLLA) coatings	To increase its cytocompatibility and also to reduce corrosion rate Mg and its alloys.
	HA (hydroxyapatite) doped PLA (polylactic acid)	To decrease the corrosion rate of AZ31 Mg alloy
	alginate and poly(L-lysine)	For the controlled delivery of growth factors
	PLA	Layers of Mg alloy AM 50 and Ca-P and/or Cap-P-Mg minimized the corrosion rate and promoted cell attachment.
Cardiovascular stents		
	PLA-poly(ethylene glycol) (PEG)	Short term long term toxicity of carbon nanotubes was reduced and also their aggregation was also prevented.
	poly(ethylene carbonate)	Stainless steel stents are spray-coated to reduce their toxicity
	PEDOT	To promote adsorption of HSA, cell adhesion and cytocompatibility,
	The copolymer of L-Lactide which is immobilized with heparin and methyl benzyloxy carbonate in THF (tetra hydro furan)	Degradation time for coated bare metal stunts was maintained for 16 weeks and only very little inflammation was observed in *in vivo* conditions.

(*Continued*)

Table 19.11 Biomedical applications of polymer coatings [72]. (*Continued*)

Area of application	Polymer used	Description
Antimicrobial surfaces		
	Antimicrobial PU coatings	Shown good adhesion strength, mechanical properties. The bacterial activity was to 95% and signifying their importance in the implants as a coating material.
	poly(1,3-*bis*-(*p*-carboxyphenoxy propane)-*co*-sebacic anhydride) 20:80	Coated films were effective against E.coli and S. aureus population and also shown CR drug delivery. This can confirm their application in developing catheters and implants.
	PCL/PLA polymers with polypropylene meshes.	Post implantation infection risk was minimized.
Drug delivery		
	hydroxypropylmethylcellulose (HPMC) and carboxymethylcellulose (CMC);	Prepared spray-coated microneedle patches can be potentially used the vaccine and intradermal delivery of pharmaceutical compounds.
	Poly methacrylic acid and chitosan	Drugs can be effectively loaded into microspheres by mesoporous silica nanoparticles.

(*Continued*)

Table 19.11 Biomedical applications of polymer coatings [72]. (*Continued*)

Area of application	Polymer used	Description
Tissue engineering		
	Bio degradable polymers-PU, PCL, PLGA. PCL.	The proliferation of chondrocytes (cartilage cells) on poly(dopamine)-coated tissues
	PGA, PLLA, PDLA along with some synthetic degradable polymers	bone tissue engineering (BTE)
	poly[2-(methyl acryloyl oxy) ethyl dimethyl-(3-sulfopropyl) ammonium hydroxide] (PMEDSAH)	Long-term growth of human embryonic stem (hES) cells are observed in different culture media
Biosensors		
	Coating of PAPBAOT on the glassy electrode of carbon. (poly(3-aminophenyl boronic acid-*co*-3-octylthiophene)	Enzyme-free glucose sensor was developed.
	Poly styerene co polyacrylic acid	Polymers imprinted with proteins were prepared. These will show the antibody-like specificity.

The droplets were impinged on to the substrates after spraying of coating formulation and spread on their surfaces, followed by drying phase. During the drying phase, dissolved materials will crystallize and forms the solid bridge between core and initial layer of drug substances and also between the successive layers. This will continues till the required layers of API is formed and the potency of pellets can be achieved.

19.7.2 Dry Coating

Apart from the coating of pellets, tableting of coated particles also gained importance from 1960s [73, 74]. Various aspects of compression of pellets such as type and amount of coating material, cushioning excipient, and compression forces are widely discussed in the literature [75, 76]. Breakage of coating, changes in quality of product in terms of dissolution are the main drawbacks of tableting of pellets. The "soft tableting" approach is used by using relatively elastic/cushioning excipients in order to minimize the damage during compression [78]. Ethylcellulose can be notably used in the compression of coated pellets because of its cushioning property; it can minimize the damage [77]. All the pharmacopoeial conditions such as uniformity of weight and content, friability, hardness, and dissolution profiles must meet during the compression of coated pellets. All these specified conditions make the formulation of coated pellets as an extremely complex process. With these specified reasons, very few compressed pellets have reached the market like a prolonged-release formulation with metoprolol [79], an enteric-coated formulation with omeprazole (Antra® MUPS), and similar theophylline (Theo-Dur®), lansoprazole (Prevacid®) and esomeprazole (Nexium®) products.

19.8 Conclusion

Broad research endeavors have been made in the past to coat various dosage forms with different natural and synthetic polymers. Film coating is an excellent choice for formulation development procedures where it is useful in playing diverse roles. Taste masking or protection forms the external environment, which are simple examples of effectively used approaches. The film coating can be achieved through the use of water-soluble, cationic, anionic, or neutral insoluble polymers from different chemical structures. Enteric polymers are used to restrict the release of drug in the upper GIT. Polymer coatings in the field of biomedicine have been used widely to improve the mechanical strength, to enhance biocompatibility, for corrosion protection, electrical conductivity, and to tailor surface chemistry.

References

1. Porter, S.C. *et al.*, Coating of pharmaceutical dosage forms, in: *Remington the Science and Practice of Pharmacy*, L.V.E.A. Allen (Ed.), pp. 977–987, Pharmaceutical Press, London, 2012.
2. Chauhan, N.P.S., Pathak, A.K., Bhanat, K., Ameta, R., Rawal, M.K., Punjabi, P.B., Pharmaceutical Polymers, in: *Encyclopedia of Biomedical Polymers and Polymeric Biomaterials*, pp. 5929–5942, Taylor and Francis, New York, 2016.

3. Bruce, H.F., Sheskey, P.J., Garcia-Todd, P., Felton, L.A., Novel low-molecular-weight hypromellose polymeric films for aqueous film coating applications. *Drug Dev. Ind. Pharm.*, 37, 12, 1439–1445, 2011.
4. Kablitz, C.D., Harder, K., Urbanetz, N.A., Dry coating in a rotary fluid bed. *Eur. J. Pharm. Sci.*, 27, 2–3, 212–219, 2006.
5. Zheng, W., Cerea, M., Sauer, D., McGinity, J.W., Properties of theophylline tablets powder-coated with (meth)acrylate ester copolymers. *J. Drug Deliv. Sci. Technol.*, 14, 4, 319–325, 2004.
6. Jarray, A., Gerbaud, V., Hemati, M., Polymer-plasticizer compatibility during coating formulation: Amulti-scale investigation. *Prog. Org. Coat.*, 101, 195–206, 2016.
7. Laboulfie, F., Hemati, M., Lamure, A., Diguet, S., Effect of the plasticizer onpermeability, mechanical resistance and thermal behaviour of composite-coating films. *Powder Technol.*, 238, 14–19, 2013.
8. Jarray, A., Gerbaud, V., Hemati, M., Prediction of solid – binder affinity in dryand aqueous systems: Work of adhesion approach vs. ideal tensile strength approach. *Powder Technol.*, 271, 61–75, 2015.
9. Ould-Chikh, S., Elaboration, mise en forme et résistance mécanique de bimatériaux sphériques: Application en catalyse, Thèse université de Lyon, Ecole Normale supérieure de Lyon, Lyon, France, 2008.
10. Cuppok, Y., Muschert, S., Marucci, M., Hjaertstam, J., Siepmann, J., Alelsson, A., Siepmann, J., Drug release mechanisms from Kollicoat SR:Eudragit NE coated pellets. *Int. J. Pharm.*, 409, 1–2, 30–37, 2011.
11. Obara, S. and Mc Ginity, J.W., Influence of processing variables on the properties of free films prepared from aqueous polymeric dispersions by a spray technique. *Int. J. Pharm.*, 126, 1–2, 1–10, 1995.
12. Porter, S.C. and Felton, L.A., Techniques to assess film coatings and evaluate film-coated products. *Drug Dev. Ind. Pharm.*, 36, 2, 128–142, 2010.
13. Joshi, S. and Petereit, H.-U., Film coatings for taste masking and moisture protection. *Int. J. Pharm.*, 457, 395–406, 2013.
14. Varum, F.J., Merchant, H.A., Basit, A.W., Oral modified-release formulations in motion: The relationship between gastrointestinal transit and drug absorption. *Int. J. Pharm.*, 395, 1–2, 26–36, 2010.
15. Freire, A.C., Basit, A.W., Choudhary, R., Piong, C.W., Merchant, H.A., Does sex matter? The influence of gender on gastrointestinal physiology and drug delivery. *Int. J. Pharm.*, 415, 1–2, 15–28, 2011.
16. Agyilirah, G.A. and Banker, G.S., Polymers for enteric coating applications, in: *Polymers for Controlled Drug Delivery*, P.J. Tarcha (Ed.), pp. 39–66, CRC Press, Boca Raton, FL, 1991.
17. Al-Gousous, J., Amidon, G.L., Langguth, P., Toward biopredictive dissolution for enteric coated dosage forms. *Mol. Pharm.*, 13, 6, 1927–1936, 2016.
18. Al-Gousous, J., Sun, K.X., McNamara, D.P., Hens, B., Salehi, N., Langguth, P., Bermejo, M., Amidon, G.E., Amidon, G.L., Mass transport analysis of

the enhanced buffer capacity of the bicarbonate-CO2 buffer in a phase-heterogenous system: Physiological and pharmaceutical significance. *Mol. Pharm.*, 15, 11, 5291–5301, 2018.

19. Dvorácková, K., Rabisková, M., Gajdziok, J., Vetchý, D., Muselík, J., Bernatoniene, J., Bajerová, M., Drottnerová, P., Coated capsules for drug targeting to proximal and distal Part of human intestine. *Acta Pol. Pharm.*, 67, 2, 191–199, 2010.
20. Takeuchi, H., Kojima, H., Yamamoto, H., Kawashima, Y., Polymer coating of liposomes with a modified polyvinyl alcohol and their systemic circulation and RES uptake in rats. *J. Control. Release*, 68, 195–205, 2000.
21. Martin, C., Somavarapu, S., Alpar, H.O., Mucosal delivery of diphtheria toxoid using polymer coated bioadhesicve liposomes as vaccine carriers. *J. Drug Deliv. Sci. Technol.*, 15, 301–306, 2005.
22. Henriksen, I., Smistad, G., Karlsen, J., Interactions between liposomes and chitosan II: Effect of selected parameters on aggregation and leakage. *Int. J. Pharm.*, 101, 227, 994.
23. Takeuchi, H., Yamamoto, H., Niwa, T., Hino, T., Kawashima, Y., Mucoadhesion of polymer-coated liposomes to rat intestine *in vitro*. *Chem. Pharm. Bull.*, 42, 1954–1956, 1994.
24. Takeuchi, H., Matsui, Y., Sugihara, H., Yamamoto, H., Kawashima, Y., Effectiveness of submicron-sized, chitosan-coated liposomes in oral administration of peptide drugs. *Int. J. Pharm.*, 303, 160, 2005.
25. Miyazaki, T., Kohno, S., Sasayama, K., Inoue, Y., Hara, K., Ogasawara, M., Sato, T., Sunamoto, J., Tohoku, Polysaccharide-coated liposomal amphotericin B for the treatment of murine pulmonary candidiasis. *J. Exp. Med.*, 168, 483, 1992.
26. Traini, D., Scalia, S., Adi, H., Marangoni, E., Young, P.M., Polymer coating of carrier excipients modify aerosol performance of adhered drugs used in dry powder inhalation therapy. *Int. J. Pharm.*, 438, 150–159, 2012.
27. Barakh Ali, S.F., Afrooz, H., Hampel, R., Mohamed, E.M., Bhattacharya, R., Cook, P., Khan, M.A., Rahman, Z., Blend of cellulose ester and enteric polymers for delayed and enteric coating of core tablets of hydrophilic and hydrophobic drugs. *Int. J. Pharm.*, 567, 1–12, 2019.
28. Shravania, D., Lakshmia, P.K., Balasubramaniam, J., Preparation and opti mization of various parameters of enteric coated pellets using the TaguchiL9 orthogonalarray design and their characterization. *Acta Pharm. Sin. B*, 1, 1, 56–63, 2011.
29. Nair, A.B., Gupta, R., Kumria, R., Jacob, S., Attimarad, M., Formulation and evaluation of enteric coated tablets of proton pump inhibitor. *J. Basic Clin. Pharm.*, 1, 4, 215–221, 2010.
30. Kuang, C., Sun, Y., Li, B., Fan, R., Zhang, J., Yao, Y., He, Z., Preparation and evaluation of duloxetine hydrochloride enteric-coated pellets with different enteric polymers. *Asian J. Pharm. Sci.*, 12, 3, 216–226, 2017.

31. Patil, A., Payghan, S., Disouza, J.I., Formulation and Evaluation of Enteric coated tablets of Azithromycin dehydrate. *Int. J. ChemTech Res.*, 3, 3, 1479–1484, 2011.
32. Raval, M.K., Ramani, R.V., Sheth, N.R., Formulation and evaluation of sustained release enteric-coated pellets of budesonide for intestinal delivery. *Int. J. Pharm. Investig.*, 3, 4, 203–111, 2013.
33. Badoni, A., Gnanarajan, G., Kothiyal, P., A Research Article on Formulation And Evaluation of Enteric Coated Tablet Loaded With Rabeprazole for Mucoadhesive Drug Delivery. *Pharma J.*, 1, 8, 50–58, 2012.
34. Zhou, M.M. *et al.*, Application and Preparation of Enteric Coating Materials. *IOP Conf. Ser.: Mater. Sci. Eng.*, 170, 012021, 2017.
35. Varum, F.J.O., Merchant, H.A., Goyanes, A., Assi, P., Zboranová, V., Basit, A.W., Accelerating the dissolution of enteric coatings in the upper small intestine: Evolution of a novel pH 5.6 bicarbonate buffer system to assess drug release. *Int. J. Pharm.*, 468, 1–2, 172–177, 2014.
36. Alotaibi, H.F., Elsamaligy, S., Mahrous, G.M., Bayomi, M.A., Mahmoud, H.A., Design of taste masked enteric orodispersible tablets of diclofenac sodium by applying fluid bed coating technology. *Saudi Pharm J.*, 27, 3, 354–362, 2019.
37. Al-Gousousa, J., Ruana, H., Blecharc, J.A., Sun, K.X., Salehid, N. *et al.*, Mechanistic analysis and experimental verification of bicarbonatecontrolled enteric coat dissolution: Potential *in vivo* implications. *Eur. J. Pharm. Biopharm.*, 139, 47–58, 2019.
38. Tribedi, S., Ananthapur, M., Sabitha, J.S., Mathappan, R., Prasanth, V.V., Formulation and Evaluation of Enteric Coated Tablets of Pantoprazole. *Int. J. Pharm. Chem. Sci.*, 2, 3, 1454–1461, 2013.
39. Rathore, S.B.S., Sharma, A., Garg, A., Sisodiya, D.S., Formulation and evaluation of enteric coated tablet of Ilaprazole. *Int. Curr. Pharm. J.*, 2, 7, 126–130, 2013.
40. Pastham, R., Reddy, S., Allenki, V., Design and *In-vitro* Evaluation of Enteric Coated Pulsatile Drug Delivery System of Zileuton Tablets. *Pharm. Chem. J.*, 4, 4, 57–65, 2017.
41. Raval, M.K., Ramani, R.V., Sheth, N.R., Formulation and evaluation of sustained release enteric-coated pellets of budesonide for intestinal delivery. *Int. J. Pharm. Investig.*, 3, 4, 203–11, 2013.
42. Alsulays, B.B., Kulkarni, V., Alshehri, S.M., Almutairy, B.K., Ashour, E.A., Morott, J.T., Alshetaili, A.S., Park, J.-B., Tiwari, R.V., Repka, M.A., Preparation and evaluation of enteric coated tablets of hot melt extruded lansoprazole. *Drug Dev. Ind. Pharm.*, 43, 5, 789–796, 2017.
43. Li, C., Zhou, K., Chen, D., Xu, W., Tao, Y., Pan, Y., Meng, K., Shabbir, M.A.B., Liu, Q., Huang, L., Xie, S., Solid lipid nanoparticles with enteric coating for improving stability, palatability, and oral bioavailability of enrofloxacin. *Int. J. Nanomedicine*, 14, 1619–1631, 2019.

44. Gupta, A., Tiwari, G., Tiwari, R., Srivastava, R., Rai, A.K., Enteric coated HPMC capsules plugged with 5-FU loaded microsponges: A potential approach for treatment of colon cancer. *Braz. J. Pharm. Sci.*, 51, 3, 591–605, 2015.
45. Dreu, R., Ilić, I., Srčič, S., Development of a multiple-unit tablet containing enteric-coated pellets. *Pharm. Dev. Technol.*, 16, 2, 118–126, 2011.
46. Setia, A., Kansal, S., Goyal, N., Development and optimization of enteric coated mucoadhesive microspheres of duloxetine hydrochloride using 32 full factorial design. *Int. J. Pharm. Investig.*, 3, 3, 141–50, 2013.
47. Khan, Z.A., Tripathi, R., Mishra, B., Design and evaluation of enteric coated microporous osmotic pump tablet (ECMOPT) of quetiapine fumarate for the treatment of psychosis. *Acta Pol. Pharm.*, 69, 6, 1125–1136, 2012.
48. Rai, G., Yadav, A.K., Jain, N.K., Agarwal, G.P., Enteric-coated epichlorohydrin crosslinked dextran microspheres for site-specific delivery to colon. *Drug Dev. Ind. Pharm.*, 41, 12, 1–11, 2018.
49. Wang, Y., Liu, H., Liu, K., Sun, J., He, Z., Design and evaluation of enteric-coated tablets for rifampicin and isoniazid combinations. *Pharm. Dev. Technol.*, 18, 2, 401–406, 2013.
50. Thomas, S., Chong, Y.N., Chaw, C.S., Preparation and characterization of enteric microparticles by coacervation. *Drug Dev. Ind. Pharm.*, 39, 7, 1142–51, 2013.
51. Hosny, K.M., Ahmed, O.A.A., Al-Abdali, R.T., Enteric-coated alendronate sodium nanoliposomes: A novel formula to overcome barriers for the treatment of osteoporosis. *Expert Opin. Drug Deliv.*, 10, 6, 741–746, 2013.
52. Sachdeva, C., Mishra, N., Sharma, S., Development and characterization of enteric-coated microparticles of biochanin A for their beneficial pharmacological potential in estrogen deficient-hypertension. *Drug Deliv.*, 23, 6, 2044–2057, 2015.
53. Swain, S., Behera, U.A., Beg, S., Sruti, J., Patro, Ch. N., Dinda, S.C., Rao, M.E.B., Design and characterization of enteric-coated controlled release mucoadhesive microcapsules of Rabeprazole sodium. *Drug Dev. Ind. Pharm.*, 39, 4, 548–560, 2013.
54. Honsy, K.M., Alendronate Sodium as Enteric Coated Solid Lipid Nanoparticles; Preparation, Optimization, and *In Vivo* Evaluation to Enhance Its OralBioavailability. *PLoS One*, 11, 5, 1–14, 2016.
55. Li, R., Chen, R., Liu, W., Qin, C., Han, J., Preparation of enteric-coated microcapsules of astaxanthin oleoresin by complex coacervation. *Pharm. Dev. Technol.*, 23, 7, 674–581, 2018.
56. Tummala, S., Kuppusamy, G., Satish Kumar, M.N., Praveen, T.K., Wadhwani, A., 5-Fluorouracil enteric-coated nanoparticles for improved apoptotic activity and therapeutic index in treating colorectal cancer. *Drug Deliv.*, 23, 8, 2902–2910, 2016.

57. Deshmukh, D., Ravis, W.R., Betageri, G.V., Delivery of didanosine from enteric-coated, sustained-release bioadhesive formulation. *Drug Deliv.*, 10, 1, 47–50, 2003.
58. Chen, K., Chang, H.H.R., Shalviri, A., Li, J., Lugtu-Pe, J.A., Kane, A., Wu, X.Y., Investigation of a New pH- responsive Nanoparticulate Pore Former for Controlled Release Enteric Coating with Improved Processability and Stability. *Eur. J. Pharm. Biopharm.*, 120, 116–125, 2017.
59. Li, P., Tan, A., Prestidge, C.A., Nielsen, H.M., Müllertz, A., Self-nanoemulsifying drug delivery systems for oral insulin delivery: *In vitro* and *in vivo* evaluations of enteric coating and drug loading. *Int. J. Pharm.*, 477, 1–2, 390–398, 2014.
60. Refai, H., Hassan, D., Abdelmonem, R., Development and characterization of polymer-coated liposomes for vaginal delivery of sildenafil citrate. *Drug Deliv.*, 24, 1, 278–288, 2017.
61. Sun, Y., Zhu, S., Lu, W., Chen, J., Sun, C., Guo, Y., Wang, B., Gao, S., Fang, W., Hu, R., A novel enteric positioning osmotic pump capsule-based controlled release system of sinomenine hydrochloride: *In vitro* and *in vivo* evaluation. *J. Drug Deliv. Sci. Technol.*, 49, 188–194, 2018.
62. Nguyen, M.N.U., Tran, P.H.L., Tran, T.T.D., A single-layer film coating for colon-targeted oral delivery. *Int. J. Pharm.*, 49, 188–194, 2019.
63. Nicholas, M.E., Panaganti, S., Prabakaran, L., Jayveera, K.N., Novel Colon Specific Drug Delivery System: A Review. *IJPSR*, 10, 2545–2561, 2011.
64. Maroni, A., Del Curto, M.D., Zema, L., Foppoli, A., Gazzaniga, A., Film coatings for oral colon delivery. *Int. J. Pharm.*, 457, 2, 372–394, 2013.
65. Hariharan, M., Wheatley, T.A., Price, J.C., Controlled-release tablet matrices from carrageenans: Compression and dissolution studies. *Pharm. Dev. Technol.*, 2, 383–393, 1997.
66. Picker, K.M., The use of carrageenan in mixture with microcrystalline cellulose and its functionality for making tablets. *Eur. J. Pharm. Biopharm.*, 48, 27–36, 1999.
67. Li, L., Zhang, X., Gu, X., Mao, S., Applications of Natural Polymeric Materials in Solid Oral Modified-Release Dosage Forms. *Curr. Pharm. Des.*, 21, 5854–5867, 2015.
68. Halliday, A.J., Moulton, S.E., Wallace, G.G., Cook, M.J., Novel methods of antiepileptic drug delivery—Polymer-based implants. *Adv. Drug Delivery Rev.*, 6, 953–964, 2012.
69. Luo, X. and Cui, X.T., Electrochemical deposition of conducting polymer coatings on magnesium surfaces in ionic liquid. *Acta Biomater.*, 7, 1, 441–446, 2011.
70. Rosiaux, Y., Muschert, S., Chokshi, R., Leclercq, B., Siepmann, F., Siepmann, J., Ethanol-resistant polymeric film coatings for controlled drug delivery. *J. Controlled Release*, 169, 1–9, 2013.
71. Smith, J.R. and Lamprou, D.A., Polymer coatings for biomedical applications: A review. *Trans. IMF*, 92, 1, 9–19, 2014.

72. Sandberg, A., Ragnarsson, G., Jonsson, U.E., Sjögren, J., Design of a new multiple-unit controlled-release formulation of metoprolol - Metoprolol CR. *Eur J. Clin. Pharmacol.*, 33, 1, S3–S7, 1988.
73. Petersen, K.-U. and Schmutzler, W., Proton pump inhibitors. Active substance release from different preparations. *Dtsch. Apoth. Ztg.*, 139, 1, 68–69, 1999.
74. Wagner, K.G., Krumme, M., Beckert, T.E., Schmidt, P.C., Development of disintegrating multiple-unit tablets on a high-speed rotary tablet press. *Eur. J. Pharm. Biopharm.*, 50, 2, 285–291, 2000.
75. Shimizu, T., Nakano, Y., Morimoto, S., Tabata, T., Hamaguchi, N., Igari, Y., Formulation study for lansoprazole fast-disintegrating tablet. I. Effect of compression on dissolution behaviour. *Chem. Pharm. Bull.*, 51, 8, 942–947, 2003.
76. Dashevsky, A., Kolter, K., Bodmeier, R., Compression of pellets coated with various aqueous polymer dispersions. *Int. J. Pharm.*, 279, 1–2, 19–26, 2004.
77. Picker, K.M., "Soft tableting:" A new concept to tablet pressure sensitive materials. *Pharm. Dev. Technol.*, 9, 1, 107–121, 2004.
78. *European Pharmacopoeia*, 6th ed., Directorate for the Quality of Medicines & HealthCare (EDQM), Council of Europe, Strasbourg, France, 07/2009.
79. Missaghi, S., Young, C., Fegely, K., Rajabi-Siahboomi, A.R., Delayed release film coating applications on ral solid dosage forms of proton pump inhibitors: Case studies. *Drug Dev. Ind. Pharm.*, 36, 2, 180–189, 2010.

Index

Made in the USA
Monee, IL
12 November 2026

0794575f-2b44-49f6-a4b5-d2096291fd9cR01